933

The ecology of animal movement

The ecology of animal movement

Edited by

Ian R. Swingland
Lecturer in Natural Sciences,
University of Kent

and

Paul J. Greenwood
Lecturer in Zoology,
University of Durham

CLARENDON PRESS · OXFORD
1983

Oxford University Press, Walton Street, Oxford OX2 6DP

London Glasgow New York Toronto
Delhi Bombay Calcutta Madras Karachi
Kuala Lumpur Singapore Hong Kong Tokyo
Nairobi Dar es Salaam Cape Town
Melbourne Auckland

and associates in
Beirut Berlin Ibadan Mexico City Nicosia

Oxford is a trade mark of Oxford University Press

Published in the United States
by Oxford University Press, New York

British Library Cataloguing in Publication Data

The Ecology of animal movement.
  1. Animal locomotion
  I. Swingland, Ian R.   II. Greenwood, Paul J.
  591.1'852   QP301
  ISBN 0-19-857575-0

Library of Congress Cataloging in Publication Data

Main entry under title:
The Ecology of animal movement.
  Bibliography: p.
  Includes index.
  Contents: Animal movement / Paul J. Greenwood and
Ian R. Swingland − Animal movements / Graham H. Pyke −
Vertebrate home-range size and energetic requirements /
Georgina M. Mace, Paul H. Harvey, and T. H. Clutton-
Brock − [etc.]
  1. Animal migration. 2. Animals−Dispersal.
3. Animal ecology. I. Swingland, Ian R. (Ian Richard),
1946−   . II. Greenwood, Paul J. (Paul John),
1952−   .
QL754.E27 1983   591.52'5   82−17807
ISBN 0−19−857575−0

Set by Thames Typesetting, Abingdon, Oxon
Printed by Thetford Press, Thetford

Foreword

T. R. E. Southwood, Department of Zoology, Oxford

Movement is, as every schoolchild learns at the outset of the biology course, a characteristic of animals. There are, as always with biological generalizations, exceptions: sessile animals and mobile organisms in other kingdoms; but this ability to change location has been a major influence in the evolution of the most diverse of the living kingdoms. Yet movement is itself the product of evolutionary pressures; whether there was a single primary cause or not is irrelevant, for movement contributes in many ways to the survival and reproduction of the animal. On the largest scale movement permits the animal to change its habitat, perhaps to string together a series of partial habitats to provide a 'life-track'. This is what is commonly understood by 'migration', though, as the chapters in this book show, a watertight definition is beset with difficulties. Movement within the habitat also enhances survival: the seeking and capturing of food, escape from enemies and from inclement environmental conditions. Animals survive to reproduce and in this process movement is again vital, the sexes must be brought together (plants too require motility for this purpose). Then movement may evolve to enhance the rate of success of the reproduction: for communication and to place the offspring in a more favourable habitat.

This simple general survey yields seven different functions by which movement may enhance the fitness, in the evolutionary sense, of the animal. The chapters in this book show how work on behaviour, biomechanics, energetics and other aspects of ecology has come together to investigate the optimization of these functions. One pitfall is to overlook the plurality of the functions of movement. This was typified by those foraging studies that neglected the role of predators.

Not only are the functions of movement multiple, so are its modes of expression: walking, swimming, flying, and the more esoteric forms such as ballooning (by some young spiders and caterpillars) and phoresy ('hitch-hiking', whereby a pseudoscorpion, for example, just attaches itself to and moves with a fly). Sometimes the individual has a choice between these modes or between variants of the same mode: an ape may be quadripedal, bipedal, or brachiate through the branches with its arms; a crocodile may crawl or 'run' with its belly off the ground; a grasshopper may walk or jump; a butterfly and a falcon may flap or glide or, perhaps, hover.

The extent of the choice, both at the individual level and in the evolutionary sense, is constrained by the animal's basic structure. Every species is a prisoner of its evolutionary history, the evolutionary pressures we measure today have been applied to a pre-existing model and not, as it were, at the drawing-board stage.

A further constraint is provided by the environment, the habitat: an ape can brachiate only where there are suitable trees. Less obvious is the interaction of habitat and structure, particularly in long-distance movement: this determines the extent to which it is the animal or the prevailing current (in air or water) that controls the overall direction of movement. A salmon may 'go where it will', but a flat fish must depend on the currents' direction. The power ratio between the animal and ambient medium fixes their relative contributions to the directionability of movement (Southwood 1981).

There are thus a great variety of functions fulfilled by movement, and a range of modes of locomotion, though this rich matrix is constrained for individuals (and individual species) by their design and their habitat. Much of the confusion and controversy surrounding the ecological interpretation of movement has arisen because of a failure to recognize this diversity and the constraints.

In this volume the editors have sought not only to explore the better known facets, but also some that have been neglected. Their intention has been to provide a conspectus of present thought and work, stressing the variety of the subject, rather than a conformity of concept. I consider they have succeeded admirably. Whilst clearly reviewing past work and achievements, new questions are raised and the diversity of movement in both function and form is highlighted.

Preface

The idea for this book was conceived in the Zoology Department at Oxford during a lull one day in March 1979. Our basic plan was to persuade active research workers in the areas of evolutionary and ecological aspects of animal movement to contribute to a book that would be useful to both undergraduates and those engaged in research. We wanted to concentrate on current and fundamental problems in movement and to stimulate and surprise by enabling the authors to put forward new ideas in a well-reasoned and coherent way. Such a project is fraught with difficulties. The disparate styles of the various authors will, we hope, add spice rather than discord. The subject areas which we have chosen are entirely eclectic, as is the choice of the fifteen authors.

Terminology is the bane of animal movement. Few people agree on the terms to be used, let alone their definitions. Migration, immigration, emigration, dispersal, philopatry, nomadism, life-time track are all terms which vary in usage depending on the topic, organism, school, and inclination of the researcher. R. R. Baker (1978) has recently tried to cut a swathe through the confusion by proposing that all forms of animal movement should be termed migration. Most people would be sympathetic to a standard, all-embracing terminology but Baker's broad definition of migration as the movement of individuals from one spatial unit to another will have few adherents. It is difficult to imagine a consensus of approval for using the word 'migration' to describe such temporal and spatial differences as the seasonal movement of terns from the Arctic to the Antarctic and back, the flight of a bumblebee from one inflorescence to another, and the brief excursion of a male grey squirrel outside its home range in search of a female. There may be general rules which underly each of these behaviours, but the sheer scale of the differences makes one label incongruous at the present time.

In this volume we have taken animal movement as our starting-point. We have viewed movement from an ecological vantage-point and, in a very loose sense, we have described spatial and temporal changes in location and distribution. Such a vantage-point involves behavioural, evolutionary, and genetic aspects. We have encouraged the contributors to use whatever terms come naturally to them, believing that 'movement' encompasses all terms and that any distinctions are purely arbitrary. Since the contributors make explicit the context in which they use their terms, we trust that confusion has been avoided.

Yet a further problem is the actual measurement of movement. Historically, ecologists have tended to use environmental units as their yardstick for measuring both plant and animal movement. This is frequently in the form of describing a population shift from one spatial unit to another, such as a change in habitat. This approach has the disadvantage that the spatial component of movement which incorporates the dispersion of individuals themselves, may be lost. Nowadays it is increasingly realized that the spatial distribution and movement of particular individuals has profound ecological as well as evolutionary ramifications. We are dealing with events which not only have implications for speciation, species, and spatial distributions, and the efficacy of group, kin, and individual selection but also life histories, spatial dynamics, population control, and predator–prey interactions. To facilitate the bringing together of these areas within the framework of animal movement, we have concentrated on two questions: 'Why move?' and 'What are the consequences?' We are concerned not with the mechanisms, such as navigation, but with the ultimate ecological and evolutionary principles. For this reason the reader will find obvious taxonomic omissions in the coverage of the book (e.g. plankton, humans) but, since we are dealing with problems rather than systematics, our apology is muted.

We have divided the book into three sections, each prefaced by a commentary chapter. The first commentary (Chapter 1) briefly considers some phylogenetic constraints on animal movement in the context of examining the gaps in research between the fine detail considered in Chapter 2, which deals with small-scale and short-term foraging behaviour, and the gross features covered by Chapter 3 which considers home-range sizes and energetics. The second commentary (Chapter 4) deals with dispersal in relation to population structure and dynamics. It is concerned with the decisions and implications of who should move, and when, in the life history, and offers a number of predictions about the consequences of the different possibilities. This area is considered in more detail in the following four chapters which concentrate on inter-specific and intraspecific comparisons; the causes and consequences of dispersal in small mammals (Chapter 5), intraspecific differences in movement (Chapter 6), mating systems, dispersal, and the evolution of altruism, disruption, and sex ratios (Chapter 7), and finally, optimal inbreeding and the evolution of philopatry (Chapter 8). The final section has an opening commentary on migration strategies (Chapter 9) introducing three chapters on large-scale movement patterns. The link between the behaviour of individual insects and the biology of their populations, namely movement, is considered in Chapter 10. In Chapter 11 the ecological correlates of colonizing ability are discussed and the section concludes with an examination of the functions of long-distance movements in vertebrates (Chapter 12).

We have been blessed with co-operative authors, understanding referees, patient and sympathetic editors at Oxford University Press, and the help and enthusiasm of those around us: to all we give thanks.

I. R. S.   P. J. G.
University of Kent   *University of Durham*   1981

Acknowledgements

Without the expert help and guidance of the staff of the Oxford University Press this book would never have been born; we are particularly grateful to them.

We would like to thank the following for: permission to quote from unpublished material or for permission to modify and use published information; help in stimulating discussions, formulating ideas; help in research; comments on chapter drafts; and funding and supporting the research. Academic Press Inc. (London) Ltd., Z. Agur, American Association for the Advancement of Science, Y. Ayal, Ballière Tindall, A. d'A. Bellairs, G. E. Belovsky, Biological Laboratory, University of Kent, C. B. Brownsmith, Blackwell Scientific Publications Ltd., H. L. Carson, E. Charnov, A. S. Cheke, R. Chesser, A. Coekburn, P. A. Colinvaux, Collins Publishers, E. F. Connor, R. Cook, R. Cowie, J. R. Crook, K. L. Crowell, M. H. Cunningham, N. Davies, A. Diamond, H. Dingle, University of Durham, Edward Grey Institute of Field Ornithology, Oxford, T. Felsenburg, E. Framstad, M. S. Gaines, L. L. Getz, A. Gilboa, T. C. Grubb Jr, I. Hanski, P. H. Harvey, W. D. Hamilton, L. Hansson, J. M. Hoff, S. Hubbard, J. R. Krebs, D. Lavee, E. G. Leigh Jr, C. M. Lessells, W. Z. Lidicker, A. Lomnicki, University of Lund, Macmillan Journals Ltd., B. G. Murray Jr, National Geographic Society, National Scientific and Engineering Research Council, Canada, Natural Environment Research Council, U.K., Nederlandse Ornithologische Unie, Norwegian Research Council for Science and Humanities, University of Oslo, A. Pashtan, C. M. Perrins, J. Phillipson, R. Prŷs-Jones, T. Reed, Royal Geographical Society, Royal Society, Royal Society Aldabra Research Station, T. Royama, I. Rubinoff, R. Shine, J. B. Slade, P. J. B. Slater, M. D. Smith, S. C. Stearns, Swedish Natural Science Research Council, R. H. Tamarin, L. R. Taylor, United States–Israel Binational Science Foundation, T. Whitham, G. C. Williams, Z. Yiftakh, M. Zimmermann, Department of Zoology, University of Oxford.

Contents

Contributors

Clutton-Brock, Timothy H., *Department of Zoology, Downing Street University of Cambridge, UK*

Greenwood, Paul J. (ed.), *Department of Adult and Continuing Education, University of Durham, 32 Old Elvet, Durham, UK*

Harvey, Paul H., *School of Biological Sciences, The University of Sussex, Falmer, Brighton, Sussex, UK*

Horn, Henry S., *Department of Biology, University of Princeton, Princeton, New Jersey, USA*

Mace, Georgina M., *Department of Zoology, The University, Newcastle upon Tyne, UK.*

Pyke, Graham H., *Department of Vertebrate Ecology, The Australian Museum, 6–8 College Street, Sydney, NSW, Australia.*

Ritte, Uzi., *Department of Genetics, The Hebrew University of Jerusalem, Jerusalem, Israel*

Rogers, David, J. *Department of Zoology, University of Oxford, South Parks Road, Oxford, UK.*

Safriel, Uriel N., *Department of Zoology, The Hebrew University of Jerusalem, Jerusalem, Israel*

Shields, William M., *College of Environmental Science and Forestry, State University of New York, Syracuse, New York, USA*

Sinclair, Anthony R. E., *Institute of Animal Resource Ecology, University of British Columbia, 2075 Westbrook Mall, Vancouver, British Columbia, Canada*

Stenseth, Nils Chr., *Zoological Institute, University of Oslo, PO Box 1050, Bindern, Oslo 3, Norway*

Swingland, Ian R. (ed.), *School of Continuing Education, Rutherford College, University of Kent, Canterbury, UK*

Taylor, L. Roy, *The Insect Survey, Rothamsted Experimental Station, Harpenden, Hertfordshire, UK*

Taylor, Robin A. J., *Entomology Department, Pennsylvania State University, University Park, Pennsylvania, USA*

1
Animal movement: approaches, adaptations, and constraints

PAUL J. GREENWOOD and IAN R. SWINGLAND

In recent years, fine-grained laboratory experiments and field observations, which attempt to understand the rules which govern the short-term foraging decisions of animals, have been at the forefront of research. One major reason why animals move is to find food. Or, in current parlance, an animal may leave one patch for another in order to maximize its net rate of energy intake. Many of the studies which have dealt with movement in relation to foraging are considered by Pyke in the next chapter.

At the other end of the spectrum, it has been assumed for a long time that the distribution of animals in space is intimately bound up with the distribution and nature of their food supplies. This idea can, in fact, be traced at least as far back in time as Aristotle. Such nebulous statements, however, provide little insight as to how animals exploit their food and how their diet and metabolic requirements might influence their movement and dispersion. Nearly twenty years ago McNab (1963) was the first to appreciate the relationships which might be expected on metabolic grounds between the size of an animal's home range or territory, its body size, and diet. A complementary line of research also emerged in studies of social organization (e.g., Crook 1964; Lack 1968) and the economics of defending an area, particularly in relation to the evolution of territorial and mating systems (e.g., J.L. Brown 1964; Orians 1969; Brown and Orians 1970). Despite a number of methodological objections, not least the cause-and-effect problem inherent in the comparative approach, the development initiated by McNab has been an active area of research. Enough studies of a quantitative nature have now been done on a wide range of taxonomic groups, particularly within birds and mammals, to allow the emergence of important generalizations about the broad patterns across species. This is the area explored in Chapter 3 by Mace, Harvey, and Clutton-Brock.

In this chapter which introduces the section on small-scale movement, we shall briefly examine to what extent there can be a meeting-ground between the fine-grained optimality approach and the coarse-grained comparative one. The origins of an optimality approach to foraging in relation to movement and dispersion owe much to the theoretical insights of MacArthur and Pianka (1966), a paper which appeared not long after McNab's. Since these papers the two lines of research have tended to run in parallel with little interchange between them. It is only recently that a merger has seemed possible.

Territories and foraging

Some of the first studies to take a cost–benefit approach to spatial patterns were those on species of birds which feed on nectar (e.g., F.B. Gill and Wolf 1975*a,b;* F.L. Carpenter and MacMillen 1975*a,b*). The advantage of dealing with such species is that most of the variables which influence the size of a territory defended by an individual can be accurately quantified. For example, in their study of the golden-winged sunbird (*Nectorinia reichenowi*), which was extracting nectar from one species of plant (*Leonotis nepetifolia*), Gill and Wolf (1975*a*) could count the number of flowers in each territory, measure the amount of nectar and its rate of production, and assess the metabolic needs of the birds from their activity states. In addition, environmental conditions were relatively constant from day to day, there was little threat to the birds from predators and the sunbirds engaged in only a small number of activities – feeding, resting, and defending. Gill and Wolf (1975*a*) found that although there was considerable variation in the sizes of the territories, the number of flowers in each territory was relatively constant and the amount of nectar available to each bird matched quite closely their daily energetic needs.

From these detailed observations on territory sizes and time budgets, Pyke (1979) went on to examine which of a set of plausible optimality models most closely predicted the sunbird's behaviour and dispersion. He considered four possibilities, that the sunbirds were

1. Maximizing their net daily gain of energy;
2. Minimizing the amount of time they spent resting;
3. Minimizing the daily energetic cost; or
4. Maximizing the ratio of gross daily energetic gain to daily energetic cost.

Two assumptions were that daily gain was in balance or positive and that the amount of time they allocated to their three main activities were fixed. The model which most closely approximated to the sunbird's behaviour was that they were minimizing their daily energetic costs. It is possible, as Pyke (1979) suggests, that under different circumstances (e.g. pre-migration, breeding) one of the alternative hypotheses would be more appropriate.

Dispersion and foraging

When a territorial system has evolved, it should pay an individual to exploit the food available to it as efficiently as possible. Clearly, a sunbird would be unlikely to obtain enough nectar if it defended a large number of flowers but spent its time visiting only a few of them. They do, apparently, avoid repeat visits to recently exploited flowers (Wolf, Hainsworth, and Gill 1975). If, as is the case with nectar and many other prey items, food is being renewed over some time period then it is possible to envisage a system whereby individuals visit their range in a systematic fashion operating on the basis of some optimal return time (Charnov, Orians, and Hyatt 1976). Such an arrangement could in fact be an important factor in the evolution of spacing patterns and territorial behaviour. It would be inappropriate for an individual or group to operate on a return time schedule if the area was also being exploited by competitors (Charnov *et al.* 1976). By defending a territory it is likely that the degree of interference is minimized.

The pattern of exploitation of a territory or home range has been an integral part of the research on the foraging behaviour of primates for some time (e.g., Clutton-Brock 1974). With the increasing use of telemetry, many more species of mammals are now being studied. For most of the species, however, the results are often of a qualitative nature because it is usually too difficult to quantify the amount of food available and its rate of renewal in relation to the ranging behaviour of the individual owner or group, let alone other possible confounding variables. Recently, there have been two studies of bird species where a more quantitative examination has been made of the pattern of visits to different parts of a territory and the disadvantages posed by an owner's systematic exploitation to an intruder's foraging success.

Kamil (1978) showed that territorial breeding pairs of Hawaiian honey-creepers (*Loxops virens*) avoid revisiting flowers where they have recently fed, allowing time for the nectar to replenish. Intruders, on the other hand, who are unfamiliar with the pattern of visits of the owners are more likely to visit flowers with little nectar. A similar picture emerges from a study of the pied wagtail (*Montacilla alba*) (N.B. Davies and Houston 1981). Pied wagtails were observed defending territories along a river bank. The owner maximized its feeding rate by systematically working its way around the territory allowing time at each spot for the food supply to renew before revisiting. Intruders, unaware of which sites had been recently visited, had a feeding rate lower than that of the owner. Although not apparently a contributory factor in the pied wagtail, it is possible in other species that a system of territorial ownership could, in itself, cause a reduced rate of intake for the intruder irrespective of its knowledge of the distribution of food. If the intruder is likely to be attacked and defeated by the resident, the need to be vigilant may divert its attention away from efficient poaching even in an area of high food density.

Dispersion, foraging, and predation

For both the pied wagtail and the golden-winged sunbird the limiting factor which affected their territorial behaviour appeared to be an energetic one; wagtails maximizing their daily energy intake on cold winter territories, sunbirds minimizing their energetic costs in warmer climes. However, both situations are perhaps unusual in that few variables seem to influence territory size. By adding other variables to the system we might find that the optimal territory size is very different. Consider just one additional variable. If an animal is at risk from predators when feeding in certain areas of its territory, a resulting change in foraging behaviour to minimize that risk may alter its territorial requirements. Milinski and Heller (1979) elegantly demonstrated in a laboratory experiment the effect that a model of a kingfisher has on the foraging behaviour of stickle-backs. In the absence of a predator, hungry sticklebacks preferentially attack water fleas at the centre of a swarm. In the presence of a predator, the fish switch their attack to the 'easier' peripheral regions. The reason, the authors conclude, is because of a greater need for vigilance at the expense of a lower feeding rate. Similarly, in great tits, handling time of prey, look-up rate, and interprey waiting-time increase after exposure to a stuffed sparrowhawk (J.R. Krebs 1980). It would seem reasonable to extrapolate from these experiments to the situation where we might expect the patterns of foraging and dispersions of individuals to be substanially altered by a high predation risk. Such changes may seriously constrain an animal in its ability to maximize its energy gain.

Some indication that such energetic constraints occur comes from a study of the sockeye salmon (*Oncorhynchus nerka*) by Brett (1971). The salmon have a daily pattern of vertical migration. During the day they stay in the deep water where the temperature is colder ($5 - 10°C$) than at the surface. At dusk they rise to the surface to feed, remain close to the surface during the night, selecting a water temperature of $15°C$, feed again at dawn, and then return to the depths. A high rate of growth is presumably important in salmon for maximizing reproductive output and the chances of survival. As ectotherms, the growth rate is dependent on temperature and is maximized at $15°C$. One interpretation of their diurnal behaviour is that they select a region of low light intensity during the day to minimize the risk of predation at the expense of a low growth rate. (They will, perhaps incidentally, also reduce their main-tenance costs at the lower temperatures.) During the night they choose a water temperature which results in an optimal growth rate when the risk from visual predators is lower. Thus, food and predators are considered to be two main factors which influence the distribution and movement of sockeye salmon. Similarly, but on a much broader scale, the assumption that feeding and pre-dation are major limiting factors has been the basis for many of the comparative studies of avian and mammalian social organizations from their inception (e.g., Crook 1964) to more recent times (e.g., Clutton-Brock and Harvey 1977*a*; Wittenberger 1979).

Home-range sizes

To what extent can the fine-grained approach to dispersion and movement be intergrated with the comparative one? During the development of quantitative comparative studies, there were a number of inherent weaknesses which tended to undermine the validity of some of the interspecific patterns which were reported and made some zoologists sceptical of their value. Most of these weaknesses are now beginning to disappear. How this has come about is best appreciated by looking in more detail at some of the criticisms.

First, the finding that home-range size is related to body size across numerous taxonomic groups is only indirect evidence that dispersion may be determined by food supply and, ultimately, metabolic needs. Such a criticism is partially offset by the consistent differences which occur between different dietetic groups (see Chapter 3), irrespective of other variables which might be important for particular species. In Chapter 3, Mace, Harvey, and Clutton-Brock take a preliminary step towards developing energetic models which investigate the first-order relationship between average daily metabolic needs and home-range size in primates. These models can be viewed as an extension of the optimality approach to the comparative field.

Second, it is clear from Chapter 3 that many of the findings of earlier studies were the result of inappropriate techniques highlighting spurious biological differences. The development of a firm analytical base enables the authors to investigate the differences which do occur between various taxonomic and dietetic groups. It also identifies more precisely those areas where the data available at the moment are inadequate and where more fine-tuned observations and experiments are needed to test predictions.

Third, when a relationship is established between home-range size and diet or some other variable such as body size, it doesn't necessarily solve the problem of which is cause and which effect (see N.B. Davies and Krebs 1978). However, the findings of some of the important, detailed studies reported above (e.g., F.B. Gill and Wolf 1975a; N.B. Davies and Houston 1981) would seem to indicate that, in evolutionary terms, the size of a home range or territory is determined by energetic requirements and not vice versa. There is a need for more studies of this kind for a synthesis of the two approaches should be a goal of future research.

Further problems

We end this introductory chapter to small-scale movement by mentioning a number of problems which have received scant attention to date. Why and when animals move has been the main focus of attention of studies on optimal foraging in relation to movement (see Chapter 2). Very little is known about how they should move when travelling from one place to another. Most species have a number of options open to them. For instance, a horse can either walk, trot, or gallop. As it changes speed it also changes to the most economical gait whilst within each gait it selects the speed that minimizes its energy consumption (Hoyt and Taylor 1981). This sort of approach now needs to be extended to investigate the mode of movement an animal should adopt in order to feed efficiently. One obvious set of experiments would entail exposing a ground-feeding bird to patches of food of differing spatial distributions. It should be possible to predict from a cost—benefit analysis whether it should fly or walk to the next patch.

These sorts of questions also raise important phylogenetic problems. Birds with the power of flight can move easily and quickly around their territories. Their time-scale of operation is much faster than, for example, most mammals which have to walk or run. But birds and also some mammals have the constraint during the breeding season of returning periodically to a fixed site to feed their young. What effect these differences between taxonomic groups have on their patterns of dispersion and movements awaits further study.

Finally, we have considered in this chapter the emphasis that zoologists have placed on the dispersion patterns of birds and mammals where food and predators are probably the major limiting factors. On the other hand, the activity of many terrestrial ectotherms is probably much more constrained by temperature than any other factor. For example, garter snakes are inactive below 15°C. In parts of their range the temperature profile may only enable them to be active for several months each year. During this period each snake will probably consume in terms of biomass what an equivalent-sized mammal eats in a day (Porter, Mitchell, Beckman, and Tracy 1975). Rules which govern the dispersion and movement of higher vertebrates may have little bearing on other groups.

2
Animal movements: an optimal foraging approach
GRAHAM H. PYKE

Introduction

Animals move from one place to another for a variety of reasons. Animals may, for example, undergo seasonal migrations related to changing environmental conditions. On a smaller scale, animals may make movements that are related to foraging, avoiding predators, finding mates, aggressive interactions, and so on. In this chapter I shall not deal with all possible aspects of animal movements but shall focus instead on movements from one place to another that occur in the context of foraging. I shall not, I might add, consider movements of limbs or mouthparts that do not involve a change in location of the animal as a whole.

Since initial studies by MacArthur and Pianka (1966) and Emlen (1966), many authors have attempted to develop a predictive theory of animal foraging behaviour. This theory, now known as optimal foraging theory, has been recently reviewed by Pyke, Pulliam, and Charnov (1977) and J. R. Krebs (1978). In the following discussion I shall examine the extent to which this theory has been developed for movements of foraging animals. In so doing I shall overlap somewhat with the above more general reviews. I shall, however, attempt to provide a more detailed and critical treatment of the present narrower subject.

In all attempts so far to understand animal foraging behaviour using the optimal foraging approach, there has been a single basic hypothesis (Pyke *et al.* 1977; J. R. Krebs 1978). This hypothesis is that animals forage in ways that maximize their Darwinian fitness (i.e., contribution to the next generation). Further hypotheses have been derived from this basic hypothesis. These derivations involve assumptions concerning the 'currency' of fitness and constraints on behaviour (Schoener 1971). It is usually assumed, for example, that the appropriate currency is net rate of energy gain (Pyke *et al.* 1977; J. R. Krebs 1978; Pyke 1980*a*). It is consequently usually hypothesized that foraging behaviour will be such that the net rate of energy gain is maximized (Pyke *et al.* 1977; J. R. Krebs 1978; Pyke 1980*a*). In all the studies discussed below this hypothesis should be a reasonable one. The animals were observed in contexts that involve foraging but not other activities such as avoiding predators or finding mates. Eventually, of course, more complicated contexts involving interactions between foraging and other activities must be considered (see Chapter 1).

As animals forage they must make three kinds of decision (i.e., choice) in terms of their movements. They must decide

1. When to move or cease moving;
2. Where to move (animals do not move randomly with respect to direction or distance – see Pyke 1978*b* for review);
3. How to move. (Several modes such as walking, running, or hopping may be possible. For any particular mode of movement a range of speeds may also be possible.)

The three decisions – when, where, and how to move – provide a logical set of categories for the following discussion.

The decision-making of foraging animals should be a continuous process. At every instant in time an animal must decide whether or not to change its present behaviour. A moving animal, in particular, must decide whether or not to continue moving or to change direction, mode, or speed of movement. In many cases, however, it is reasonable to view animal decisions as discrete events. Suppose, for example, that an animal is foraging for nectar at flowers, several of which occur on each plant. In this case it seems reasonable to assume that each decision, concerning whether or not to move to another flower on the same plant and to remove nectar from that flower, is a discrete event that occurs just as the animal finishes feeding at each flower (e.g., Pyke 1978*c*, 1980). In such cases an animal is assumed to commit itself to certain behaviours for certain lengths of time. The animal makes a new choice when either the behaviour or the time is completed. Such discretization of animal decision-making should always, however, be done with care.

Animal decision-making can be considered over a range of spatial and temporal scales. For example, for an animal that is foraging in a highly patchy environment, either within-patch or between-patch movements could be examined. Not all scales will be considered below. Instead attention will be focused only on those scales for which sufficient information is available to generate worthwhile discussion.

When to move

Theory

There are two reasons why a foraging animal might move from one place to another. First, as an animal spends more and more time at one place it may deplete the food supply there and suffer a decreasing rate of food (or energy) gain. Eventually the animal would do better in terms of food (or energy) if it left its present location than if it stayed. Second, the future quality in terms of food of an animal's present location may depend in a statistical sense on the past quality there. Hence, based on past experience or sampling, an animal may be able to predict the future rate of food gain in its present location. If the expected rate is sufficiently low the animal should change locations.

Of the two potential reasons for leaving one location and moving to another, depletion of food availability has received more attention. A theory applicable to this situation was derived by Charnov (J. R. Krebs, Ryan, and Charnov 1974; E. L. Charnov 1976). His model visualizes an animal that accumulates energy from patches (i.e., locations) in a continuous manner but at decreasing rate (Fig. 2.1). In addition there is a certain average length of time required for

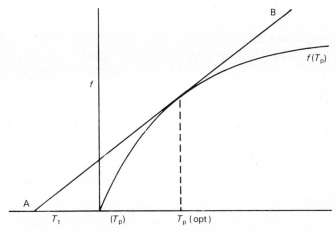

Fig. 2.1. Graphical representation of Charnov's (1976) model of optimal patch use for the special case where all patches are the same and the energetic costs are constant. The curve $f(T_p)$ represents the average energy gain (f) as a function of the time (T_p) spent in each patch. The abscissa is extended in the negative direction to a distance T_t which is the mean time taken to travel between patches. The maximum average rate of energy intake that is obtainable can be determined by constructing the tangent to $f(T_p)$ from point A. The instantaneous rate of energy intake in the patch is then equal to the over-all rate in the habitat. The predator will maximize its average rate of energy intake if it leaves each patch after time T_p (opt). T_p (opt) will decrease as T_t decreases. (Redrawn from Cowie (1977).)

an animal to move from one patch to another (i.e. to change locations). Revisitation of patches is assumed to be of negligible importance. From this model Charnov showed that such an animal will maximize its over-all net rate of energy gain (i.e. forage optimally) if it remains in each patch until its net rate of energy gain in that patch (i.e., its marginal net rate of energy gain) has decreased to the over-all rate in the habitat (see Fig. 2.1). In other words if an animal can do better elsewhere it should leave; otherwise it should stay where it is. E. L. Charnov (1976) termed this result the marginal value theorem. Essentially the same theory was derived by Cook and Hubbard (1977).

Examples

Foraging by great tits The best test so far of Charnov's marginal value theorem has been carried out by Cowie (1977). He studied great tits (*Parus major*) foraging in the lab for mealworms. The mealworms were mixed with sawdust in small containers (patches) placed on artificial trees. (Each patch had the same number of mealworms.) The travel time taken by the birds to move between these patches was varied by adding two kinds of covers to the containers. The covers varied in terms of difficulty of removal. Cowie (1977) determined the mean number of prey caught as a function of the time spent searching in a patch. As expected, the relationship was positive with a decreasing slope. Cowie (1977) was then able to determine the optimal time spent per patch for each of six birds and for both kinds of covers (i.e., for twelve different between-patch travel times). The observed average times spent per patch agreed closely with the optimal times predicted by Charnov's theory (Fig. 2.2).

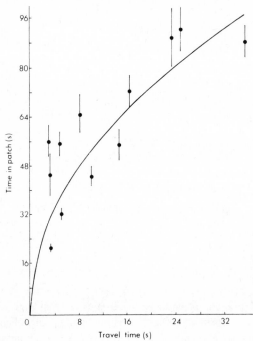

Fig. 2.2. Graphical summary of results of Cowie's (1977) study of great tits. Shown is the relationship between mean travel time and the mean time spent in each patch. The vertical bars represent the standard errors. The curve shows the optimal solution. (Redrawn from Cowie (1977).)

Foraging by bumblebees A similar test is provided by Whitham's (1977) field study of bumblebees. In this case Bumblebees (*Bombus sonorus*) were observed collecting nectar from the flowers of desert willow (*Chilopsis linearis*). Each flower consists of five outwardly radiating grooves which hold nectar by capillary action and a central pool which receives nectar when the grooves are full.

The bumblebees remove the pool nectar at a higher rate than the groove nectar. Not surprisingly they begin with the pool nectar and then feed on the groove nectar. The rates of removal of pool and groove nectar were found to be constant (i.e., independent of the nectar volume present). The relationship between nectar (i.e. energy) gain and time spent in a patch therefore consists of a sequence of linear segments with progressively lower slopes (Fig. 2.3). Consequently, assuming energetic costs are constant, a bumblebee should, according

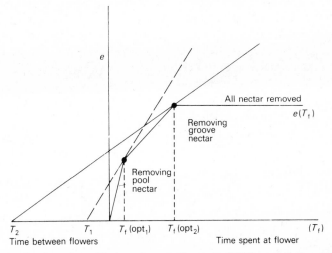

Fig. 2.3. Application of graphical version of Charnov's (1976) model of optimal patch use to Whitham's (1977) study of bumblebees. Tangents to the 'curve' $e(T_f)$ of energy gained from a flower versus time spent at the flower intersect points on the curve corresponding either to the removal of just the pool of nectar or to the removal of all the nectar. The bumblebees should remove just the pool nectar if the between-flower travel time is less than some threshold.

to the marginal value theorem, remove either all the nectar from a flower or just the pool nectar (Fig. 2.3). Allowing for differing costs of feeding and flying the same result holds (Whitham 1977). In fact the bumblebees should remove all the nectar from each flower whenever the average nectar volume per flower is less than 2.0 μl but should only remove the pool nectar if the average volume exceeds this amount (Whitham 1977). Such threshold behaviour was not, however, exhibited by the bumblebees. Instead the average amount of nectar left behind by the bumblebees decreased steadily as the average amount available decreased through time (Fig. 2.4).

The most likely explanation for the discrepancy between observation and prediction in Whitham's (1977) study is that the theory is not completely appropriate. The theory summarized above deals only with a declining rate of energy gain in a patch. It does not consider the possibility of sampling. Just as the nectar volumes in flowers from the one plant should be positively

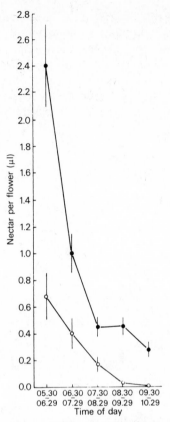

Fig. 2.4. Summary of results of Whitham's (1977) study of bumblebees. The amount of nectar in flowers is plotted as a function of the time of day. Solid circles represent the mean standing crop of nectar in available flowers, and open circles represent the mean amount of nectar remaining immediately after a bumblebee had visited a flower. Vertical bars indicate one standard error (Redrawn from Whitham 1977).

correlated (see below), so too should be the nectar volumes in the five grooves of a desert willow flower. Consequently a bumblebee might 'reject' a flower after obtaining small nectar volumes from some of the grooves. Such sampling and rejection could lead to the observed pattern shown in Fig. 2.4. If the nectar volumes in grooves of the flowers are independent, the bumblebees should remove the nectar from either none or all of the grooves of each flower. Optimal foraging theory typically predicts simple threshold behaviour. When such behaviour is not observed, sampling or the acquisition of information is commonly invoked as an explanation for the discrepancy. Few attempts have been made, however, to evaluate this possibility (see below).

Foraging by chickadees The first test of Charnov's marginal value theorem was carried out by J. R. Krebs *et al.* (1974). Krebs *et al.* (1974) observed chickadees

(*Parus atricapillus*) foraging on artificial pine cones. Each pine cone was a small piece of wood with six holes drilled in it. Single small pieces of mealworm were placed in some of these holes. A sticky label which the birds learned to peel off was placed over each hole. The cones were hung on five artificial trees each of which had three groups of four cones. In this case a patch or location could be a single cone, a group of four cones, or a tree. In Experiment 1 the birds were first given a training period in which each cone contained one larva. Then under the same conditions the behaviour of the birds was recorded (six tests per bird). Finally, three randomly chosen trees were provided with three larvae per cone (high-intensity trees) while the other two remained as low-intensity and the behaviour of the birds was once again recorded (six tests per bird). In Experiment 2 the birds were first tested for two days in a 'poor environment' in which there were five of each of three types of cone group with one, three, and six larvae respectively. Then the birds were tested for two days in a 'rich environment' which had cone groups containing three, six, and twelve larvae. In Experiment 2, the trees were indentical on average.

J. R. Krebs *et al.* (1974) tested the marginal value theorem in a different manner to those described above. They assumed (implicitly) that the chickadees base their decision as to whether or not to leave a patch and move to another on the time since the last prey-capture. This time, they argue, should be inversely proportional to the instantaneous rate of food gain (Krebs *et al.* 1974). It follows from the marginal value theorem that

(a). The birds should have a constant 'giving-up time' for all patch types within a habitat;
(b). Giving-up time should be shorter in habitats where the average capture rate is higher;
(c). The ratio of giving-up time in two habitats should be equal to the ratio of the mean capture rates for the two habitats (Krebs *et al.* 1974).

Giving-up time is the time interval between the last prey-capture and departure from the patch. Krebs *et al.* (1974) tested the first of these predictions using the results of both Experiments 1 and 2. In the case of Experiment 1 they compared the average giving-up times in the high- and low-intensity trees of the second set of tests. As predicted, they found no significant difference. In the case of Experiment 2 they compared the average giving-up times in the three types of cone group. For each of the two environments (i.e., rich and poor) there were no significant differences between the three types of cone group. This was as predicted. J. R. Krebs *et al.* (1974) also tested the second prediction using the results of both experiments. In Experiment 1, they compared the average giving-up time in the first set of tests with that in the second set. In the first set the habitat was poor, the rate of prey capture being 2.8 prey per minute. In the second rich habitat the rate of prey capture was 3.5 prey per minute. As predicted, the average giving-up time was shorter in the richer habitat. In the case of Experiment 2 they found, as predicted, that the average giving-up time was shorter in the rich environment than in the poor environment. Krebs *et al.* (1974) tested the third of the above predictions by combining the results of Experiments 1 and 2 to give four environments. The relative capture rates were, from richest to poorest, 2.04 : 1.51 : 1.34 : 1.00. The relative reciprocals of the giving-up times in the four experiments were 2.07 : 1.42 : 1.37 : 1.00. As predicted, these two relationships are extremely similar.

The study of J. R. Krebs *et al.* (1974) appears at first glance to provide strong support for the optimal foraging hypothesis. There are, however, several problems. The major one is that the observations can be explained without recourse to optimal foraging theory. This theory assumes that an animal 'knows' what sort of patch it is in. If an animal does not know at any time the quality of the patch it is in, its behaviour should be basically the same in all types of patch (within a habitat or environment). If the chickadees of Krebs *et al.* (1974) were unable to tell what sort of patch they were in (i.e. what sort of tree in Experiment 1 and cone group in Experiment 1) their average giving-up times should therefore have been the same for all patch-types within a habitat or environment. Thus optimal foraging is not necessary to generate the first of the above predictions.

The second of the above predictions is more difficult to explain without optimal foraging theory. Suppose, however, that the departure of an animal from a patch is determined by some factor which occurs independently of prey capture. Then the more often prey are encountered (i.e., the richer the habitat), the shorter must be the time between departure and the last prey-capture. The third of the above predictions could be explained in the same manner as the giving-up time would then be inversely proportional to the rate of prey encounter. In the case of the chickadees of Krebs *et al.* (1974) departure might have been caused, for example, by background noise (Cowie, personal communication).

Another problem, as pointed out by J. R. Krebs and Cowie (1976), is that J. R. Krebs *et al.* (1974) did not demonstrate that any depletion of rate of prey capture occurred during the time an animal spent in a patch. In fact 'the tests lasted about 5 min each, a short enough time for depletion effects not to be a major consideration' (Krebs *et al.* 1974, p. 956). Only for the poor habitat in Experiment 1 were many of the prey removed during a test. In this case about 14 out of 60 (23 per cent) were removed during each test. By comparison about 16 out of 132 (12 per cent) were removed during each test in the rich habitat of Experiment 1. Still lower fractions were removed in Experiment 2. It is therefore possible that a theory assuming depletion in a patch through time is inappropriate for the study of Krebs *et al.* (1974), at least for the three richer environments.

In the study of Krebs *et al.* (1974), sampling could have been important in the three richer environments. In these cases it is possible, as mentioned above, that the chickadees could not tell, except possibly by sampling, what type of patch they were in. A sampling scheme suitable for such a situation might consist of a giving-up time although more complicated sampling schemes are also possible (Pyke *et al.* 1977). The employment of a giving-up time would result in more time being spent in better patches and consequently in an increased over-all rate of prey capture. In addition such a giving-up time should clearly decrease as the over-all rate of prey capture (i.e., habitat quality) increases. In other words the same predictions as above might be made if animals employed a giving-up time as part of a sampling regime established to allocate time to patches which are of differing and initially unknown quality.

Although depletion was small in the study of J. R. Krebs *et al.* (1974), it could still have been important. The travel time between patches would, for all habitats, have been the time required to move from one tree or cone group to another. As the distances were of the order of one metre, this time would have been only a second or two on average. This is short compared with the average time to find a single prey ($> 10s$). Consequently little depletion of a patch is necessary for departure to be optimal (see Fig. 2.1). The general problem of separating the effects of depletion and sampling will be a very difficult one to solve (see below).

Foraging by a parasitic wasp Another study of interest was carried out by Cook and Hubbard (1977). They reanalysed data from Hassell (1971), who had studied a parasitic wasp (*Nemeritis canescens*) searching for its host the almond moth (*Ephestia cautella*). The hosts were arranged in a grid of patches of six different densities. The locations of the patch-types were chosen at random. A number of different parasite densities were used. Cook and Hubbard (1977) assumed that the wasps know the quality of each patch and that the number of hosts attacked in a patch is given by the Random Parasite Equation (i.e., $N_{par} = N_i [1 - \exp(-aT_{ip}/(1 + abN_i))]$ where N_{par} is the number of hosts parasitized, N_i the host density in the ith patch-type, T_i the time spent in the ith patch-type, p the parasite density, b the handling time per host, and a the rate of sucessful parasite search). They obtained the constants a and b, the average between-patch travel time, and the total available time from Hassell (1971) and then calculated the optimal amounts of time for each patch-type as a function of parasite density. They assumed that (energetic) costs are constant and their theory is essentially the same as that of Charnov (1976). The predicted and observed time budgets are shown in Fig. 2.5. Agreement between the two is encouraging (Cook and Hubbard 1977).

Cook and Hubbard (1977) did however find one clear discrepancy between observed and predicted behaviour. They predicted that especially at very low parasite densities, some patches should not be visited at all. They found, however, that the wasps always spent some time in each patch type (Fig. 2.5). They

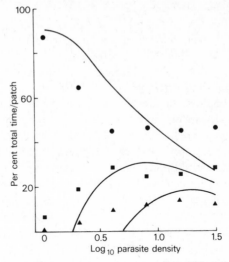

Fig. 2.5. Summary of Cook and Hubbard's (1977) analysis of Hassell's (1971) data on foraging by the parasite wasp *Nemeritis*. The percentage total time spent by the wasps in the different patches is plotted against parasite density. Circles (●), squares (■), and triangles (▲) correspond to patch-types of progressively lower initial prey densities. The lines are, from highest to lowest, the corresponding predicted time budgets. (Redrawn from Cook and Hubbard (1977).)

suggested that this is not really surprising since 'it would be necessary for the parasites to monitor the environment in some way in order to decide what the optimal solution was' (Cook and Hubbard 1977, page 120). Pursuing this suggestion it seems likely that the wasps, at least initially, did not know the quality of each patch. Consequently, they should have spent at least a small amount of time in each patch before restricting their attention to the better patches. This problem should have been relatively short-lived, however, as the wasps were allowed to forage for 24 hours in a small area. Sampling should therefore have played a minor role in the foraging of the wasps and the depletion model should have been a good approximation. The good agreement between observation and prediction supports this idea.

Foraging by parasitic wasp Hubbard and Cook (1978) extended the study of Cook and Hubbard (1977) with experiments of their own with the ichneumon *Nemeritis canescens*. In accordance with optimal foraging theory, they found that the terminal rates of encounter between parasites and healthy prey, calculated for a 10-min period at the end of each visit, were approximately the same for each of five different patch-types. Like J. R. Krebs *et al.* (1974), Hubbard and Cook (1978) assumed that their predator departs from patches as the result of exceeding a giving-up time. Consequently, they predicted that the giving-up times would be the same for each patch-type. They found in fact no significant differences between these giving-up times. As a final test of the theory, they considered the results of varying the total time spent by the parasites in their laboratory environment. For each of four total times they calculated, using the methods of Cook and Hubbard (1977), the optimal allocation of time to each of the five patch-types. They found that the observed changes in time budgets with increases in the total time were similar to and in the same directions as the predicted trends. They also found, however, that especially at low total times, the parasites spent more time than predicted in low-quality patches and conversely less time than predicted in the best patch. As the total time increased, however, agreement between observed and predicted time budgets increased. Hubbard and Cook (1978) attributed the discrepancy between observed and predicted time budgets at low total times to the acquisition by the parasites of information about the qualities of the patches.

Foraging by great tits and ovenbirds J. R. Krebs and Cowie (1976) have applied Cook and Hubbard's (1977) model to the data of J. N. M. Smith and Sweatman (1974) and of Zach and Falls (1976a). J. N. M. Smith and Sweatman (1974) observed foraging by great tits in an aviary. The birds obtained mealworms from six patches, each consisting of a 16 X 16 array of food cups covered with foil caps. Each patch contained a different prey density, created by placing a half mealworm in 0, 4, 8, 12, 16, or 20 cups per patch and leaving the other cups empty. The experimental design of Zach and Falls (1976a) was very similar. They observed ovenbirds foraging for half mealworms in a field arena. In this case the prey were scattered on the ground in four patches of different density (i.e., 2, 4, 8, or 16 prey per patch). In both studies the birds were tested several times without any changes in the locations of each patch-type. In both cases the birds quickly learned, as expected, to allocate more time to better patches. J. R. Krebs and Cowie (1976) compared the behaviour of both groups of birds in later trials with the behaviour predicted by Cook and Hubbard's (1977) model. They found in both cases that the birds spent more time than expected in the poor-quality patches (Fig. 2.6).

In these cases sampling of the sort described above would not seem to provide a convincing explanation for the observed behaviour. The ovenbirds had apparently learned the location of the different patch-types for they made more visits to the better patches (Zach and Falls 1976a). The same was apparently true for the great tits which continued to allocate considerable search effort to the location of the best patch when its density and the density of the worst patch was reversed (J. N. M. Smith and Sweatman 1974). Sampling of another kind, however, could perhaps explain the observed behaviour (Smith and Sweatman 1974). It is probable not only that in the wild these birds encounter food that is patchily distributed but also that the spatial distribution of food changes continuously. In other words a good patch now may be a bad patch later. If such is the case, the birds should constantly sample patches so as to

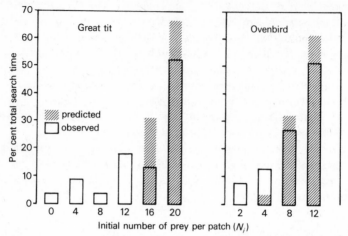

Fig. 2.6. Application by Krebs and Cowie (1976) of Hubbard and Cook's (1977) model to Smith and Sweatman's (1974) study of great tits and Zach and Fall's (1976a) study of ovenbirds. The per cent total search time per patch-type, both observed and theoretical, is plotted against the initial number of prey per patch. (Redrawn from Krebs and Cowie (1976).)

be able to respond appropriately to changes in the food distribution. Thus they might be expected to exhibit such sampling in artificial situations even if these do not change through time. Hence it is important that experiments of animals in artificial situations be run long enough for the behaviour of the animals to have reached an equilibrium. Perhaps in a constant environment the sampling described above would decrease very slowly to a negligible level.

Foraging by ovenbirds The final test of Charnov's marginal value theorem that I shall discuss was carried out by Zach and Falls (1976b). They observed ovenbirds foraging in a laboratory arena for freeze-killed adult flies. The arena contained six patches which were always identical in terms of the initial number of prey. Two experiments were run. In each of these the birds were deprived of food for 30 min and then each bird was allowed to forage in the arena until all patches had been visited or until searching ceased. In Experiment 1 the birds were exposed to patches with four prey each on five occasions and then once to patches with two prey each. The behaviour of the birds was observed on the fifth (E 1/1) and sixth (E 1/2) exposure. In Experiment 2 the same procedure was carried out except the birds (a different group) were exposed five times to two-prey patches and then once to four-prey patches. Zach and Falls (1976b) measured the giving-up times per patch for the four test runs. They found no significant differences between the average giving-up times. That there is no difference between giving-up times between the two tests in each experiment is, they point out, to be expected. At the times of the sixth exposure in each experiment (E 2/1 and E 2/2) the birds had experienced only the conditions of the first five exposures. As their experience with the conditions of the sixth exposure was very limited, little or no change in the birds' rules of behaviour should therefore have occured. Zach and Falls (1976*b*) argue, on the other hand, that according to optimal foraging theory, the average giving-up time should have been shorter during the fifth exposure of Experiment 1 (E 1/1) than during the fifth exposure of Experiment 2 (E 2/1), the over-all search-time per prey being lower in Experiment 1 (see Fig. 2.1). This is not necessarily so, (see below).

The basic problem with the argument of Zach and Fall (1976*b*) is that it assumes that the animals' rule of departure from a patch consists of a giving-up time. This assumption, as pointed out above, has also been made by several other authors (J. R. Krebs *et al.* 1974; J. R. Krebs and Cowie 1976; Hubbard and Cook 1978; N. B. Davies 1977; E. L. Charnov 1976). In the present case it seems particularly doubtful. Zach and Falls (1976*b*) found that, in all of the test exposures, the ovenbirds found 86–90 per cent of the prey in each patch. They also found that the ovenbirds systematically search the patches. The ovenbirds should not therefore have experienced a steady decline in their rate of prey encounter within a patch. Instead the rate would probably have been almost constant until the search of each patch was completed. The search, though systematic, could easily have missed about 10 per cent of the prey in a patch. It seems likely therefore that the departure rule of the ovenbirds would have involved the fraction of the patch so far searched or the time so far spent in the patch. It could also have involved the number of prey so far found and the time since the last prey capture (i.e., giving-up time). Zach and Falls (1976*b*) found in fact that if the birds found prey in a given patch quickly they subsequently tended to search longer in that patch. Since all the above factors could be involved it is also possible that the departure rule could change between habitats while leaving the *average* giving-up time unaltered. More data are necessary to settle the issue.

Foraging by spotted flycatchers By comparison with the attention given to resource depletion as the factor promoting departure from a patch (see above plus G. A. Parker and Stuart 1976), very few studies of sampling have been carried out (N. B. Davies 1977; Pyke 1978*c*, 1980). Davies (1977) observed spotted flycatchers (*Muscicapa striata*) hawking insects in a garden. The birds had several regularly-used perches at which they employed a 'sit-and-wait' foraging strategy. After capturing a prey they often but not always returned to the same perch as before. In addition they sometimes flew directly from one perch to another. Davies (1977) found that there was no significant change in the time interval between successive captures made from the same perch. Thus the flycatchers were not depleting their prey. Davies (1977) suggested that the insect prey of the flycatchers probably come along in batches and that the birds employ a rule of patch departure that is adapted to dealing with this patchy situation. Whenever the prey density near a perch is too low, a bird should move from that perch to another. This idea is supported by the observation that the longer the waiting time between prey capture and the next the less likely is a bird to return to the same perch after the last capture (N. B. Davies 1977). A long waiting time probably indicates a low prey density.

Davies (1977) attempted to develop and test a more precise model of this sampling regime. Here he restricted his attention to perch departures that occur at times other than at the time of a prey capture. He assumed, like so many other authors (see above), that the departure rule of the flycatchers consists of a giving-up time. If no prey is captured during this amount of time, a bird is assumed to leave its present perch for another. The importance of waiting time suggests, however, that the departure rule of the birds might also involve the time between the last and second last prey capture and possibly even previous capture–capture intervals. Davies (1977) then carried out the following computer simulation. He first calculated the frequency distribution of capture–capture intervals. He then presented prey to a computer flycatcher at intervals chosen at random from this distribution. The probability of the bird leaving its perch at the time of a prey capture was assumed to depend in a constant fashion on the waiting time between the last and second last captures (see above). Handling time per prey capture was estimated to be 4 s and the between-perch travel time was estimated to be 5 s. The number of captures in a 30-min foraging bout was determined for various values of giving-up time. The number of captures was found to increase with increases in giving-up time, possibly being close to an asymptote for giving-up times in excess of about 20 s (Fig. 2.7).

Davies (1977) concluded from this that 'the giving-up time of 29 sec chosen

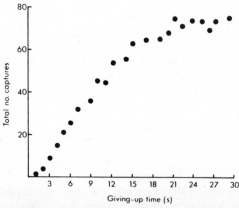

Fig. 2.7. Results of Davies's (1977) computer simulation of a foraging flycatcher. The total number of prey captured in a 30-min foraging bout is plotted against patch giving-up-time. (Redrawn from Davies (1977).)

by the flycatchers in the wild is just long enough to maximize the number of prey caught'. Such a conclusion seems premature, however, for the following reasons. First Davies' reasoning should apply equally to any times greater than about 20 s (Fig. 2.7). Second, and more importantly, Davies' simulation was not a completely realistic representation of the flycatchers' world. He assumed that successive capture–capture intervals are independent. Consequently, in his model, a bird would be unable to estimate future capture success on the basis of past experience. It follows, that the optimal solution in Davies' model is an infinite giving-up time (see Fig. 2.7). In other words a bird should never have any reason to leave a perch. A more realistic simulation would include some sort of correlation betweem successive capture–capture interals. There should then be a finite optimal giving-up time.

Foraging by hummingbirds In an earlier study of mine (Pyke 1978c), I considered both depletion and sampling as reasons for departure from one location and movement to another (see Pyke 1980 for a detailed critical discussion). The locations in this study were inflorescences of scarlet gilia (*Ipomopsis aggregata*). The animal subjects were hummingbirds (*Selasphorus platycercus* and *S. rufus*) which foraged for nectar at the gilia flowers. Each inflorescence had several flowers and so the hummingbirds must have decided after probing each flower whether or not to leave the present inflorescence and fly to another. I found evidence that this decision depended on the number of flowers available on an inflorescence, the number of flowers already visited, and the amount of nectar obtained at the last flower. In other words, the departure rule of the hummingbirds appeared to be much more complicated than a simple giving-up time. That the amount of nectar obtained at the last flower should influence the hummingbirds is expected since a positive correlation was found to exist between any two flowers from the same inflorescence (Fig. 2.8). The humming-

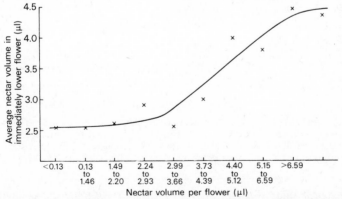

Fig. 2.8. Plotted against the nectar volume in a flower of scarlet gilia (grouped into intervals) is the average nectar in the immediately lower flower. The line is fitted to these data by eye. This relationship was found to be independent of the relative positions of the two flowers. (From Pyke (1978c).)

birds should therefore be more likely to leave an inflorescence with decreases in the nectar volume obtained at the last flower. This would be the sampling part of the birds' foraging strategy. Depletion of the nectar resource also occurs. As either the number of flowers already visited increases or the number of flowers available decreases, the probability that the next flower is a revisit increases (Fig. 2.9). Consequently, the likelihood of departure from an inflorescence should increase with increases in the number of flowers already visited and with decreases in the number of flowers available. It is possible that the departure rule of the hummingbirds could involve still more factors than the three mentioned above. I assumed, however, that it did not. The next step was to determine the optimal departure rule.

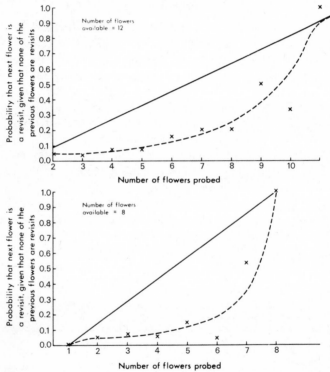

Fig. 2.9. The probability that the next scarlet gilia flower probed by a hummingbird is a revisit is plotted against the number of flowers already probed for inflorescences with 8 and 12 flowers available. The dashed lines are fitted by eye. The solid lines show the re-visitation probabilities if the birds chose each flower at random. (From Pyke (1978c).)

The marginal value theorem of E. L. Charnov (1976) applies, as mentioned above, only to situations where there is depletion of the rate of food or energy gain as time is spent in a patch. It is not applicable when sampling is also occurring. The theorem was also derived for a continuous and deterministic situation and not for one that is discrete and stochastic. I therefore proposed (without proof) the following discrete, stochastic analogue of Charnov's marginal value theorem. The optimal departure rule (i.e., that which maximizes the net rate of energy gain) of the hummingbirds is the rule such that

1. Given a bird chooses to remain on its present inflorescence, the ratio of the expected net energy gain at the next flower to the expected time required to visit that flower is equal to the over-all net rate of energy gain in the habitat;
2. Given a bird chooses to leave its present inflorescence and fly to another, this ratio is less than the over-all net rate of energy gain.

Originally this theorem seemed likely to be a good approximation to the foraging problem. It now seems less so (see below).

I then used the above theorem to derive optimal nectar thresholds. If the amount of nectar obtained at the last flower is greater (less) than the appropriate threshold, the bird should remain at (leave) the present inflorescence. As one would expect, because of flower revisitation, these thresholds increase with increases in the number of flowers already visited and with decreases in the number of flowers available (Table 2.1).

TABLE 2.1. *Optimal nectar thresholds for different numbers of flowers probed on an inflorescence and for two different numbers of flowers available per inflorescence. (Adapted from Pyke 1978c)*

Number of flowers available per inflor- escence	Optimal nectar thresholds (μl)									
	Number of flowers so far probed per inflorescence									
	1	2	3	4	5	6	7	8	9	10
8	-0.00	0.35	0.57	1.00	1.80	2.92	∞			
12	-0.00	0.25	0.43	0.66	1.22	2.15	2.83	4.75	∞	

Finally, to test the theory, I determined what the frequency distributions of numbers of flowers probed per inflorescence for two inflorescence sizes, namely eight flowers and twelve flowers, would be if the birds employed the optimal nectar thresholds (Table 2.1). These optimal frequency distributions were then compared to the observed behaviour of the hummingbirds. Agreement between the two was good both in terms of the shapes of the distributions and their means (Fig. 2.10). In the light of more recent work, however, it seems likely that such agreement was coincidental (see below).

Foraging by bumblebees In two recent field studies I have found clear indication that my discrete stochastic analogue of Charnov's marginal value theorem

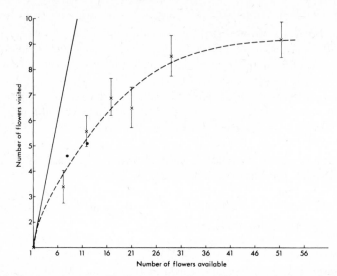

Fig. 2.10. Number of scarlet gilia flowers visited per inflorescence by hummingbirds versus number of flowers available on each inflorescence. The crosses are observed points, the dots are theoretical. The vertical bars represent the standard errors. The dashed line is fitted by eye to the observed data points. The continuous line is the line for equal numbers of flowers visited and flowers available. (From Pyke (1978*c*).)

(Pyke 1978*c*) is at times not even a good approximation to the optimal departure rule (Pyke 1982). In the first of these studies I observed bumblebee workers (*Bombus appositus* and *B. flavifrons*) foraging for nectar from flowers of monkshood (*Aconitum columbianum*). The flowers occurred in varying numbers on vertical inflorescences. I found, just as in the previous study, that there was a positive correlation between the nectar volumes of flowers on the same inflorescence. Hence sampling should have been a part of the bumblebees' departure rule. If, on the other hand, we focus on the behaviour of the bumblebees after the first flower they probe on each inflorescence, depletion should not have been a factor affecting departure. For the second flower is never a revisit and was found to have the same distribution of nectar volumes as the first flower. In addition I found that there was no relationship between the frequency distribution of nectar volumes per flower and the number of flowers available on inflorescence (Pyke 1982). Consequently, according to my analogue of the marginal value theorem (see above), the departure rule of the bumblebees should have been independent of the number of flowers available. I found, however, that there was a strong tendency for the likelihood of departure from an inflorescence to decrease with increases in the number of flowers available (Fig 2.11).

Foraging by honeyeaters (Pyke 1981) In the second of the two studies mentioned above I observed honeyeaters (*Phylidonyris novaehollandiae* and *P. nigra*)

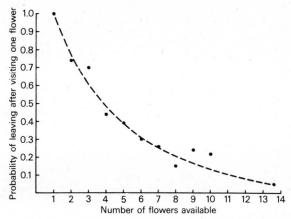

Fig. 2.11. The probability of a bumblebee leaving an *Aconitum* inflorescence after probing just one flower versus the number of flowers available on the inflorescence.

foraging for nectar amongst inflorescences of *Lambertia formosa*. In this case there are almost invariably seven flowers per inflorescence, arranged in a circle of six plus one in the centre. Once again I found a positive correlation between the nectar volumes of flowers on the same inflorescence. In addition, laboratory data from another honeyeater species (*Acanthorhynchus tenuirostris*) indicate that virtually no flower revisitation occurs up to about the fifth flower probe. After that the probability of a revisit begins to increase markedly. It follows that up to about the fifth flower probed sampling but not depetion should have determined the departure rule of the birds. Furthermore, since there is no spatial pattern to the flower nectar volumes in each inflorescence, the departure rule should, by my discrete, stochastic analogue of the marginal value theorem, be independent of the number of flowers already probed up to about the fifth flower. The likelihood of departure varied significantly, however, as a function of the number of flowers already visited (Fig 2.12). The departure rule must have varied at the same time.

To determine the exact optimal departure rule for this foraging, I have commenced some computer simulations of the problem. In these simulations a bird is assumed to forage at each of 274 inflorescences. The nectar volumes encountered by the bird are those found in a field sample of this number of *Lambertia* inflorescnces. The bird is assumed to take certain amounts of time to probe each flower and to move between inflorescences. These times were estimated from field observations. The bird is also assumed to make departure decisions after probing each flower. These decisions are assumed to depend on the average nectar volumes so far obtained from flowers on the present inflorescence and on the number of flowers already probed on the inflorescence.

One important result has so far been produced by these simulations. If there is no revisitation at all up to the seventh flower probed, the optimal departure

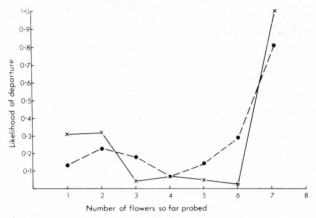

Fig. 2.12. Observed (X) and theoretical (●) likelihoods of departure of a honeyeater from a *Lambertia* inflorescence versus the number of flowers so far probed.

rule is such that the likelihood of departure should vary with the number of flowers already probed in a manner similar to that observed (Fig 2.12). The optimal departure rule, allowing for the observed extent of flower revisitation, has also been determined. The optimal and observed likelihoods of departure are similar but not identical (Pyke 1981).

General conclusions

The above discussion leads to several conclusions. First, while there is much qualitative support for the hypothesis that foraging animals choose when to move in ways that maximize their net rate of energy gain, there is very little support of a precise, quantitative nature. Only the study of Cowie (1977) survived unblemished the admittedly pedantic analysis above. There is, however, an important aspect of the foraging process that Cowie does not consider. He does not deal at all with the actual departure rule of the tits, only its consequences in terms of time spent in each patch. Second, very little attention has been paid to the actual departure rules employed by animals. Instead most authors have assumed that their animal subjects use very simple departure rules such as a giving-up time. In order to develop an appropriate theory for this aspect of foraging it will be necessary to incorporate realistic assumptions concerning departure rules. Much more exploration of these rules are therefore necessary. Finally, existing theory concerning when foraging animals should change locations is inappropriate in most situations. Charnov's marginal value theorem is appropriate only when depletion occurs but not when sampling also occurs. In addition, my discrete stochastic analogue of this theorem is sometimes a poor approximation to the optimal departure rule. Realistic computer simulations seem likely to provide useful solutions to these problems.

Where to move

Foraging animals should use information not only in deciding when to change locations but also in choosing where to move. In the latter case three kinds of information are potentially important. Information about the spatial and temporal distribution of food clearly should be important. Information about the past patch of the animal and about past food encounters would also be important. From such information an animal could estimate aspects of future foraging. In addition, if the rate of food renewal is sufficiently low, it may be advantageous for an animal to avoid recrossing its previous path. Finally, an animal might make use of information concerning potential future locations. For example, a nectar-feeding bee may be able to see a number of plants it could visit next and to judge aspects of their quality such as number of flowers and distance. Alternatively, an animal might possess such information from earlier experience in the same area.

The use by foraging animals of information about the spatial and temporal distribution of food has received scant attention. Many authors have observed animals increasing the amount of turning in their paths following encounter with prey (see Pyke *et al.* 1977; Pyke 1978*b* for a review). Some of these authors have suggested that such behaviour represents an adaptation for exploiting a patchily-distributed food resource (e.g. Laing 1937, 1938; Flanders 1947; Dixon 1959; Kaddou 1960; Mitchell 1963). No one, however, has incorporated a patchy food-distribution into a foraging model and determined the optimal response to food encounter in terms of movement behaviour. Instead all optimal foraging models that have considered where animals move have assumed either a uniform food distribution or that movements are independent of food encounter (Cody 1971, 1974; Pyke 1978*a,b*).

In terms of use of information by foraging animals about possible locations, studies have so far focused on two extreme situations. Two studies assumed that such information had no effect on where animals move (Cody 1971, 1974; Pyke 1978*b*). These studies both visualized an animal moving about a bounded grid of regularly arranged points. Food was assumed to occur at these points and to be non-renewable. Movements were assumed to occur only between a point and its four equally closest neighbours. The direction of each movement was assumed to depend on the direction of the previous movement. In both studies computer simulations were used to determine the relationship between direction of successive movements that resulted in the maximum rate of food gain.

The two studies came to quite different conclusions as a result essentially of different assumptions concerning movement behaviour at the boundaries of the grid. Cody (1971, 1974) assumed a reflecting boundary with movements to a boundary followed by about-turns. He found good agreement between predicted movement behaviour and that of finch flocks. I argued that this assumption is unrealistic and incorporated in my simulations boundary behaviour observed in hummingbirds foraging at a grid of artificial flowers (Pyke 1978*b*.) I also explored the consequences of varying the size of the grid and the length of the foraging bout. As a measure of the relationship between directions of successive movements I used D. A. Levin, Kerster, and Niedzlek's (1971) directionality. It provides an estimate of the ongoing tendency of an animal's movements, being 1 when movement is completely linear and 0 when movement is random (i.e., all directions of each movement equally likely). I found that over the range of grid areas and foraging bout lengths the optimal directionality was always between 0.8 and 1.0. A survey of the literature, however, yielded observed directionalities between 0.0 and 0.6 (Pyke 1978*b*). I concluded that it is unrealistic to ignore use by foraging animals of information about future possible locations. Such information could be obtained through sight, sound, odour, or memory. I also suggested that Cody's (1971, 1974) agreement between observed and predicted could have been coincidental.

At the other end of the spectrum in terms of use of information about future locations is another study of mine (Pyke 1978*a*). In this study I developed a model of an animal moving between randomly scattered resource points. I assumed that from any resource point the animal could see many others and could tell their distance and direction. In terms of use of information about past movements, I considered several different memory levels. In the simplest case an animal was assumed to remember only its direction of arrival at the present resource point. In the most complex case an animal was assumed to also remember the change in direction at the previous resource point and the amount of food obtained at the present point. Finally I assumed that an animal chooses the next resource point in the following manner (see Fig 2.13). It aims its departure in some direction relative to the direction of arrival at the present resource point. It then scans visually a sector of angular width *w* on either side of the aimed departure direction. Finally, it chooses the closest resource point within this 'scanning sector'. The aim of the model was to determine the optimal relationship between the arrival and aimed departure directions and the optimal width of the scanning sector.

Several qualitative predictions emerged from the model. First, whatever the memory capabilities of an animal, the arrival and aimed departure directions should be very similar. Consequently there should be a strong positive correlation between the directions of successive movements. Second, if the animal knows the amount of food it obtained at its present resource point, then it should increase the width of the scanning sector employed at that point with increases in the amount of food. The correlation between arrival and departure directions should therefore decrease with increases in the amount of food obtained in between. At the same time the average distance moved to the next resource point should decrease. Finally, if the animal is also able to remember the change in direction at the previous resource point, then it should

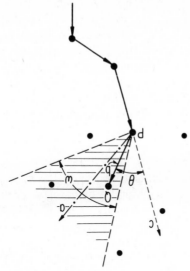

Fig. 2.13. After reaching the resource point P, the animal aims its direction of departure from P in the direction 'a', then visually scans the shaded sector of angular width ω and finally moves to the closest resource point within this sector. The direction of arrival at P is indicated by 'c', the direction of actual departure from P is indicated by 'b', and θ is the actual change in direction at the resource point P. (From Pyke (1978*a*).)

tend to alternate left and right turns.

I compared each of these predictions with observations made on foraging bumblebees (Pyke 1978*a*). In this case the resource points were inflorescences, either of *Delphinium nelsoni* or *Aconitum columbianum*. The behaviour of the bumblebees (*Bombus flavifrons* and *B. appositus*) was consistent with all of the above predictions. Thus the bumblebees are apparently able to remember all of the potential information considered above. It is possible that they make use of even more information.

There are three problems with the above model. First, it does not lead to precise quantitative predictions but only to qualitative ones. This problem appears to arise from the analytic nature of the model and will be remedied only when it is replaced by appropriate computer simulations. Another problem with the model is that it omits a factor of much importance in the foraging of many animals. It does not consider the possibility that an animal may be able to tell not only the distances and directions of possible future resource points but also some aspect of their qualities in terms of food yield. Nectar-feeding animals, for example, can apparently judge the number of flowers on nearby inflorescences or plants, for they tend to move to those with many flowers (Pyke 1978*c*, 1980). A final problem is that the model makes some assumptions about the animal's movement rule that may not be correct. The model assumes that the animal always chooses the closest resource point within a scanning sector. It is possible, however, that the directions of possible points affect the animal's choice in a more continuous manner. It is also possible that the animal does not always choose the closest resource point (of a given quality). Suppose, for example, that the animal does not know its direction of arrival at the present resource point and that, consequently, it must choose the next point solely on the basis of distance (assuming all resource points of equal quality). Then, if it always chose the closest resource point, it would eventually spend all its time moving between the same two points. The optimal strategy in this situation might be to choose the closest, second closest, etc. with different probabilities. This might also be the case in some situations if the animal can remember its arrival directions. Once again, a computer simulation seems neccessary to determine the optimal strategy in this case. It is also possible, as implied above, that an animal's choice of the next resource point is determined not just by the absolute properties of the possible points but also, or instead, by the properties of each relative to the properties of other points.

To develop realistic models of where foraging animals move it is therefore necessary to know much more about the nature of movement choices made by animals. This can be achieved only by following animals, mapping their movements, and also mapping the spatial locations and qualities of all possible resource points that the animal could have chosen. Only when we know approximately what information an animal uses in making its foraging decisions can we ask what the optimal use of this information is.

How to move

A foraging animal that is in the process of moving may do so using one of several modes of movement, such as hopping, flying, etc. and may do so at any of a range of possible speeds. Mode of movement of foraging animals has, however, received no attention so far. Perhaps this is because the situation is straightforward and trivial in most cases? It seems hardly surprising, for example, that a bird or a bee will fly from one inflorescence to another rather than hop or crawl along connecting stems or earth. Though either of the latter modes of movement would be energetically less expensive than flying the time required to travel the distance would probably be considerably lower if the animal flew. The speed of movement of foraging animals has, on the other hand, received a little attention (Ware 1975).

Ware (1975) reanalysed laboratory data collected by Ivlev (1960) on the foraging swimming speed of bleak, a planktivorous fish. He calculated the relationship between swimming speed and rate of food (energy) gain. He then combined this with the relationship, determined by Ivlev (1960), between swimming speed and rate of energy expenditure, to yield the relationship between swimming speed and net rate of energy gain. He found that the observed average swimming speed of the bleak (107 m/h) was very close to the swimming speed that maximizes the net rate of energy gain (i.e., optimal speed − 111m/h).

Most foraging situations are likely to be more complicated than that considered by Ivlev (1960) and Ware (1975). The fish observed by Ivlev (1960) moved continuously while foraging. Most animals, however, stop and start fairly frequently. Nectar-feeding birds and bees, for example, spend time at inflorescences or plants between flights from one to another. Most movements will therefore involve acceleration and deceleration rather than a constant speed. Hence it may be insufficient in general to predict and measure *average* movement speeds of foraging animals.

Conclusion

While the optimal foraging approach to movements of foraging animals seems to be a profitable one, it can hardly be said that there is a wealth of precise evidence in favour of it. Much more work is clearly needed in all areas of the development and testing of optimal foraging theory. This work should start with the assumptions that must be made. Any models must include realistic assumptions about information use and behavioural limitations or constraints. The theory is also in need of much attention. Simple models of limited applicability and realism must be replaced by more complicated and realistic models. Only when sufficient time and effort are devoted to dealing with these difficulties will the true usefulness of optimal foraging theory be clear. I do not expect this theory to emerge as a panacea for explaining animal foraging behaviour. I expect, however, that it will continue to provide much insight into such behaviour in many situations.

3
Vertebrate home-range size and energetic requirements

GEORGINA M. MACE, PAUL H. HARVEY and T.H. CLUTTON-BROCK

Introduction

Within a population, individual animals use distinctive areas, called home ranges or territories, in pursuit of their routine activities. Both the nature of these areas and the pattern of dispersion of the population vary widely between different species, but each is determined primarily by the underlying resource distribution (J. L. Brown 1975). In this chapter we investigate some of the factors which determine the sizes of home ranges or territories over which individuals move during everyday activities.

The area that an individual occupies, exclusive of large-scale migrations or uncharacteristic erratic wanderings, is generally referred to as its home range (Burt 1943; Jewell 1966). This is defined and measured without reference to other animals, or any particular kinds of display or aggressive behaviours. In contrast, territories are non-overlapping areas from which other individuals are actively excluded and the area becomes available for the sole use of an individual or group.

The degree of territoriality varies enormously between species. In some there is almost complete overlap of home ranges (e.g. Steller's jay, J. L. Brown 1963) while in others defence of areas and exclusion of conspecifics is common. Although territoriality may be limited to a particular part of the home range, such as a mating area, feeding area, or nest (e.g. coati, Kaufmann 1962; antbird, Willis 1967, 1968) it can also encompass the entire home range (e.g. typical songbird territories such as those of the song sparrow, Nice 1937).

Selective forces favouring the evolution of territoriality have been discussed extensively. Wynne-Edwards (1962) considered territorial behaviour to be a consequence of interpopulation selection: populations that did not over exploit their resources would be favoured over those that did. However, subsequent theoretical studies have shown that individual selection, which can favour immediate reproductive success at the cost of efficient resource exploitation, is a much stronger force than interpopulation selection (see Wiens 1966; Maynard Smith 1976a). Most recent studies view the evolution of territoriality as a result of individuals actively defending food and other resources needed for survival and reproduction (Lack 1954, 1966; Hinde 1956; J. L. Brown 1964, 1975).

Defence of an area clearly involves some cost in time and energy, and for territoriality to have evolved there must be compensatory advantages to a territory holder so that its inclusive fitness is increased. Such advantages may result from exclusive access to food resources whose distribution the animal knows and can therefore exploit prudently and economically, access to mates or nest sites, or benefits from reduced rates of predation, parasites, or disease which may result directly from increased inter-individual distances (J. L. Brown and Orians 1970; Waser and Wiley 1979). Each of these, including an area free from disease or parasites, may be viewed as a resource that the territory holder is defending. Some species will have access to defensible resources while others will not (J. L. Brown 1964) and it is the defensibility of a resource (largely a function of its spatial or temporal stability) together with its value to the individual, offset against the costs of defence, that will determine the extent to which the species is territorial. For example, aerial food resources are generally not defensible while many species with more spatially and temporally predictable food resources, such as seeds, can and do defend territories (see Lack 1966; J. L. Brown 1975). Within species, home ranges may overlap more extensively in areas where resources are more abundant and where the net benefits of territory defence are reduced (e.g. bears, Jonkel and McCowan 1971).

What then are the factors that determine the size of individual's home ranges or territories? An upper limit in the case of territory holders must be set by the size of an area that the individuals can effectively defend (Crook 1970, 1972). Among non-territorial species, some parts of a home range are rarely visited and the upper limits in this case are probably set by disadvantages the animal may experience if it frequents areas with which it is unfamiliar. In either case the area is unlikely to exceed the minimum space necessary to provide key resources (but see Knowlton and Parker 1979).

Although the nature and distribution of these resources may determine territory or home-range size, the key resources may differ between segments of a population. For example, in small rodents, female home ranges tend to be determined by food, shelter, and water (L. E. Brown 1966) while males are subject to sexual selection for increased access to females and their home ranges are almost invariably larger since they must encompass those of several females (Eisenberg 1966). Home-range size can also vary with age of individuals (Ralph and Pearson 1971; Fitch and Shirer 1970) and with season (L. E. Brown 1966).

Among adult female vertebrates, however, the primary determinant of home-range or territory size seems to be access to food (Belovsky and Slade in preparation). If this is so, the animals' metabolic requirements, food quality, and food density should be the major factors determining home-range or territory size. Since McNab (1963) first demonstrated a relationship between home-range size, body weight, and diet across a sample of small mammal species, numerous studies have investigated the form and validity of similar relationships in other taxonomic groups.

In this chapter we have three aims. First, we review these studies and show

how their diverse conclusions are often attributable to different methods being employed on different taxa. Second, we undertake a systematic consideration of the associations between body size, home-range size, and metabolic requirements in natural populations of vertebrates. Finally, we make a preliminary study of data collected from one group (the primates) in order to test predictive models which make explicit assumptions about the relationships between metabolic needs, home-range size, and diet. Other aspects of resource distribution among individuals have been reviewed recently (optimal foraging, J. R. Krebs 1978; Pyke *et al.* 1977; behavioural mechanisms, Waser and Wiley 1979; establishment of ranging areas, R. R. Baker 1978), or are discussed elsewhere in this volume. (see Chapters 1 and 2).

Review and critique

McNab's small-mammal study demonstrated that home-range size (*HR*) was related to body weight (*W*) by a function of the form

$$HR = a \cdot W^b$$

On a double logarithmic plot *b* measures the slope of the line of best fit and *a* the elevation, or the value of *HR* when *W* is unity. McNab estimated the value of *b* as 0.63 and, since this did not differ significantly from the value of 0.75 which relates basal metabolic rate to body weight (Kleiber 1961), he suggested that home-range sizes were determined by the animals' metabolic needs. McNab (1963) also showed that 'hunters' used larger home ranges for their body weights than 'croppers' and suggested that this was a result of differences in the relative densities of the two kinds of food.

Subsequently, Armstrong (1965) and Schoener (1968) reported similar studies across samples of bird species. They established slopes of 1.23 and 1.16 respectively, both considerably greater than that obtained by McNab. The exponent relating basal metabolic rate to body weight in birds is between 0.65 and 0.75 (J. R. King and Farner 1961; Lasiewcki and Dawson 1967), which is similar to that for mammals. Armed with this information, Armstrong (1965) concluded that avian home ranges generally exceed energetic requirements, all the more so in larger species, and that relationships between energetic needs and home-range size differ fundamentally between birds and mammals.

Schoener (1968), however, divided his species into feeding categories according to the proportion of animal food in the diet of adult birds. In agreement with McNab's mammals, he found that territory size for a given body weight was highest for species feeding on the least abundant food items (carnivores) and lowest for those with more abundant food (herbivores). Interestingly, when relationships were established within feeding categories, the slopes were more similar to those found by McNab (1963); omnivores had a slope of 0.51 and herbivores (with data from only three species) had a slope of 0.70. The slope through predators (1.39), however, was considerably higher than that for any other dietetic group. Schoener suggested that this was a consequence of relatively high hunting costs associated with animal prey, which result in the density of economically acceptable prey items per unit area decreasing as body size increases. Reconciling this with McNab's (1963) lower slopes within dietetic categories, Schoener (1968) calculated a slope of 1.41 through nine species of terrestrial predatory mammals.

In order to re-examine relationships between size, metabolic needs, and home-range size, Turner, Jeinrich, and Weintraub (1969) undertook a careful study of relationships across thirteen species of terrestrial lizards. Because of difficulties in comparing home-range sizes calculated by different methods, and because of uncertainty about the way in which the energy expenditure of a wild animal relates to basal metabolism measured in the laboratory, they doubted the accuracy of previous studies. Using carefully controlled measures of home-range size, they found a slope of 0.95 across lizard species, which differs significantly from a slope of 0.62 which relates lizard metabolic rates to body size (Bartholomew and Tucker 1964).

More recent studies have either demonstrated or confirmed that home-range size is related to body size and diet in primates (Milton and May 1976; Clutton-Brock and Harvey 1977a, b), North American mammals (Harestad and Bunnell 1979), and a variety of other vertebrate and invertebrate taxa (Belovsky and Slade, in preparation). In spite of the abundant information, no conclusion has yet been reached on McNab's (1963) prediction that energetic requirements will determine home-range size. Indeed, in their consideration of this problem, many studies are open to criticism based on both methodology and data interpretation.

First, most discussion has centred around the numerical values of exponents relating either home-range size or metabolic rate to body weight. These exponents describe the slope of best-fit lines (on logarithmically transformed data) which are generally set by regression analysis. Regression is inappropriate for this kind of data because it assumes that the values on the abscissa (body-weight axis) are measured without error. The effect of errors in this measurement is to reduce the value of the slope to an extent related to the variance of those errors (see Harvey and Mace 1982). A more suitable method for establishing the best-fit line is the reduced major axis or (when the data are logarithmically transformed) the major axis technique (Kermack and Haldane 1950; Harvey and Mace 1982).

A second difficulty is non-independence of data points. Most studies use species as independent points although some have assigned equal weighting to subspecific or population measures (McNab 1963; Turner *et al.* 1969). Large groups of congeneric species which have similar diets, body size, and home-range size can strongly bias the slope of the line. Nested analysis of variance was used by Clutton-Brock and Harvey (1977a) (see also Harvey and Mace 1982) to identify the correct level at which to perform an analysis. In that case, genera were found to be valid as independent points.

Care must also be taken over the composition of the entire sample. Part of the reason that both Armstrong (1965) and Schoener (1968) found such high slopes for birds (1.23 and 1.16 respectively) was that their samples comprised a group from the order Passeriformes – all small omnivores or invertebrate predators, and another group from the Falconiformes – all large predators specializing on vertebrate prey. Since the two groups are so widely separated on the body-weight axis, the slope of a line through all predators, or through all birds, largely measures differences in the diets of the two groups rather than saying anything about changes in home-range size with weight.

Finally, although the purpose of many of these studies was to disentangle the roles of metabolism and body weight as determinants of home-range size, few have correctly addressed this problem. As several authors have pointed out (Turner *et al.* 1969; Martin 1981) the correct way to do this is to use measures of daily energy expenditure, rather than laboratory estimates of basal metabolic rate which are calculated on resting and postabsorptive animals. Harestad and Bunnell (1979) state that 'increases in metabolic rate with activity will appear in the proportionality quotient and not the exponent'. It is not clear that this will necessarily be the case since most studies of the energetic costs of various forms of activity have shown that they do not increase isometrically with body weight or with basal metabolic rate (C. R. Taylor, Schmidt-Nielsen, and Raab 1970; Schmidt-Neilsen 1972). In nearly all cases relative costs (i.e. energy used per unit body weight) are greater for smaller animals. Direct measurements of energetic costs incurred by active animals in their natural habitats are ideally needed, but such measures are unfortunately hard to obtain for most vertebrate groups.

A related criticism of studies using estimated weights of groups of animals occupying a common home range (e.g. Clutton-Brock and Harvey 1977*a,b*) was made by Martin (in press): a group consisting of many smaller animals has a larger metabolic requirement than one with a similar group weight consisting of fewer, larger individuals.

In the remainder of this chapter we will re-examine the relationship between body size, metabolism, and home-range size, taking into account the criticisms outlined above. Since this method is largely investigative and descriptive, and therefore open to the criticism that it does not actually test any hypothesis, we will give an account of a predictive study involving average daily metabolic needs and ranging patterns in primates.

An investigative approach – home range and metabolic needs across three orders of vertebrates

Introduction

This study attempts to answer three questions

(a). Do taxa of birds and mammals differ from one another in the exponent relating home-range size to body size?
(b). Does this exponent differ between dietetic groups?
(c). Does daily energy expenditure determine home-range size?

Three orders of vertebrates were selected for this study – the Passeriformes, the Rodentia, and the Squamata (lizards and snakes). These three groups were appropriate because data are readily available on their home ranges and energetics, because they each comprise species with a variety of diets, and because none of them contain species where groups of animals share a common home range.

Methods

Data on home-range size and body weight were extracted from published sources as follows: Passeriformes (Schoener 1968); Galliformes (Belovsky and Slade, in preparation); Rodentia (Mace 1979 and sources therein); Squamata (Turner *et al.* 1969; Belovsky and Slade, in preparation). The analysis treated genera as independent points. Home-range size and body-size measures for each genus were calculated as the median of component species values. Data are reproduced in Table 3.1.

TABLE 3.1. *Values of home-range size, body weight, and daily energy expenditure used in the analysis. The median value for each genus is calculated from component-species values extracted from sources in the literature (see text). Diet categories are as follows:* C=*Carnivore,* O=*omnivore,* H=*herbivore,* G=*granivore*

Genus	Weight (g)	Home range (ha)	Daily energy expend- iture (kcal/day)	Diet
Rodentia				
Aplodontia	1135.0	10.00	––	H
Sciurus	760.5	7.50	––	G
Tamiasciurus	252.0	1.82	41.16	G
Marmota	1868.0	1.03	––	H
Citellus	726.0	1.48	––	H
Citellus	197.8	1.95	16.43	G
Tamias	107.0	0.91	––	G
Eutamias	48.2	1.06	––	G
Glaucomys	72.5	1.91	24.19	G
Xerus	580.0	4.19	––	O
Perognathus	30.0	0.35	12.34	G

Table 3.1 (*cont.*)

Genus	Weight (g)	Home range (ha)	Daily energy expend- iture (kcal/day)	Diet
Dipodomys	36.3	0.59	––	G
Microdipodops	15.0	––	8.78	G
Liomys	44.0	0.33	––	G
Reithrodontomys	9.0	0.47	6.55	G
Peromyscus	21.0	0.28	10.59	G
Baiomys	7.0	0.07	––	O
Sigmodon	80.4	0.29	24.73	H
Neotoma	290.0	0.64	––	H
Dicrostonyx	52.0	0.20	––	H
Clethrionomys	23.25	0.09	10.58	H
Ondatra	1130.0	0.29	––	H
Microtus	23.0	0.10	11.42	H
Synaptomys	27.0	0.13	––	H
Arvicola	89.0	––	21.43	H
Psammomys	140.0	1.64	––	G
Spalax	204.0	––	32.20	G
Micromys	8.7	––	5.76	G
Mus	24.85	0.13	14.26	O
Apodemus	32.0	0.20	12.42	G
Muscardinus	3.05	––	1.44	G
Glis	5.08	––	0.95	G
Zapus	22.0	0.27	10.64	G
Napoezapus	24.0	––	11.61	G
Microcavia	220.0	0.44	––	H
Cavia	550.0	0.16	––	H
Galea	350.0	0.43	––	H
Proechimys	498.0	0.35	––	H
Squamata				
Uta	2.8	0.014	––	–
Sceloporous	13.0	0.347	––	C
Sauromelas	188.0	0.631	––	H
Tropidurus	6.0	0.024	––	–
Ameira	12.0	0.032	––	–
Cnemidophorous	11.0	0.045	––	–
Eumeces	6.8	0.033	––	O
Lygosoma	1.9	0.003	––	–
Leiolopisma	1.2	0.002	––	–
Crotophytus	40.0	0.112	––	–
Ophisaurus	120.0	0.223	––	C
Basiliscus	70.0	0.017	––	O
Dasia	15.0	0.034	––	–
Mabuga	65.0	0.033	––	O
Draco	8.0	0.032	––	–
Agama	28.0	0.050	––	–
Anolis	2.3	0.001	––	C
Natrix	461.0	3.24	––	C
Agkistrodon	423.0	3.14	––	C
Flaphe	727.0	13.87	––	C
Coluber	531.0	11.98	––	C

Genus	Weight (g)	Home range (ha)	Daily energy expend- iture (kcal/day)	Diet
Thamnophis	507.0	6.80	--	C
Carphophis	6.3	0.03	--	C
Heterodon	136.0	9.46	--	C
Masticophus	388.0	2.20	--	C
Crotalis	671.0	2.83	--	C
Lampropeltis	388.0	2.91	--	C
Haldea	0.4	0.01	--	C
Pituophis	1139.0	23.70	--	C
Galliformes				
Lagopus	510.0	2.57	--	H
Colinus	200.0	1.10	--	H
Meleagris	6023.0	54.96	--	H
Dendragapus	510.0	1.21	--	H
Bonasa	908.0	10.00	--	H
Centrocercus	1135.0	14.60	--	H
Tympanuchus	908.0	5.32	--	H
Phasianus	1136.0	12.50	--	H
Other bird orders				
Asio	252.0	55.04	159.0	C
Zenaida	120.0	--	49.7	-
Aegolious	96.0	--	59.0	-
Calypte	4.8	--	6.7	-
Passeriformes				
Empidonax	11.1	0.88	--	C
Tyrannus	40.4	8.38	--	C
Contopus	13.8	4.37	--	C
Eremophila	40.8	0.49	--	O
Lullula	30.0	8.30	--	C
Corvus	1410.0	938.70	--	C
Nucifraga	130.0	13.23	--	O
Aphelocoma	72.4	2.14	--	O
Parus	11.0	1.89	22.5	C
Sitta	10.4	1.86	--	G
Chamea	14.8	0.32	--	C
Troglodytes	10.35	0.71	--	C
Thyromanes	11.0	0.49	--	C
Thryotherus	18.5	0.12	--	C
Mimus	50.1	0.40	28.8	C
Dumatella	35.9	0.11	--	O
Turdus	86.1	0.43	--	O
Oenanthe	25.2	1.54	--	C
Sialia	30.8	1.01	--	C
Montacilla	17.5	1.01	--	C
Lanius	39.1	4.58	--	C
Vireo	10.7	0.95	--	C

Table 3.1 (*cont.*)

Genus	Weight (g)	Home range (ha)	Daily energy expenditure (kcal/day)	Diet
Dendroica	9.5	0.63	— —	C
Setophaga	9.0	0.19	— —	C
Protonotaria	16.1	1.50	— —	C
Oporonis	11.3	0.77	— —	C
Geothypis	9.8	0.53	— —	C
Wilsonia	9.3	1.01	— —	C
Seiurus	18.9	1.01	— —	C
Icteria	27.0	0.13	— —	C
Sturnella	89.0	3.04	— —	C
Habia	35.25	5.50	— —	C
Melospiza	21.6	0.16	— —	O
Giuraca	27.9	6.19	— —	O
Spizella	15.1	2.39	— —	C
Ammodramus	16.7	1.09	— —	C
Richmondena	41.2	0.15	— —	O
Passerina	14.3	0.11	— —	C
Pipilio	45.5	2.11	— —	O
Progne	48.95	— —	39.0	—
Spiza	35.0	— —	24.4	—
Passer	29.0	— —	23.9	—
Calcarius	28.6	— —	40.0	—
Parus	18.0	— —	22.5	—
Anthus	12.0	— —	12.0	—

For analysis, the data were normalized by logarithmic transformation and the reduced major axis was used to establish lines of best fit through bivariate plots. Lines relating home range to body size were set within each dietetic category in each order. Since there are no herbivorous genera in the Passeriformes, data from the Galliformes were used for comparison with rodent herbivores. Daily energy expenditures for birds were taken from J. R. King (1974) and those for rodents from N. R. French *et al.* (1972) together with other sources given in Mace (1979). Metabolic costs in ccO_2 were converted to kcal by the equality 1 litre oxygen = 4.8 kcal (Dawson 1974) and all data were standardized to represent kcal consumed per 24 hours.

Results

The best-fit relationships between home range and body weight for each diet category within each order are given in Table 3.2. Fig. 3.1 shows bivariate plots for each diet type with the different orders represented by different symbols.

1. *Exponents.* Across orders exponents are highest in the passerines and galliforms – two orders of birds. The value for lizard and snake carnivores lies below the 95 per cent confidence limits for passerine carnivores, and the

TABLE 3.2. *Statistics describing best-fit relationships between home-range size (H) and body weight (W) set by reduced major axis (see text). For functions of the form* $H = aW^b$ *where a is the elevation of the best-fit line, b its slope, r the correlation coefficient and N is the number of genera*

Order	Feeding category	a	b	$L1_b - L2_b$	r	N
Passeriformes	carnivores	−5.09	1.75	1.29−2.21	0.75	29
	omnivores	−7.15	1.72	0.59−2.85	0.44	10
Galliformes	herbivores	−7.59	1.39	0.95−1.83	0.93	8
Squamata	carnivores	−5.70	1.21	0.93−1.49	0.91	15
	omnivores	−2.63	−0.29	−−	−0.52	3
Rodentia	omnivores	−4.82	0.97	−−	0.99	3
	herbivores	−5.31	0.81	0.48−1.14	0.71	14
	granivores	−3.86	0.88	0.64−1.12	0.87	14

exponent for rodent herbivores lies below the 95 per cent confidence limits for galliform herbivores (Table 3.2). A poor correlation in passerine omnivores and few data for rodents precludes comparisons of omnivores across orders.

Comparing dietetic categories, within both rodents and birds, the lowest exponents are for herbivores, although differences are not statistically significant. Omnivores tend to be intermediate and carnivores have the highest exponents.

2. *Elevations.* For all diet types, passerine and galliform genera have larger home ranges for their body sizes than do either the Squamata or the rodents (Fig. 3.1, Table 3.2). There are also elevation differences attributable to diet. In passerines the elevation of the best-fit line for carnivores is above that for omnivores; in rodents it is higher in the granivores than in the omnivores and higher in the omnivores than in the herbivores (Fig. 3.1, Table 3.2).

3. *Daily energy expenditure.* Fig. 3.2 shows daily energy expenditure (DEE in kcal/day) plotted against body weight for rodents and birds. The best-fit relationships are:

$$\text{Rodents: } DEE = 0.80W^{0.77}; r = 0.89; n = 19$$
$$\text{Birds} \quad : DEE = 2.03W^{0.75}; r = 0.96; n = 11$$

The exponents are similar for birds and rodents but elevations differ significantly so that at a given body weight, birds have higher energy demands. Basal metabolic rates for rodent genera were also plotted against body size in order to compare the scaling of basal and active rates. Only genera for which both measures were known were used so as to ensure equivalent samples. Basal rates were found to have the higher exponent, reinforcing the view that energy needs of active animals will increase at a lower rate with body size than basal rates.

Home-range size is plotted on daily energy expenditure for both birds and rodents in Fig. 3.3. If DEE determines home-range size, we would expect there

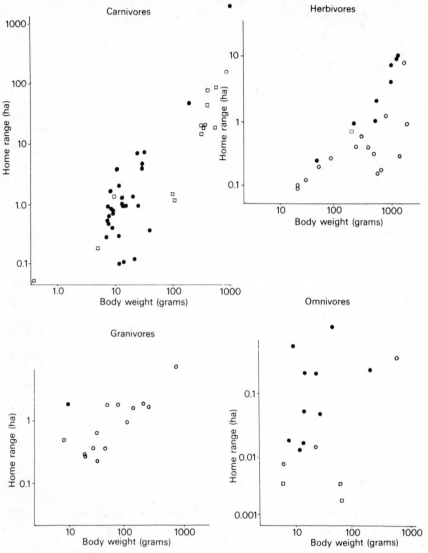

Fig. 3.1. Home range (ha) plotted against body weight (g) in four different feeding categories. ● = Passeriformes (Galliformes in herbivore graph); ○ = Rodentia; □ = Squamata.

to be a strong positive relationship between the two variables with the line of best fit having a slope of 1.0. Unfortunately, there are data available for only a few species on both home-range size and daily energy expenditure, but the best-fit relationships are:

$$\text{Rodents: } HR = 0.001 DEE^{2.26}; r = 0.60; n = 12$$
$$\text{Birds : } HR = 0.0004 DEE^{2.35}; r = 0.88; n = 13$$

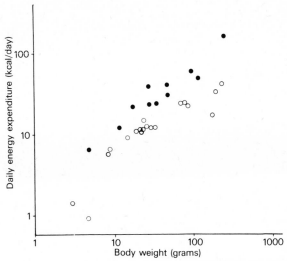

Fig. 3.2. Daily energy expenditure (kcal/24 hours) plotted versus body weight (g) for birds (●) and rodents (○).

Clearly, the exponent for rodents at least differs from a predicted value of 1.0 and the relationship does not differ between birds and rodents. Inspection of Fig. 3.3 does indicate differences in elevation due to diet: herbivores tend to have smaller home ranges for their daily energy requirements than do equivalently sized granivores.

Sex differences in home-range size

Among mammals, males and females generally use different-sized home ranges, and in most cases it is the male range that is the larger. Large male ranges are usually attributed to breeding patterns and increased access to females (Eisenberg 1966) and differences are greatest in highly promiscuous forms such as the root vole (*Microtus oeconomus*) (Tast 1966) where males traverse many female ranges. It is among females, therefore, that the closest relationships between home-range size and energetic needs should be found. Unfortunately, there are too few studies where male and female ranges have been calculated separately, and where comparable methods have been used, to examine scaling factors closely in males and females. However, data from ten herbivorous and sixteen insectivorous or granivorous genera indicated that there is more variation among male home-range sizes than female, when the effect of body size is taken into account. Much of this may be due to differences in breeding systems, though the small samples and the highly variable nature of the data render such a hypothesis untestable at present.

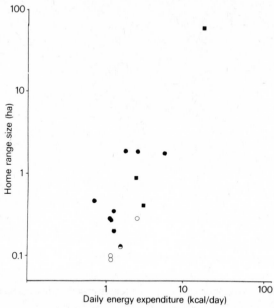

Fig. 3.3. Home range (ha) plotted versus daily energy expenditure for avian carnivores (■), rodent granivores (●), rodent herbivores (○), and rodent omnivores (○).

Sexual dimorphism in body size is to some extent a product of the breeding system: monogamous species tend to show little dimorphism but among polygynous forms the degree of dimorphism is highly variable (Ralls 1977). Thus it is unsurprising that the degree of sexual dimorphism in body size is not correlated with the degree of sexual dimorphism in home-range size ($r = 0.198$; 24 d.f.). The most size-dimorphic forms among the sample of rodents are some ground squirrels where males defend territories containing several females (e.g., *Citellus columbianus*, Murie and Harris 1978). However, highly promiscuous species such as the root vole: (Tast 1966) where males wander widely and tend to have ranges several times larger than females show little size dimorphism. Such differences may arise through varying levels of intermale conflict which could select for increased male size. Also, although there are several genera where females are larger than males (e.g. *Glaucomys, Eutamias*), in no cases are female ranges larger.

More precise home-range data should enable a closer examination of its relationship to breeding systems. At present, it is wise to note that this constitutes an additional source of variation, and that energetic models for home-range size should ideally be based on female data when there are marked sex differences in ranging patterns. There is also some evidence now emerging that metabolic requirements of males and females of similar body sizes may vary (Coelho 1974).

Discussion

As in most previous studies, this analysis has revealed differences in home-range size that are attributable to some extent to differences in body size and the relative densities of preferred food items. Three further effects are found — elevations of best-fit lines are highest in the bird taxa, and slopes are higher in both the bird taxa and the predatory dietetic groups. The elevation difference is attributed to birds having higher energy requirements than similar-sized rodents (Fig. 3.2) and this explanation is further supported by the finding that home range relative to metabolic needs seems to be equivalent in the two groups (Fig. 3.3).

We have already mentioned Schoener's (1968) hypothesis as to why carnivores might be expected to have higher rates of increase of home-range size with body size than other dietetic groups, and this may be relevant to the differences observed between the Passeriformes and Galliformes. In addition, among herbivores, larger forms tend to be less selective over the items they choose (Clutton-Brock and Harvey In press) and so home-range size might be expected to increase at a lower rate. There are no significant differences between dietetic categories in the rodents, and insufficient data to look for them in the Squamata. Consistent differences in slope between taxa present a more difficult problem: the slope for herbivorous birds is significantly greater than that for herbivorous rodents. One explanation might rest on fine dietetic differences between the two orders. The larger rodent herbivores (e.g. *Marmota*) are mostly folivorous and forage over areas of high foliage productivity, with predictably distributed food items. The galliform herbivores, which frequently include quite large amounts of fruit and seed in their diet, may have to expand their home ranges more rapidly with increasing body size in order to include enough of their more patchily distributed food items. It is likely that differences in relationships between home range and body size between groups depend on both habitat productivity and both large- and small-scale distribution of food items.

Daily energy expenditure increases with body weight with approximately the same exponent in rodents and passerines but the elevation is consistently lower in rodents (see fig. 3.2). Fig. 3.3 shows home-range plotted against energetic expenditure; if home-range sizes were determined by energetic requirements we would predict a slope of 1.0, with some elevation differences due to diet. In fact, the slope is greater than 2.0 so that home-range size increases far more quickly than energetic needs. However, there are indications of the predicted differences resulting from diet. There is no evidence of differences between taxa.

Contrary to McNab's (1963) findings, we have shown that within both birds and rodents, body size predicts home-range size far more precisely than do energetic needs. There are, therefore, no 'economies of scale' due to increases in body size. Two factors have generally been considered important as determinants of home-range size — habitat productivity and the animals' energetic needs, the latter being determined largely by body size. We add a third to this list; an interaction between body size and productivity of the habitat. For resources that are patchily distributed, size increase results in an increase in patchiness of resource distribution so that an animal has to range more widely in order to gather the same amount of energy per unit area. Among some prey types (e.g. vertebrates) increasing patchiness is manifested in lower benefits of hunting small prey (see Schoener 1968, 1969); for other items (e.g. fruit, seeds) the effect is due to their pattern of distribution.

A predictive approach – home-range size, metabolic needs, and diet among primates

Introduction

We have shown how home-range depends on both body size and diet within and among several vertebrate groups, and we have suggested how aspects of diet such as the relative density of food items, their pattern of dispersion, and energy content could have profound influence on relationships between home-range size and metabolic requirements (or body size). Little further progress can be made with the investigative approach outlined in the last section until much more detailed information is available on the microdistribution of resources.

In this section we shall use some of the extensive field data available from primates to demonstrate how simple predictive models might be formulated and tested for group-living species, in order to discover more about underlying relationships between available resources and metabolic requirements.

Theory

Primates usually live in groups, the members of which share a common home-range. If the home-range area depends on the metabolic needs of the group, then we might expect

$$HR = ADMN/a.$$

HR is the home-range size in hectares, $ADMN$ is the group's metabolic need in kcal/24 h (Average Daily Metabolic Needs), and a is a measure of the amount of food available, or productivity of the habitat (kcal/ha 24 h). If a was constant across species and habitats, a double logarithmic plot of home range on average daily metabolic needs would have an expected slope of 1.0 and an intercept of $-a$. As we showed above this is not the case and we suggested (as have others before) that a varies with diet, habitat, and the ecological stratum which the species exploits (among primates this would be arboreal versus terrestrial). If, for the moment, we concentrate on diet and simplify primates' food intake into foliage and non-foliage ('fruit' though with some species this includes other components such as gum and insects) and further assume that our measures of proportion of foliage in the diet represent *energetic* proportions, then we can construct various predictive models that relate home-range size to metabolic needs and diet.

Average daily metabolic needs

Coelho (1974) developed a method for estimating the *ADMN* of primate groups. From detailed observations of the amounts of time animals in a group spend on various activities, and using published data on the metabolic costs of those activities, he was able to estimate the total amount of energy needed by a group over a typical 24-hour period. Coelho, however, had more satisfactory data than those available to us because our aim is a cross-species comparison; in addition to the assumptions made for his analysis, we have made others about the body sizes of individuals and the energetic costs of various behaviour patterns. In order to test the robustness of our conclusions, we altered those assumptions in a series of sensitivity analyses (Harvey and Clutton-Brock 1981); the conclusions were not affected. The data were taken from sources quoted in Clutton-Brock and Harvey (1977*b*) and only included those species for which sufficient data was available to calculate *ADMN*.

Three models

We consider three models. The first two assume that dietetic or physiological constraints determine the proportion of foliage in the diet, and that for all primate species the productivity of different habitats for both foliage and 'fruit' is the same. If this is so, then one of the two components will be limiting in density and thus dictate the home-range. It could well be that for species with more than a certain proportion of foliage in the diet, foliage is the limiting component while for other species 'fruit' limits the home-range size.

Foliage-limiting model The expected functional relationship is

$$HR = (ADMN \cdot p)/c \tag{1a}$$

where *p*, which lies between 0.0 and 1.0, is the proportion of foliage in the diet and *c* is a measure of the habitat's productivity of foliage in kcal/ha 24 h. ln (*HR*/*ADMN*) plotted on the ordinate against ln (*p*) should yield a best-fit line with a slope of 1.0 and an intercept of −ln(*c*).

'Fruit'-limiting model Here we would expect

$$HR = (ADMN(1-p))/d \tag{1b}$$

where *d* is a measure of the habitat productivity of non-foliage. ln(*HR*/*ADMN*) plotted against ln(1−*p*) should give a slope of 1.0 and an intercept of −ln(*d*)

As we have pointed out above, there may be some critical proportion of foliage in the diet below which the first (foliage-limiting) model is no longer relevant and the second ('fruit'-limiting) model applies. If this were so, we would expect minima to occur on the plots of ln $(HR/ADMN)$ (a measure of relative home-range size) against diet. Such minima would denote the critical proportions of the different dietary components at the points where each limited home-range size.

Ability to digest foliage-limiting model   Perhaps a more realistic model for 'foliage limitation' might be that the amount of available foliage differs between species. Some species are able to deal with appreciably lower-grade foliage intake than others, or to detoxify secondary plant compounds. In this case

$$ADMN = (HR \times c') + (HR \times d) \tag{2}$$

which is

$$HR = ADMN/(c' + d) \tag{3}$$

where c' is a variable across species (rather than the constant c in the first 'foliage-limiting' model), and:

$$p = c'/(c' + d) \tag{4}$$

But this model produces the same predictions as the second 'fruit'-limiting model since inserting (4) into (3) gives

$$ADMN = HR \times d(p/(1-p) + 1)$$

which is equivalent to (1b) because

$$(1/1-p) = (p/(1-p) + 1)$$

This means that, using the data presented here, we cannot distinguish between the latter two models, since we do not know if any (supposed) excess foliage is not eaten because it cannot be digested (contains poisonous compounds for example) or because non-foliage is a dietetic requirement. It is also possible that we have left out an important cost from our calculation of average daily metabolic needs, such as the added energetic costs of moving in a three-dimensional environment for arboreal species.

Results

Our estimates of average daily metabolic needs for 20 primate species are given in Table 3.3. Fig. 3.4 is a double logarithmic plot of home range on average daily metabolic needs for the 20 species; a line of slope 1.0 is superimposed through

TABLE 3.3. *Data used for the analysis of primate ranging. (Taken from collected material in Clutton-Brock and Harvey 1977a, b)*

	A	D	HRS	p	ADMN	Abb.
Lemuridae						
Lemur catta	T	FR	7.4	0.34	3329	Lc
Lemur fulvus	A	FO	0.8	0.71	1423	Lf
Propithecus verreauxi	A	FO	7.6	0.41	1615	Pv
Cebidae						
Callicebus torquatus	A	FR	20.0	0.14	279	Ct
Alouatta villosa	A	FO	38.0	0.54	6615	Av
Cercopithecidae						
Macaca fascicularis		FR	32.0	0.16	7724	Mf
Cercocebus albigena	A	FR	410.0	0.05	7097	Cl
Cercocebus galeritus	T	FR	55.0	0.14	13434	Cg
Papio anubis	T	FR	2430.0	0.08	29325	Pa
Cercopithecus aethiops	T	FR	27.0	0.12	4151	Ca
Presbytis melalophos	A	FO	21.0	0.37	4165	Pm
Presbytis obscura	A	FO	29.0	0.48	4063	Po
Colobus satanas	A	FO	60.0	0.37	7771	Cs
Colobus guereza	A	FO	15.0	0.82	6433	Cu
Colobus badius	A	FO	75.0	0.78	24463	Cb
Hylobatidae						
Hylobates lar	A	FR	54.0	0.34	1307	H1
Symphalangus syndactylus	A	FO	23.0	0.45	2128	Ss
Pongidae						
Pongo pygmaeus	A	FR	500.0	0.22	3379	Pp
Pan troglodytes	T	FR	700.0	0.28	3144	Pt
Gorilla gorilla	T	FO	620.0	0.36	34640	Gg

A = arboreal (A) or terrestrial (T); D = diet: folivore (FO) or frugivore (FR); *HRS* = home-range size *p* = proportion of foliage in the diet; *ADMN* = average daily metabolic needs (kcal/24h); Abb. = abbreviations used in Figs. 3.5 and 3.6.

the mean on both axes. There is a positive and significant relationship between the two variables which changes little under the extreme assumptions of the five sensitivity analyses (see Harvey and Clutton-Brock 1981). Although the slope of 1.0 lies outside the 95 per cent confidence limits of the principal axis, the observed value of 2.28 is very similar to that found for the equivalent relationship from field data on birds and rodents (see 'Investigative approach'). There is also a clear dietary effect with folivores having relatively small home ranges.

The models, as outlined above are tested in Table 3.4, and it is clear that they are not sufficient to account for the data. The simple foliage-limiting model

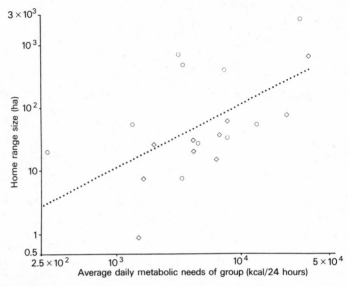

Fig. 3.4. Home range (ha) plotted on average daily metabolic needs for primate folivore (◇) and frugivore (○) groups.

TABLE 3.4. *Relationship between* (a) ln(*HR*) *and* ln(*ADMN*); (b) ln(*HR*/*ADMN*) *and* ln(*p*); *and* (c) ln(*HR*/*ADMN*) *and* ln(1−*p*). *The correlation is the product-moment coefficient, and the best-fit lines are the results of major axis analysis* (*Sokal and Rohlf 1969*)

	Correlation coefficient	Elevation (*a*)	Slope (*b*)	95 per cent confidence limits of *b*
(a)	0.54	−15.30	2.28	1.27 to 6.15
(b)	−0.50	−8.41	−3.21	−1.75 to −11.64
(c)	0.43	−1.02	5.78	2.97 to 56.03

predicts a positive linear relationship between ln(*HR*/*ADMN*) and ln(*p*), whereas the relationship is significantly negative. The other two models predict a positive linear relationship between ln(*HR*/*ADMN*) and ln (1−*p*) with a slope of 1.0; the relationship is positive but the slope is 5.78 and 1.0 does not fall within the 95 per cent confidence limits. There is some evidence of the relationship not being linear with possible minima at about 70 per cent foliage in the diet (see Fig. 3.5) or 30 per cent 'fruit' (see Fig. 3.6). However, removal of points below about 30 per cent 'fruit' would lead to an increase in slope above 5.78 and away from the predicted value of 1.0.

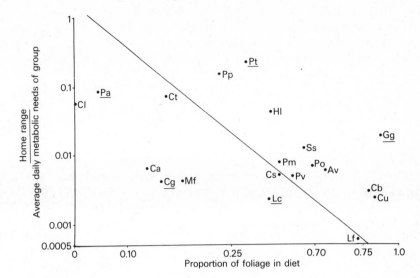

Fig. 3.5. Relative home-range size (*HR/ADMN*) plotted against proportion of foliage in the diet (*p*) for primate species. Both axes logarithmically transformed; for abbreviations see Table 3.3. Terrestrial species underlined.

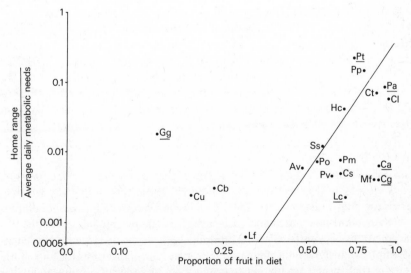

Fig. 3.6. Relative home range size (*HR/ADMN*) plotted against proportion of non-foliage in the diet (1−*p*) for primate species. Both axes logarithmically transformed; for abbreviations see Table 3.3. Terrestrial species underlined.

There is some indication from Fig. 3.6 that the terrestrial species lie away from the line of best fit (are over-dispersed about the line). Nevertheless, if the expected value of -1 is superimposed through the mean on both axes then the variance of points for arboreal species around the line (measured orthogonally to the diet axis) is not significantly less than the variance of points for arboreal species (variance ratio $F_{5,12} = 2.62$). However, removal of terrestrial species from the analysis does improve the correlation somewhat (from 0.43 to 0.54; see Table 3.5), although the slope of 1.0 still lies outside the 95 per cent confidence limits of the data.

TABLE 3.5. *Relationship between* ln(HR/$ADMN$) *on the ordinate and* ln(diet) *on the abscissa for arboreal species only. The product-moment correlation coefficient is quoted together with the results from a major axis analysis*

Abscissa	Correlation coefficient	Elevation	Slope(b)	95 per cent confidence limits of b
ln(p)	-0.71	-6.67	-2.06	-1.28 to -4.07
ln($1-p$)	0.54	-1.88	3.57	1.85 to 19.59

Without doubt, there is a relationship between relative home-range size and diet. Since the models outlined here do not correctly predict the form of that relationship, we have sought an empirical solution after trying various transformations on the two axes. In fact, the negative correlation between ln(HR/$ADMN$) and ln(p) already demonstrated (Fig. 3.5; Table 3.4) is as good as any that we can find using simple transformations. It also distinguishes effectively between the arboreal and terrestrial species by the weighting on the minor axis (orthogonal to the diet axis); the variance of points for terrestrial species is significantly greater than those of arboreal species ($F_{5,12} = 4.06$, $p < 0.05$); and the increased correlation (in absolute terms) of 0.71 after terrestrial species have been removed is appreciably higher than that with terrestrial species included (Table 3.5).

Discussion

This analysis should be regarded as a first step and it is reported in more detail elsewhere (Harvey and Clutton-Brock 1981), but it is included here because it does indicate a relationship between average daily metabolic needs and both home-range size and diet. Even though the models presented are inadequate to account for the relationship between home-range size and diet, the empirical analysis indicates that the relative home-range sizes for four of the six terrestrial species are unusually small (about an order of magnitude smaller than equivalent arboreal species), while those for the other two terrestrial species (*Pan* and *Gorilla*) are correspondingly large. It seems to us unlikely that the small home ranges of the other terrestrial species are a consequence of arboreal species incurring additional costs while climbing; such costs are unlikely to increase energetic demands by an order of magnitude. The large home range and low population densities of *Pan* and *Gorilla* have been discussed elsewhere (Clutton-Brock and Harvey 1979). It is important that we emphasize the empirical nature of this last relationship. Any interpretations arising from it should be regarded as hypotheses awaiting test by independent means.

Concluding discussion

After pursuing both sets of analyses outlined in this chapter, we are only too well aware of their preliminary nature; both the investigative and predictive approaches hold considerable promise, but the data are sparse. In addition, both studies revealed our ignorance of patterns of food distribution and differences in aspects of food choice among the taxa involved. Foraging strategies and the influence of secondary plant compounds on food choice are both topics that are being widely researched across a variety of species. Sometimes the question posed relate to the information needed before studies relating the influence of resources distribution on home-range size can proceed.

In addition, we have ignored a variety of factors that might also be important: the presence of competitors, differences in primary productivity between habitats, the availability of 'safe' nesting sites, and the element of choice inherent in deciding upon a study population (unusually high density). Perhaps we should be surprised that our analyses take us this far?

4
Some theories about dispersal

HENRY S. HORN

Introduction

This chapter is a sketchy review of an idiosyncratic selection of theorectical ideas about dispersal that are related to broad aspects of population structure and dynamics. I note especially those notions that are seemingly paradoxical or counterintuitive, arguing each case intuitively without reference to the particular mathematical formalism whence it arose. I specifically exclude more than casual mention of such important topics as island biogeography, population genetics, and epidemiology. Theories of dispersal are badly in need of a definitive review, but that would take more extensive reading, more mathematical acumen, and more hubris than I can summon just now. Partial reviews and entries into the literature can be found in Wright (1969, Chapter 12; 1978, Chapter 2), Bailey (1975), Wiens (1976), S. A. Levin (1976), L. R. Taylor and Taylor (1977), Okubo (1980), and Hassell (1980). The classic paper by Skellam (1951) still rewards study.

My review is divided into three sections, with the following topics. First, a number of adaptive advantages have been cited for dispersal. Dispersers can colonize vacant land in a harsh and variable environment, but they may also be able to take several chances at winning sites that are occupied in a healthy and stable environment. Under certain conditions, dispersal by offspring is advantageous to their parents, even though dispersers incur such heavy mortality that staying at home is preferable to dispersal for the young themselves; thus there may be a conflict of interest between parent and offspring with respect to the adaptive advantage of dispersal. Dispersal can also regulate an optimal balance between inbreeding and outbreeding. This function is discussed fully by Greenwood (Chapter 7) and by Shields (Chapter 8).

Second, dispersal can have seemingly paradoxical effects on simple models of population dynamics, depending upon the arcane details of the way that it is added to the models. It can either increase or decrease the equilibrial abundance of a single species in a patchy environment. It can allow competitors to coexist in a shifting spatial mosaic or it can bring them into mutual confrontation. It can help prey to escape their predators or it can bring more predators to an already precarious population of prey. Further effects of aggregation and dispersal on population dynamics are discussed by Taylor and Taylor (Chapter 10) and Rogers (Chapter 9).

Third, in populations that are growing, selection favours profligate reproduction and dispersal to new sites; in crowded populations, selection favours tolerance of further crowding. These extreme forms of selection are called respectively *r*-selection and *K*-selection, named after the traditional parameters of growth and crowding tolerance of the logistic equation for population growth. Many aspects of life history are often assumed to place a species on a continuum between *r*- and *K*-selection (Horn 1978), but dispersal from crowded populations cuts across this simple classification. The pattern of age-specificity of dispersal has profound and far-reaching implications for the genetic structure of populations, which in turn affects the mechanisms of natural selection that are important in the evolution of many aspects of life history and behaviour, including dispersal itself. These features are summarized in Table 4.1 near the end of this chapter.

Adaptive significance of dispersal

Dispersal and colonization of populations

Natural selection within populations is usually against dispersal, but only dispersers can found new populations. Therefore if local populations are exterminated, leaving openings, there is a selective advantage for individuals who disperse to those openings. This idea underlies the ingenious but technically flawed work of Levins (1970), and the independently derived model in a compact paper by Van Valen (1971). Further models with more complex population structure are reviewed by E. O. Wilson (1973). Gilpin (1975) gives a more robustly nonlinear treatment, discussing how the required local extinctions can be caused by unstable interactions between predators and their prey.

How realistic these models are is arguable, but D. E. Gill (1978*b*) presents a case history that allows him to extend the models. The North American red-spotted newt (*Notophthalmus viridescens*) inhabits patchily distributed ponds from which juveniles disperse, but to which adults repeatedly return to breed. Most of the time, most of the ponds produce no young at all; in any given period, a few ponds produce many. Colonization of the non-productive ponds is entirely from those few ponds that produce dispersive young. As long as a few of the normally non-productive ponds experience occasional periods of productivity, the young that colonize those ponds will found new populations, and there will be a selective mechanism favouring dispersive young. Adults are selected for homing behaviour at all those ponds that produce young; and these adults have an advantage of at least one year's reproduction over adults that disperse from non-productive ponds. It remains to be seen how many natural populations have the structure of Gill's newt population, namely local patchiness and very high inter-patch variation in reproductive success, but these essential ingredients are at least realistic and plausible. Furthermore, Gill's model predicts not only an adaptive significance for dispersal, but its pattern in relation to life history as well.

Dispersal in stable habitats

Dispersal is obviously favoured when the local environment is deteriorating while better conditions persist elsewhere, but Hamilton and May (1977) argue that it is adaptive for parents to enforce dispersal of some of their offspring even from a stable and favourable environment. In a crowded population of sedentary adults, non-dispersive offspring compete among themselves for a limited number of openings, as well as competing with dispersive offspring of other parents. By dispersing its offspring, a given parent has a chance of establishing progeny in several openings, and this benefit may exceed the cost in mortality due to dispersal itself and the numerical advantage that non-dispersers might have in competing for the few local openings near their parents (Horn 1978).

Both Roff (1975) and Hamilton and May (1977) find that the optimal pattern of dispersal, as enforced by parents on their offspring, provides less complete average occupancy of the environment than other patterns of dispersal; thus what is optimal for individuals is suboptimal for the population as a whole. There is a conflict between individual selection and group selection, as is the case with selection for altruistic traits (see, for example, Horn 1981).

A conflict of interest occurs between parent and offspring in the models of Hamilton and May (1977) and Comins, Hamilton, and May (1980). It is in the parent's best interest to disperse some of its offspring, even in the face of heavy mortality during dispersal. This is so even when the mortality is heavy enough to make successful dispersal and establishment less likely for a given offspring than successful settlement near its parent. The adaptive pattern in this case must be a compromise whose resolution varies in favour of parent or of offspring as one or the other is morphologically and behaviourally equipped to disperse.

Spatial and temporal variation in environment

Hamilton and May's (1977) results have been extended analytically by Comins *et al.* (1980) to include the case where openings occur because of the exogenous imposition of frequent deaths in an otherwise crowded population of sedentary adults. Such a model had previously been explored in extensive simulations by Roff (1975). Both Roff and Comins *et al.* (1980) find that the provision of new openings augments the selective advantage to dispersal, because the presence of openings at some distance from a parent makes competition for distant openings more important in comparison to awaiting competition to succeed the parent. Comins *et al.* also find that, with exogenous extinctions imposed on a small population, the optimal proportion of offspring that should disperse actually increases even as migratory mortality increases. This seemingly paradoxical effect is apparently caused by the importance of preserving a certain level of potential colonization by dispersers which survive in the few available openings; to preserve a given number of surviving dispersers, the number of dispersers must go up as their survival decreases. However, a complete interpretation of this

effect is complex, and should be sought in the original paper by Comins *et al.*
(1980).

More broadly viewed, the importance of dispersal depends on the pattern of
variation in the environment. Gadgil (1971) argues that dispersal from an
unfavourable site is worthwhile when the temporal pattern of environmental
fluctuation varies from place to place, because there will often be another place
where conditions are better. However, he also notes that, if environmental
changes are highly correlated from place to place, unfavourable conditions in
one place mean unfavourable conditions in any other, and endurance of hard
times may be favoured over dispersal.

Inbreeding

In an elegant model, Bengtsson (1978) has presented the dilemma: should an
individual remain at home and run the risk of inbreeding, with consequent
depression of viability or fecundity of its offspring, or should it disperse and
ensure outbreeding even though dispersal contributes to mortality and takes
time from potential breeding? Bengtsson provides simple recipes for calculating
such costs and benefits of dispersal; and the resolution of the dilemma obviously
depends on whether or not costs exceed benefits. Shields (Chapter 8) develops
these points in detail, noting further that extensive outbreeding may be deleterious, and arguing that appropriate dispersal may provide an optimal balance
between inbreeding and outbreeding. From a purely genetic point of view the
optimal balance can be achieved by migration of either males or females. Where
dispersal incurs mortality or loss of some opportunity for breeding, there can be
a conflict of interest between the sexes about dispersal. Social influences on the
resolution of this conflict are discussed by P. J. Greenwood (1980, and Chapter
7 this volume) who argues that individuals of a sex that defends breeding resources should be sedentary and their mates dispersive, while defenders of access
to mates should be dispersive and their mates philopatric.

Effects of dispersal in simple population models

Explicit consideration of the patchiness and spatial structure of populations
has only recently become fashionable, though Wright (1969 for a review) started
laying the groundwork more than half a century ago. Adding dispersal to models
of patchy populations has a number of seemingly paradoxical effects, which I
review below. For other effects, see Taylor and Taylor (Chapter 10), Rogers
(Chapter 9), and Lomnicki (1978). Okubo (1980) engagingly and lucidly reviews
the effects of dispersal in a number of mathematically difficult, but powerful
and comprehensive models, in general supporting the following interpretations
of simpler models.

Models for a single species

The effect of dispersal in fragmented populations is either to increase their mean density (Vance 1980) or to decrease it (Gadgil 1971). This ostensible paradox is instantly resolved by noting that Vance's models allow most dispersers to survive to reach another population, thus removing surplus individuals who might have died in overcrowded populations and transferring them to less crowded populations, while Gadgil's models impose additional mortality on dispersers. The mortality in Gadgil's models is imposed indirectly; isolated populations receive little immigration, while accessible populations receive sufficient immigration to balance emigration; most of the emigrants from isolated populations die.

The effect of dispersal on interpopulation variance in density is generally to decrease it by homogenizing the populations. Dispersal also tends to reduce temporal fluctuations. However in discrete-time models, the regulatory effect of dispersal may be applied out of phase with the need for regulation, and thus dispersal may exacerbate fluctuations (Vance 1980), just as other mechanisms that regulate populations in continuous models may be destabilizing in discrete models (May 1973; Maynard Smith 1974*a*.)

Comparisons have been made between models of populations in which all patches are equally accessible to each other, and models in which space is explicitly taken into account and dispersal is either effected by 'stepping stones' or limited to nearest neighbours (Wright 1969; Gurney and Nisbet 1978; Comins *et al.* 1980). It is not surprising that the usual effect of spatial restrictions is to reduce the effectiveness of a given bulk level of dispersal.

Models for interspecific competition

S. A. Levin (1974) reviews simple models of interspecific competition in patchy environments. He finds that patchiness allows the possibility of coexistence of several species because of differences in initial specific compositions among the patches. Dispersal may decrease diversity by homogenizing the composition of different patches, but it may alternatively promote heterogeneity by moving fugitive species to newly vacant patches if local populations fluctuate widely.

Models for prey and predator

Hassell and May (1973, 1974) argue that locally unstable prey—predator interactions may be regionally stabilized by patchiness and contagious distribution of prey, and by aggregation of predators at sites of high prey-density. Their argument (after Hassell 1980) is basically that drawing predators to areas of prey abundance allows them to feed efficiently and to preserve their abundance, while it draws predators away from precariously small prey populations and thus provides a refuge in which prey at low density can escape predation. This powerful result allows classification of the effects of dispersal, which otherwise would be a hodgepodge. Prey may disperse to homogenize their populations, decreasing regional stability and possibly lowering diversity; or prey may disperse to colonize areas recently denuded and vacated by predators, promoting regional stability and heterogeneity. The dispersal of predators can either homogenize their populations and cause regional instabilities, or allow predators to converge on areas of high prey-abundance and stabilize regional interactions between predator and prey. The differences in effects of dispersal are not due to different levels of dispersal, but rather to the pattern of dispersive responses of prey and predator to their own and each other's local abundances.

Dispersal in the context of life history

I ended a recent review of adaptive life histories (Horn 1978) with a list of tactical questions about dispersal. Should adults disperse? Should young disperse? Should the tactics of parents and offspring agree? How should tactics of dispersal vary with the level of crowding within a local population? Further questions may be added to the list. Are there conditions that favour profligate dispersal in an otherwise conservative life history? What are the consequences of different patterns of age-specificity of dispersal for the genetic structure of populations, or for the spatial structure and dynamics of populations? How does dispersal interact with social behaviour?

I wish that I could end this chapter with a list of answers, but I cannot. However, I can review partial answers to some of the questions, and I can provide a vague road map for further explorations.

Dispersal and the 'r−K continuum'

Southwood (1977) emphasizes that the nature and timing of optimal dispersal is determined by the relation between the time course of decay of a habitat and the characteristic lifespan of its inhabitants. Frequent and copious dispersal is usually associated with r-selected species, species that are subject to extreme fluctuations in numbers, with high mortality but also high rates of population growth via fecundity, early breeding, rapid development, and small body size. This point is not entirely new, but Southwood's argument is concise and compelling in the context of modern theories of population dynamics.

However, Hamilton and May (1977) argue that dispersal may be highly adaptive for K−selected populations, populations that are crowded and constant in density, with high survival, competitive vigour, and ability to escape predators, all of which are made easier by large body size. Many of the properties of r−selected species are the opposite of those of K−selected species, and different species may be usefully compared on an $r−K$ continuum. However, the adaptive significance of dispersal cuts right across this continuum, as I have argued elsewhere (Horn 1978). Dispersal is one of the features of r−selection, but it does not fit tidily into K−selection. Similar considerattions lead Wilbur, Tinkle, and Collins (1974) to question the utility of the concept of the $r−K$ continuum. Fame and glory await whoever presents a tidy, realistic, and complete classification of mechanisms of population dynamics if that person also invents catchy jargon to name the mechanisms.

Consequences of age-specificity in dispersal

I have presented several selective mechanisms that favour the dispersal of young before reproduction, and that may allow adults to be sedentary. D. E. Gill's (1978*b*; see 'Dispersal and colonization of populations', this chapter) models favours sedentary adults and dispersive young. Hamilton and May's (1977, see 'Dispersal in stable habitats', this chapter) model predicts parental enforcement of dispersal by some offspring. Avoidance of inbreeding (see 'Inbreeding', this chapter) necessitates dispersal of pre-reproductive individuals so that they will mate at some distance from their place of birth. Two different ways of achieving pre-reproductive dispersal are for young to disperse by themselves or for parents to disperse while dribbling out new young into the environment. The feasibilies of these alternative methods depend on such things as: whether young are capable of dispersal, either actively or passively by wind or flowing water, whether parental care after birth is necessary, and whether the ecological role of parent or offspring requires a sedentary existence. Wilbur (1977) reviews a host of other ecological parameters that affect the relative advantages of producing young in batches versus dribbling them out singly. Endler (1979) describes the pattern of egg-laying by wild *Drosophila* throughout life and dispersal, and discusses the general effect of such patterns on gene flow and hence on the genetically effective size of local populations. Other factors affecting relative dispersal by adults and young are discussed by Stenseth (Chapter 5), Swingland (Chapter 6), Greenwood (Chapter 7), and further genetic consequences are discussed by Safriel and Ritte (Chapter 11).

Parental strategies of forcing young to disperse need not remain invariant throughout the parent's life. Hamilton and May (1977) and Comins *et al.* (1980) describe the conflict of interest between parents and offspring over dispersal of young in crowded and sedentary populations. As long as a parent is capable of holding its own site, it should enforce dispersal of some offspring to other sites, even if for a given offspring the chances of settling near the parent are higher than the chance of successful dispersal to a new site. However, when a parent becomes unable to hold its own site due to age or infirmity, it becomes adaptive to sow the site with many offspring to increase the likelihood that one of them will win the site in competition with dispersive offspring from parents at other sites. Thus Hamilton and May and Comins *et al.* predict a pattern of parental enforcement of dispersal through much of the parent's life, but a relaxation of this enforcement, perhaps even retention and nourishment of the young at home, late in a parent's life. This pattern is consistent with the pattern of optimal parental behaviour suggested by Trivers (1974). In species with extensive parental care, Trivers argues, there is a period when young can gain from further investment by a parent, but the parent would do better to drive away the current batch of young and invest in another batch. When the parent reaches an age of lowered future reproductive value, it may better invest further in the current young and delay future reproduction that may not be extensive anyway. Thus Trivers also predicts parentally enforced dispersal, but a lessening of the enforcement with advancing age of the parent.

These considerations are summarized in the axes of Table 4.1, which shows the population structures that result from different patterns of dispersal over life history. Table 4.1 feeds back on itself because the genetic structure of the local population is the very setting in which selection for a pattern of dispersal takes place. Different degrees of competition and crowding at various stages of life are important in determining the likelihood of an individual's establishing itself locally versus at a distance. This in turn determines the adaptive value of dispersal.

TABLE 4.1. *Genetic structure of local population for various mixes of dispersal by adults and pre-reproductive young*

ADULTS	PRE-REPRODUCTIVE YOUNG	
	Sedentary or philopatric	**Dispersive**
Sedentary or philopatric	Inbred sibling associations	Associations of outbred adults of mixed parentage
Dispersive with batched young	Associations of young siblings, casual mixture of adults, both outbred*	Casual mixture of outbred individuals
Dispersive with dribbled young	Youthful associations of mixed parentage, casual mixture of adults, both outbred	Casual mixture of outbred individuals

* See text for discussion and clarification of this entry.

Interpretation of Table 4.1 is straightforward with the exception of one entry. If adults are dispersive with batched young and pre-reproductive young are sedentary, then the population consists of associations of pre-reproductive siblings who are outbred and casual mixtures of adults. This population structure provides the opportunity for incest among siblings, simply by the attainment of reproductive maturity prior to dispersal of the sibling associations. Such incest would carry this entry in Table 4.1 into a separate category of inbred sibling associations of youth, and incestuously inbred adults of mixed parentage. Hamilton (1967) discusses the potential advantages of such incestuous behaviour, and the consequences for the evolution of extreme sex ratios and seemingly bizarre social behaviour.

The degree of local inbreeding influences the advantages to be gained from sexual recombination (Horn 1981). It also determines the amount of additional dispersal necessary optimally to balance inbreeding and outbreeding, which Greenwood (Chapter 7) and Shields (Chapter 8) discuss in detail.

Each of the types of population structure in Table 4.1 has a different implication for the form that natural selection takes, particularly selection for social
and altruistic traits (Horn 1981). Inbred sibling associations favour co-operative
behaviour toward neighbors and competition with strangers and immigrants.
Associations of the outbred siblings favour altruistic behaviour within the
sibling group and coherence of the associated siblings until reproductive
maturity, to take advantage of each other's co-operation. Associations of stable
composition, whether individuals are related or not, favour the evolution of
reciprocal altruism (Trivers 1971), where altruistic behaviour is differentially
expressed in behalf of those individuals who are likely to reciprocate in the
future. The casual mixture of outbred individuals is the only setting in which
purely Darwinian selection for maximial individual reproductive success is
relatively unaffected by other selective machinery.

Thus the detailed pattern of dispersal in life history is of critical importance
to nearly all aspects of a species' ecology and behaviour, from the dynamics
of its population to the nature of social interactions.

5
Causes and consequences of dispersal in small mammals

NILS CHR. STENSETH

> 'I don't understand', said the scientist, 'why you lemmings all rush down to the sea and drown yourselves.' 'How curious', said the lemming. 'The one thing I don't understand is why you human beings don't.' [From *Interview with a lemming* by James Thurber; cited from Gaines *et al.* (1979*a*).]

Introduction

Fairy-tales about lemmings rushing down to the sea and drowning themselves, stem from the Scandinavian naturalists at the turn of the century, and in particular from the famous, but often not sufficiently critical, Robert Collett (1911–12) (e.g., Elton 1942; Marsden 1964; Clough 1965, 1968; Curry-Lindahl 1975; for a critical discussion of this earlier literature, see C. J. Krebs 1964, pp. 55–7). The Norwegian lemming (*Lemmus lemmus*) is in fact observed moving around in great numbers (e.g., Kalela 1961; Koponen, Kokkonen, and Kalela 1961; Koshkina 1962; Myllymäki, Aho, Lind, and Tast 1962). However, all reports by laymen about lemmings purposely marching to the sea in order to commit suicide, particularly during peak densities, are nothing but fairy-tales. Unfortunately, these stories are often referred to in the popular press and even textbooks. As adopted in the scientific literature (e.g., Allee, Emerson, Park, Park, and Schmidt 1949; Wynne-Edwards 1962), the extensive lemming dispersal or migrations are often described in group-selective terms (see G. C. Williams 1966*a*, p. 244; Alexander 1979, p. 42); the lemmings are presumed to drown themselves to save their species from overpopulation. Undoubtedly, dispersal benefits the population or species (e.g., Wynne-Edwards 1962; Van Valen 1971) and in most cases begets increased stability of the ecosystem (e.g., Horn and MacArthur 1972; Maynard Smith 1974*a*; Chewning 1975; Roff 1974*a, b*, 1975; Hastings 1977, 1978; Lomnicki 1978, 1980; Ugland and Stenseth 1982; Hoff 1980; but see pp. 93–100). However, dispersal and migration may easily be understood as individually favourable (e.g., Cohen 1967; Hamilton and May 1977; Horn 1978; Comins *et al.* 1980; see review below). Therefore, *as a consequence* of lemmings living in habitats characterized by many (often large) waterbodies such as lakes, rivers, and fjords, it is not surprising to see lemmings swimming across water and as a result often dying in large numbers due to exhaustion. Neither does the observation of many migrants during peak densities demonstrate that the dispersal tendency is on the average particularly high during this period; when there are many individuals in the area there will necessarily be many migrants unless the dispersal rate drops to zero at peak density.

Except for the interest in lemming movements, migration and dispersal were for a long time ignored or considered to be of little or no significance by population ecologists working on small mammals. Often this neglect was due to the methodological difficulties in identifying dispersers and in measuring the amount of dispersal. The theoretical justification for this neglect is the assumption that the number of emigrants from a local population is equal to the number of immigrants into that population. Recently, the validity of this assumption has been questioned (e.g., Lidicker 1962, 1975; C. J. Krebs 1979; Gaines and McClenaghan 1980); thus, great attention is now being devoted to the evolution of dispersal and its consequences in small mammals (e.g., P. K. Anderson 1970; Petrusewicz 1966; C. J. Krebs, Gaines, Keller, Myers, and Tamarin 1973; C. J. Krebs and Myers 1974; Garten and Smith 1974; Lidicker 1975; Bekoff 1977; Tamarin 1977, 1978*a*; Rosenzweig and Abramsky 1980; Mihok 1981). Furthermore, conceptual models are also being developed which relate population genetics to dispersal in fluctuating small mammal populations in order to explain their regular density cycles (e.g., C. J. Krebs *et al.* 1973; C. J. Krebs 1979; Tamarin 1978*a*, 1980*a,b*; Smith, Garten, and Ramsey 1975; M. H. Smith, Manlove, and Joule 1978; Gaines 1981; Gaines and McClenaghan 1980).

In this chapter I review some of the literature that has recently accumulated on dispersal in small mammals (operationally defined as a mammal with adult weight up to 5 kg; Bourliere 1975). Although the theoretical sections are general, emphasis is placed on data for small rodents, particularly of the family *Cricetidae* and the subfamily *Microtinae* (Arata 1967). I will first discuss the evolution of dispersal by means of natural selection or kin selection (Maynard Smith 1964); I will primarily centre this section around the concept of evolutionarily stable strategies, or ESS's (Maynard Smith and Price 1973; Lawlor and Maynard Smith 1976).

Second, I discuss some important consequences of dispersal. This latter discussion will mainly relate to demographic aspects such as optimal sex ratios, and consequences of dispersal with respect to population stability. In particular, I will discuss whether or not dispersal may cause regular density cycles as commonly observed in many microtine rodent species (e.g., C. J. Krebs and Myers 1974).

From the beginning, it should be realized that I am using the terms dispersal and migration to refer to the movement of individuals in which they leave their home range (as defined by Burt 1940). Sometimes these vagrants may establish a new home range, sometimes they may die during the dispersal phase (e.g., as a result of exhaustion or being taken by a predator). Following Endler (1977, his Table 2.1) I will, particularly when analysing data, distinguish between dispersal and migration; dispersal is defined as short non-directional movements away from a home site, whereas migration is defined as directional movements by a single individual or by groups of individuals.

Why not stay home rather than disperse?

Preamble

In the following I assume a spatially heterogeneous (or patchy) environment (Fig. 5.1). The entire habitat complex or landscape may be divided into a *transition habitat* unsuitable for the survival and reproduction of rodents, and

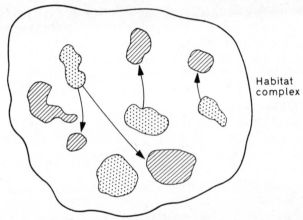

Habitat complex

Fig. 5.1. A metapopulation (Levins 1970) within a habitat complex or landscape. Shaded areas represent rodent habitats or patches (terminology as in Hansson 1977a); the remainder is transition habitat where survival and reproduction is impossible. Hatched areas are empty but suitable for rodents whereas stippled areas represent inhabited patches suitable for rodents; these are also the donor habitats (subsection 'Conclusions' in section 'Evolution of dispersal'). Movements between rodent patches are possible and indicated by solid lines. Local populations may go extinct. (Redrawn from Stenseth and Hansson 1981; their Fig. 1.).

various types of *rodent-habitats* (or patches) (Hansson 1977a; M. H. Smith *et al.* 1978; M. H. Smith, Chesser, Gaines, Stenseth, and Tamarin 1980b). In the latter types of habitat, a rodent population may, in a deterministic model, survive for an infinite length of time. (What is a rodent patch and what is a transition habitat are, of course, defined by the preferences of the species being studied.)

For small rodents, as for most other species, such a description of their habitat is indeed appropriate (see Emlen 1973; Levin 1976; Wiens 1976; Southwood 1977; Solbreck 1978, for a general justification of this way of considering the environment). For specific justifications of this point of view with respect to small mammals, see Anderson (1970), Naumov (1972), Lidicker (1973, 1975), Hansson (1977a,b, 1979a), Stenseth (1977b, 1978a,b), Stenseth, Hansson, and Myllymaki (1977a), Stenseth, Hansson, Myllymaki, Andersson, and Katila (1977b), Stenseth and Hansson (1980), Getz (1978), Getz and Carter (1980), Smith, Baccus, Chesser, Johns, Manlove, Ryman, and Straney (1980a), Smith et al. (1978, 1980b), Fleming (1979), and Rosenzweig and Abramsky (1980). However, see Roughgarden (1977, 1979, pp. 400–8) who argues that patchiness in distribution often may be only a consequence of dispersal in a nonrandomly fluctuating environment. Thus, we are faced with the problem of which comes first, dispersal or patchiness; in the following I assume the latter.

In such a habitat complex, it is by no means obvious which is the better strategy for an individual; to remain in its place of birth or to undertake a perilous dispersal through a transition habitat in search of a new home. The patch in which the individual was born and grew up is known to be a relatively good place and the individual also knows its way around. By leaving the current home patch, the risk of being taken by a predator increases greatly (e.g., Errington 1946, 1963; Metzgar 1967; Carl 1971; Kalela and Koponen 1971; Ambrose 1972; Armitage and Downhower 1974; Slade and Balph 1974; Svendsen 1974; Kozakiewicz 1976; Windberg and Keith 1976; Hansson 1977a; Beacham 1979c). As pointed out by Lidicker (1975) (see also Tamarin 1980a), dispersal may, on the other hand, represent a gain in inclusive fitness (Hamilton 1964) owing to increased access to critical resources and mates, decreased exposure to predators and disease in the current patch, and an opportunity to produce more fit offspring by avoiding inbreeding depression (Falconer 1960). However, in order to compare the advantages and disadvantages of the resident and dispersing strategy, the above-mentioned costs and gains must, ultimately, be translated into a decreased or increased number of progeny now or at some future time. By so doing I am treating the number of successful offspring as the only currency of the evolutionary process.

In this section I review some attempts to explain why natural selection, under certain conditions, favours dispersal behaviour. In addition I analyse a much simpler, but more general model that may replace the more complex (and difficult to understand) models under review.

Genetics and dispersal

Differences in dispersal tendencies must correspond to genetic differences if natural selection is to operate; otherwise, dispersal cannot be interpreted as an adaptation. Unfortunately, mainly circumstantial evidence seems to be available for this in small mammals (Bekoff 1977). Rasmuson, Rasmuson, and Nygren (1977) were able to demonstrate, in a laboratory experiment, a genetic basis for dispersal tendencies in the field vole (*Microtus agrestis*). However, the best available data of this type are probably those of J. H. Anderson (1975) (see also C. J. Krebs 1979) on the Townsend vole (*M. townsendii*) with an estimated heritability of dispersal of 62±11 per cent. P. J. Greenwood, Harvey, and Perrins (1979*a*) similarly found in a study on great tits (*Parus major*) an estimated heritability of dispersal ranging from 56 to 62 per cent.

The main difficulty in studies attempting to estimate the genetic component to dispersal is the difficulty of establishing parental lineages and the dispersal tendencies of related individuals (C. J. Krebs, Wingate, LeDuc, Redfield, Taitt, and Hilborn 1976). There is, however, a suggestion that siblings in four species of *Microtus* exhibit a similar tendency to disperse (Hilborn 1975; Beacham 1979*b*). A genetic basis for different dispersal tendencies is further indicated by studies on allelic frequencies of dispersing and sedentary individuals of the meadow vole (*Microtus pennsylvanicus*) (Fig. 5.2) and prairie vole (*M. ochrogaster*) (Myers and Krebs 1971; Gaines and Krebs 1971; C. J. Krebs *et al.* 1973; Pickering, Getz, and Whitt 1974); and deer mouse (*Peromyscus maniculatus*) and old-field mouse (*P. polionotus*) (M. H. Smith, Carmon, and Gentry 1972; Garten 1976, 1977; Massey 1977). However, no such differences in allelic frequencies could be found in the old-field mouse by Blackwell and Ramsey (1972), nor in Richardson's ground squirrel (*Spermophilus richardsonii*) by Michener and Michener (1977). Obviously, further studies are necessary.

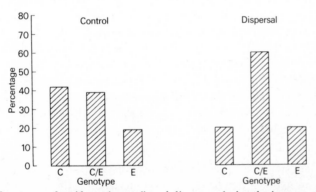

Fig. 5.2. Genotypes of residents (control) and dispersers during the increase phase of the meadow vole (*Microtus pennsylvanicus*). Transferine (Tf) genotype frequencies of dispersing females on a trapped grid compared with those of resident females on control grids in adjacent grassland. As can be seen dispersing voles in the increase phase are not a random sample from the control population. C, C/E, and E represent the three transferine genotypes Tf^C/Tf^C, Tf^C/Tf^E, and Tf^E/Tf^E respectively. (Redrawn from C. J. Krebs *et al.* (1973); their Fig. 6.)

Evolution of dispersal

Habitat suitability

Gadgil (1971) developed a model for analysing under what conditions dispersal behaviour is favoured by natural selection in a habitat complex like that depicted in Fig. 5.1. The model is based on the concept of *crowding*, which may be defined as $\rho = N/K$, where N is the current density of the species in the particular habitat patch and K is its carrying capacity. *Habitat suitability* as defined by Fretwell and Lucas (1970) and Fretwell (1972), p. 82) is then measured by 1-ρ. Note, however, that, for a sexually reproducing species, a mate must exist in the patch in order for it to be considered a suitable patch. Furthermore, it should be realized that the carrying capacity, K, of a patch, besides depending on the size of the patch and the quantity and quality of the food resources in it, is also influenced by the density of competing species in that patch; see P. R. Grant (1972, 1978*a*).

Based on the ideas developed by Findley (1951, 1954) and Svärdson (1949), P. R. Grant (1969, 1970, 1971*a,b*, 1972, 1975, 1978*a*) (see also Morris and Grant 1972) developed his model for interspecific competition with reference to *Microtus*, *Peromyscus*, and *Clethrionomys*. These general ideas are also applicable to other species constellations (Hoffmeyer 1973; Myllymäki 1977*b*; Henttonen, Kaikusalo, Tast, and Viitala 1977; Gliwicz 1980, 1981; see Fleming 1979 for review), and were formalized into a mathematical model by Stenseth *et al.* (1977*b*). A similar approach to the same problem, although different in details, was taken by Rosenzweig (1974, 1979, 1981) and Pimm and Rosenzweig (1981).

By applying this theoretical framework of Fretwell and Grant, Gadgil's model and the generalized model developed in this chapter [see eqn (3)] may easily be extended to incorporate the effect of interspecific competition as well as applying to a habitat complex consisting of patches with varying qualities. This general framework was assumed in the conceptual model developed by Hansson (1977*a*) and M. H. Smith *et al.* (1980*b*); see in this connection Lidicker's (1978) recommendation of adopting such a community approach for understanding small-rodent population dynamics.

Asexual reproduction
Gadgil's model Generally speaking, dispersal is favoured when there is a chance of colonizing another habitat-patch more favourable than the one that is presently inhabited. Grinnell (1904, 1922) originally suggested this idea which was further developed by Gadgil (1971) and MacArthur (1972). In Gadgil's formulation, a more favourable patch is defined as one which is less crowded, i.e. a patch with lower ρ than the patch presently inhabited.

A pronounced tendency to disperse will be favoured under the following environmental conditions:

1. When habitats have low duration stability relative to the species' generation time (cf. Southwood, May, Hassell, and Conway 1974);
2. When highly suitable habitat patches are relatively close together (a conclusion also following from the model discussed by G. A. Parker and Stuart

1976 and G. A. Parker 1978);
3. When catastrophic mortality creates temporarily vacant habitat patches.

Dispersal is most favourable when the pattern of environmental fluctuations (affecting patch suitability) varies from place to place; in such cases there will often be other places where conditions are better. However, if the changes in suitability are highly correlated from place to place, then hard times in one patch mean hard times in another, and site or patch tenacity is favoured over dispersal. Note that Gadgil's model suggests that the dispersal rate is expected to be *low* in habitat complexes with, on average, high densities in the patches.

Andersson (1980) has recently extended this analysis using a somewhat different model. His model predicts that the relative merit of dispersal is higher when resources changes cyclicly rather than randomly over time. In cases of cyclic food production, dispersal is favoured by large litter size, high juvenile survival, and low adult survival.

Lomnicki's model The analysis of dispersal in a patchy environment due to Gadgil (1971) was extended by *L*omnicki (1978, 1980). *L*omnicki analysed cases where there are differences among the individuals in the population (e.g., social hierarchy). Individual differences are assumed to be a result of unequal resource-partitioning among individuals in the population. Consider, for example, an unequal resource-partitioning among individuals as depicted in Fig. Fig. 5.3. There are $N, N < K$, individuals in the patch. While L individuals may obtain sufficient resources to reproduce and leave progeny, the remaining

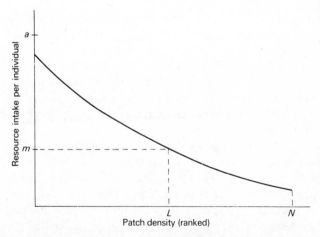

Fig. 5.3. Hypothetical resource intake of N individuals as a function of their rank. Of the N individuals, only L receive m units of resources or more (i.e., the threshold value below which reproduction is impossible). The parameter a represents the maximum possible resource intake per individual. See text for further discussion. (Redrawn from *L*omnicki (1980); his Fig. 2.)

(N–L) individuals that do not survive to the time of reproduction, will have their fitness in this patch equal to zero. Therefore, their emigration, even with a very low probability of finding a better place, is of selective advantage; there will always be a finite probability that a dispersing individual will be successful in finding a better habitat patch. Hence, there will be selection for dispersal, particularly in species exhibiting some kind of social hierarchy. Essentially this model is more mechanistically appropriate than the Gadgil model.

The Hamilton–May–Comins model It is no surprise that dispersal should be advantageous when the local environment is deteriorating; nevertheless, Gadgil's model was essential a decade ago in specifying and formulating our ideas relating to dispersal in varying environments. However, our understanding of why organisms disperse was greatly improved by the inclusive fitness model developed by Hamilton and May (1977). They argued that it may be adaptive for parents to enforce dispersal of some of their young even in *stable* environments. Hamilton and May (1977) assumed a habitat complex such as that depicted in Fig. 5.1; each patch is inhabited by resident adults whose offspring compete within their own generation for the space vacated by the death of adults. For the asexual population, Hamilton and May derived, in the limiting case of a large number of habitat patches and a large number of offspring, that the optimal (or ESS) dispersal rate, v^*, may be expressed analytically as

$$v^* = 1/(2-p), \tag{1}$$

where p is the probability of a dispersing individual being able to find another patch in which to reproduce. From this equation it follows that parents ought to enforce some dispersal of their offspring (half or more of them) even when the local environment is not deteriorating and even when the process of dispersal itself incurs considerable mortality. As in the case of Gadgil's model, the one by Hamilton and May suggests that dispersal is not at its maximum in habitat complexes with, on average, high population densities in the patches; see further discussion of this below.

Comins *et al.* (1980) extended this model of Hamilton and May. The most general model discussed by Comins *et al.* (1980) is a discrete generation island-type model (Wright 1969) in which reproduction, migration, and competition are stochastic processes; in each patch K individuals may live. Environmental variability or uncertainty is introduced in the form of a fixed probability E that a patch is destroyed by extrinsic forces during a one-generation time interval. Reproduction is, as in Gadgil's model, asexual and individuals of each genotype have a certain probability v of being born with a predisposition to migrate.

The results of this model are depicted in Fig. 5.4. From this we see that the following features favour increasing tendencies of dispersal.

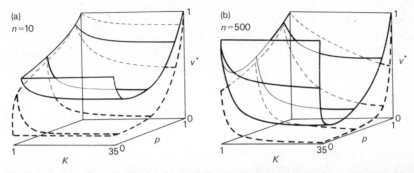

Fig. 5.4. A three-dimensional plot of the ESS-migration probability v^* as a function of the size of the patch K and the probability p of surviving migration through the hostile transient habitat. Broken lines represent the planes derived for situations without any extrinsically imposed extinction rate E; i.e., $E = 0$. The solid lines represent the planes derived for unstable environments; i.e. $E = 0.1$. (a) represents an organism with an average number of offspring per site (n) equal to 10. (b) represents an organism with $n = 500$. (Redrawn from Comins *et al.* (1980) who give reasons for disregarding the 'rim' dose to $p = 0$ as an artefact.)

1. When the degree of extrinsic disturbances E increases (i.e., when habitats have short duration stability);
2. When the hostility of the transition habitat $1-p$ decreases;
3. When the patch size K decreases.

These are essentially the same conclusions as derived from Gadgil's (1971) model. Remember in particular Gadgil's definition of crowdedness, $\rho=N/K$. Contrary to Andersson's (1980) model, this model by Comins and co-authors predicts a higher dispersal rate to be optimal with decreasing litter size. However, contrary to Comins and co-workers, Andersson did not analyse a genetic model. In the Comins and co-workers model, for a fixed number of patches in the habitat complex, a decreasing litter size (n) implies an increased variability in gene frequency from patch to patch. This may result in an increased probability of a dispersing individual finding individuals of a genetically different type in their patch of arrival and thereby increase the probability of competing with dissimilar individuals in that patch. Under these conditions it will be worthwhile attempting to outcompete other individuals in this patch. Overall, it may lead to an increased dispersal rate with decreasing litter size. Such effects are, by assumption disregarded in Andersson's model. In his model increased litter size contributes to the crowding as well as the probability of reaching another more suitable patch; both imply increased dispersal as a result of increased litter size. Thus, comparing these two models it becomes obvious how important it is to analyse genetic models. The inclusive fitness concept due to Hamilton (1964) is indeed appropriate for such analyses.

A generalized model The general feature of these models may all be simplified into the following general two-age-class Leslie-matrix model (a formally similar model was discussed by Stenseth 1978c; see that paper for technical details).

Let the density of juveniles and adults at time t in a particular patch be denoted by $n_{0,t}$ and $n_{1,t}$, respectively. Then we have

$$\begin{bmatrix} 0 & m(1-v)+m\,g\,v \\ s(1-v)+s\,p\,v & 0 \end{bmatrix} \begin{bmatrix} n_{0,t} \\ n_{1,t} \end{bmatrix} \begin{bmatrix} n_{0,t+1} \\ n_{1,t+1} \end{bmatrix}$$

where the Leslie matrix may be written as

$$\begin{bmatrix} 0 & m[1-v(1-g)] \\ s[1-v(1-p)] & 0 \end{bmatrix}$$

where m is the reproductive output for adults; s is the survival from juveniles to adults; v is, as in the Hamilton–May model, the probability of migration of juveniles, i.e. $1-v$ of the juveniles remain in the patch where they were born; those juveniles that disperse, are subjected to a mortality rate equal to $1-p$ in the transition habitat; and $g>1$ is the gain in reproductive output obtained by reproducing in another patch. The tendency to disperse is assumed to be genetically fixed. No other genetic aspects are explicitly considered; genetic relatedness is, however, considered when interpreting the results of the analysis of the model.

If a strategy is to be an ESS, the largest positive eigenvalue λ_1 must be at its local maximum with respect to the evolutionary variable v. This optimum value is denoted v^*. For $v = v^*$, we have $\lambda_1 = 1$. A value v^* is sought such that for any $v \neq v^*$, $\lambda_1 < 1$ (see Charlesworth 1974).

The eigenvalues of the Leslie matrix are the roots of an equation

$$\varphi(\lambda, v) = \lambda^2 + A \tag{2}$$

where A is a known function of $v,p,g,s,$ and m.

If v^* is to be an ESS, then (in the generic case) $(d\lambda/dv)_v{}^*=0$ and $(d^2\lambda/dv^2)_v{}^*<0$. To find $d\lambda/dv$ it is noted that

$$d\varphi = \frac{\partial\varphi}{\partial\lambda} \cdot d\lambda + \frac{\partial\varphi}{\partial v} \cdot dv = 0.$$

Hence $d\lambda/dv=0$ if and only if $\partial\varphi/\partial v=0$. From this we have that the condition for v^* to be an ESS is

$$\left(\frac{\partial\varphi}{\partial v}\right)_{v^*} = 0$$

which implies that

$$\left(\frac{\partial A}{\partial v}\right)_{v^*} = 0$$

After derivation and minor rearrangements, this gives

$$v^* = \tfrac{1}{2}\left[\frac{1}{1-p} - \frac{1}{g-1}\right]. \qquad (3)$$

We require that $0 < v^* < 1$; if the values of p and g are such that v^* as found by eqn (3) yields a v^*–value outside this interval, v^* is assumed equal to 0 or 1, respectively. For v^* to be greater than zero; that is, for dispersal to be optimal at all, we must have that

$$g > 2-p. \qquad (4)$$

This condition is indeed intuitively meaningful; it represents a formalization of the assumption that the gain in fitness must, at least, be greater than the loss in fitness due to dispersal.

The following conclusions may be drawn from this Leslie-matrix model; Eqn (4) implies that the optimal rate of dispersal will increase with decreasing hostility of the transient habitat. In addition, since density-dependent factors operates on m within each patch, g will increase with increasing rate of extinction due to extrinsic factors E, that is, v^* will increase with increasing value of E. The relation between g and K, the carrying capacity, is more difficult to appreciate. However, from the basic assumption of crowdedness p discussed in connection with Gadgil's model, we have that, if K decreases, the gain by leaving the patch will increase, and particularly so if the litter size n is small; this is so because if n is small, many other patches in the habitat complex are also likely to be unoccupied. Thus, v^* will increase with decreasing K and decreasing n. In cases where it is important for the animal to know its way around in the habitat patch where it reproduces, g will be small (and most likely less than one). Thus, this model also incorporates indirectly the fact that an individual staying in its place of birth may take advantage later in life of familiarity with its original area or patch (P. J. Greenwood and Harvey 1977).

There is great consistency when these conclusions are compared with those deduced from the models reviewed above, and in particular with the Hamilton–May–Comins model (Fig. 5.4). For instance, if $p=0$, eqn (1) yields $v^*=\tfrac{1}{2}$. The assumption underlying its derivation is, however, that there is a large number of patches in the habitat complex. Irrespective of the litter size in such a system, the advantage of dispersing and attempting to find a vacant place (preferably with individuals genetically different from the dispersing individual) will become large. Consequently, g is implicitly assumed to be large. Then, if $p=0$ and $g=\infty$, eqn (3) yields $v^*=\frac{1}{2}$, i.e., the result derived by Hamilton and May. Interpreting my g within the framework of Comins and co-workers' model would, I believe, improve our appreciation and understanding of their model. Obviously further theoretical work is necessary here; this further work should also attempt to incorporate the general ideas of habitat suitability discussed earlier.

A life-history model: the demographic consequences of cyclic and stable densities Elsewhere I have discussed an extension of the model developed by Pianka and Parker (1975) (Stenseth 1978a; Stenseth and Framstad 1980; Stenseth and Ugland 1982; Ugland and Stenseth 1982). As originally presented, my argument was related to cyclic small rodent populations (Stenseth 1978a). These cycles, whatever their causes, may in the majority of cases be schematized as in Fig. 5.5; that is, a long increase phase with positive specific growth rate

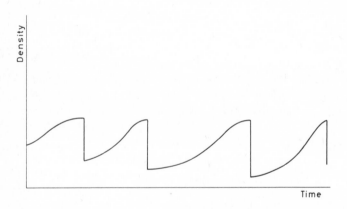

Fig. 5.5. An idealized, but nevertheless typical, small-rodent multiannual density cycle as seen in regions with pronounced cycles. Three phases are depicted; the increase phase with $dN/(dt \cdot N) \gg 0$, the stable phase with $dN/(dt \cdot N) \approx 0$, and the decrease or crash phase with $dN/(dt \cdot N) \ll 0$; only the increase and stable (or peak) phases are assumed to last for any significant length of time.

followed by a relatively short peak phase and then the rapid crash (Stenseth 1977a, 1978a, 1981). The predictions derived from the following model depend on this assumption. Other patterns of density changes are often found for temperate populations (C. J. Krebs and Myers 1974). My predictions are likely to apply mainly to arctic and sub-arctic regions.

    The argument in this model relates to the optimal allocation of resources between current reproductive rate m_x and residual reproductive value V^*_x (G. C. Williams 1966b; Pianka and Parker 1975). The age-specific residual reproductive value is defined as

$$V_x^* = \frac{l_{x+1}}{l_x} \cdot V_{x+1} \qquad (5)$$

where V_{x+1} is the age-specific reproductive value (R. A. Fisher 1930) measuring the expected number of offspring born from (and including) age $x+1$, until death or the occurrence of sterility. Here l_x (and l_{x+1}) is the survivorship function commonly used in demographic analyses.

Only certain combinations of V_x^* and m_x are possible under prevailing conditions: these possible combinations constitute the fitness set (Levins 1968). Such fitness sets are depicted in Fig. 5.6(d) and (e). They are constructed from the two component functions depicted in Fig. 5.6(a)–(c) as described by Stenseth and Ugland (1982). The resulting fitness sets are depicted in Fig. 5.6(d), (e). Notice in particular that costs due to gestation as well as lactation are

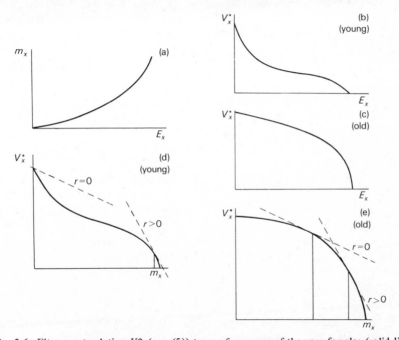

Fig. 5.6. Fitness set relating V_x^* (eqn (5)) to m_x for young-of-the-year females (solid line in (d) and over-wintered females (solid line in (e)). These are constructed by applying the component function, $m_x = f(E_x)$ (where E_x is reproductive effort; see Stenseth and Framstad 1980), depicted in (a) and deduced by Stenseth and Framstad (1980) on the basis of data on the European common vole (*M. arvalis*) (this is assumed to be typical for microtine rodents in general) and the component function, $v_x^* = g(E_x)$, assumed by Schaffer and Tamarin (1973) to be generally valid for microtine rodents, depicted in (b) for young-of-the-year and in (c) for overwintered females. The broken lines in (d) and (e) represent the adaptive functions picking out the optimal demographic strategy in a population with a particular specific growth rate, r; this optimal value is given by the m_x-value for which the adaptive function touches the fitness set (cf. Stenseth 1978*a*, 1980*a,b*; Stenseth and Framstad 1980).

included in the fitness sets (Stenseth, Framstad, Migula, Trojan, and Wojciechowska-Trojan 1980; Stenseth and Framstad 1980).

The corresponding adaptive function (Levins 1968) picking out the optimal combination of m_x and V_x^* values from the fitness set is

$$V_x^* = e^r V_x - e^r m_x \qquad (6)$$

(Stenseth 1978*a*; Stenseth and Ugland 1982; Ugland and Stenseth 1982; see also Pianka and Parker 1975; Stenseth and Framstad 1980). Following Hairston, Tinkle, and Wilbur (1970) and Stenseth (1978*a*; particularly the discussion about his Fig. 4; Stenseth and Ugland 1982; Ugland and Stenseth 1982), r is assumed to be a parameter determined by the state of the environment where any given individual finds itself. For a cyclic population with long increase phase (Fig. 5.5) (i.e. a mainly growing population, $r>0$) the more cyclic the population is, the greater is r (Stenseth 1978*a*; Ugland and Stenseth 1982). Furthermore, since the small-rodent populations tend to have synchronized density variations over large areas (C. J. Krebs and Myers 1974), r may be assumed to be approximately the same for most of the patches in the habitat complex; r is therefore a characteristic of the habitat complex. Thus, the adaptive function will have a steeper slope in areas where the populations are more cyclic (Stenseth 1978*a*; Ugland and Stenseth 1982; see also Pianka and Parker 1975, for a less formal argument).

The basic idea then, is that the evolving population attempts to allocate resources to m_x and V_x^* so that V_x is maximized under the prevailing constraints set by the environment, i.e., the constant r. Essentially, this is a generalization of the ESS-argument. In a specific environment with a given r, we can ask what strategy is the optimal one favoured by natural selection. Of course, the environmental situation will be altered, i.e. r will change, if invasion occurs; however, that is *not* what is being studied as the ESS-argument relates to non-invasion.

From this it follows that

1. More resources should be invested in current reproductive output in the increase phase of cycles than in the peak phase (Stenseth 1978*a*);
2. Current reproductive output is expected to be higher when the population is more cyclic (Stenseth, Hansson, and Ugland 1982).

From these conclusions, it follows that, during the increase phase, a female who has recently reproduced ought to deliver her next litter elsewhere (e.g., in another patch). By not forcing her previous litter (i.e. her current reproductive output) to leave their area of birth, she will increase their chance of survival and thereby increase her own current reproductive output; note that this is independent of, but consistent with, inclusive fitness arguments. At the same time, she may (according to previously reviewed models) easily find a vacant place where she can deliver her next litter. Presumably, the juveniles might have much more difficulties in finding another patch. Furthermore, if the female had her next litter in the same nest as her previous one, particularly if she did not force her previous litter to leave, the chance of being taken by a predator might increase greatly (MacLean, Fitzgerald, and Pitelka 1974; Fitzgerald 1977).

Therefore, leaving the patch during periods of great expansion will lead to an extensive gain in the dispersing female's inclusive fitness (through her previous litter remaining in their patch of birth) at a negligible loss in her immediate future survival and reproductive output. Again, these conclusions are consistent with and complementary to those reached earlier in this section.

All the models reviewed thus far suggest that dispersal is *not* expected to be at its maximum during periods of peak density in the habitat complex.

Sexual reproduction

The Hamilton–May–Comins model According to Comins *et al.* (1980), the effect of introducing sexual reproduction is only slight. Essentially, sexual reproduction is found to increase the population size by one-half of an individual; therefore, the effect of sex is only important in the case of low population densities.

Roff (1975) developed a model for the evolution of dispersal in a sexually reproducing species. He found an optimal dispersal rate which was always smaller than the one predicted by Hamilton and May (1977); however, in view of the result of Comins *et al.* (1980) that the population size needs to be increased by one-half, this difference between the results of Hamilton and May on the one hand, and Roff, on the other hand, is as expected.

Bengtsson's model Bengtsson (1978) (see also Chesser and Ryman 1980 for a more general model) has discussed a model for analysing whether to stay at the place of birth (and then mate with a close relative) or disperse (see Chapter 7). Although the general conclusions reached above are still valid, Bengtsson finds more pronounced effects of sexuality than Comins *et al.* (1980). For sexually reproducing populations a dispersing individual must (unless it is a pregnant female) find a mate in order to avoid effective sterility. Hence, it is expected that the optimal rate of dispersal should be relatively low during periods of low population densities in the habitat complex.

Bengtsson considers a species where adults may move to areas where they can mate with unrelated mates. There is a cost $(1-p$, where p is as defined above) associated with trying to find an unrelated mate; this is the exposure to predation in the transition habitat, difficulties in entering established demes (e.g., Getz 1978), and the difficulties in finding a mate in another patch. Furthermore, there is another cost associated with the resident behaviour (or strategy); this is the effects of inbreeding depression (Falconer 1960). It is in specifying that both these costs exist that Bengtsson's model differs from the sexual version of the Comins model. Nor were such aspects of sexuality considered in the models of Roff, and Hamilton and May. However, such effects are relatively easily incorporated within the framework of eqn (3).

Asexual–sexual comparison From the discussion in this section it follows that most of the general conclusions derived from the models for asexually reproducing organisms are also valid for species with sexual reproduction. The complicating factors of sexual reproduction are the effects of inbreeding and the difficulties of finding a mate. Overall, this suggests that the optimal dispersal rate is low during periods of sparse *as well as* peak densities; only at intermediate densities when the population in the habitat complex is expanding the most, is the optimal dispersal rate expected to be high.

Conclusions
Predictions Within a habitat complex or a large region, cyclical small-rodent populations are commonly observed to exhibit synchronized density changes in different patches (C. J. Krebs and Myers 1974). This implies a low spatial density-variation within rodent habitats. During the low phase, this spatial density-variation is likely to be at its maximum as small rodents are often observed to be very unevenly distributed (e.g., Lidicker 1973; Emlen 1973). Therefore, although low-density (essentially empty) patches are easy to find during this phase, mates outside the birth or home patch may be difficult to find. This implies small p and g; thus, *a low dispersal rate (but greater than zero) is expected during periods of extreme low densities.*

At peak densities, on the other hand, most patches are inhabited by many individuals. This can support a high predator density in the entire habitat complex (O. P. Pearson 1966, 1971; Hansson 1979*b*); under such conditions, dispersal through transition habitats is very perilous. Furthermore, well established demes are known to resist invasion of strangers (Andrejewski, Petrusewicz, and Walkova 1963; Sadlier 1965; Healey 1967; Lidicker 1975; Boonstra 1978; Getz 1978; Getz and Carter 1980). Both features result in low p-values *implying low dispersal rates during peak densities.* During such periods of general crowding, natural selection is more likely to favour highly competitive strategies (D. E. Gill 1974, 1978*a*; Stenseth 1978*a*). For similar reasons, dispersal should also be low, if not lower, during crash periods. This is because both N and K (in the expression for crowding $\rho = N/K$) will decrease greatly during crash phases.

The optimal dispersal rate will be at its maximum during the most intensive increase phase. Overall, a positive correlation between the optimal specific dispersal rate, v^*, and the specific growth rate of the population, $(\mathrm{d}N/\mathrm{d}t)1/N$ (Fig. 5.7(a)) is expected to be found when analysing data from a given region (see also C. J. Krebs (1978*b*). Correlating v^* and the density N should yield no significant correlation although a clockwise rotation over time is expected (Fig. 5.7(b)). As can be seen from Fig. 5.7, a non-zero dispersal rate is always expected to be optimal.

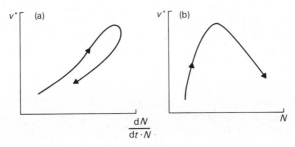

Fig. 5.7. Predicted changes in the optimal dispersal rate v^* (a) as a function of the specific growth rate, $dN/(dt \cdot N)$ and (b) density N. The optimal dispersal rate is predicted to be slightly higher in the early increase phase as compared with the late increase phase; thus, the depicted clockwise rotation (a).

If populations exhibiting high degrees of density variations are also characterized by more extensive and intense increase phases (Fig. 5.5), a positive correlation between v^* and a cyclicity index is expected (for possible measures of such density variations, see Tamarin 1978*a*; Stenseth and Franstad 1980; Stenseth *et al.* 1982; see also Fig. 5.15). This prediction can be tested by comparing populations of different species or populations of the same species in different geographical areas.

Adaptive (or ESS) and non-adaptive dispersal, rather than presaturation and saturation dispersal The observation that dispersal is high during periods of extensive population expansion and before the carrying capacity of the patch is reached, made Lidicker (1975) invent the term 'presaturation' dispersal; the underlying idea was anticipated by him in an earlier paper (Lidicker 1962) when he, somewhat misleadingly, called it 'density responsive' dispersal, and by Howard (1960, 1965) who called it 'innate' dispersal. Lidicker (1975) introduced the concept of presaturation dispersal as a contrast to 'saturation' dispersal; saturation dispersal is indeed the orthodox view and earlier authors have often accepted its existence without further argument (e.g., MacArthur 1972; Gaines, Vivas, and Baker 1979*b*). Lidicker (1975, p. 106) states that 'presaturation emigrants may be characterized by possesing a particular sensitivity to increasing densities'. On the other hand, he states on p. 105 that

saturation emigration is the outward movement of surplus individuals from a population living at or near its carrying capacity. Such individuals (exhibiting saturation dispersal behaviour) are faced with the immediate choice of staying in the population and almost certainly dying. They would represent more often than not, social outcasts, juveniles, and very old individuals, those in poor conditions, and in general those least able to cope with local conditions.

In earlier sections I have found theoretical evidence for expecting presaturation dispersal to be a real phenomenon. However, there is no theoretical reason for expecting a particularly high 'saturation' dispersal rate to be an *adaptive* phenomenon. Rather, there are reasons to expect the optimal dispersal rate to have comparable magnitudes during high and low densities. This led M. H. Smith *et al.* (1980*b*) to suggest the term 'ambient' dispersal which in general will be lower than presaturation dispersal.

In this respect it is indeed interesting to observe that Lidicker (1975, p. 107) states that 'in general, it can be claimed that where conditions favour the evolution of dispersal behaviour, it will be the presaturation type that will appear'. Thus, Lidicker seems to have anticipated the theoretical arguments given above in favour of presaturation as opposed to saturation dispersal. Indeed, it is worth noting that Lidicker (1975) primarily, and possibly exclusively, was able to find examples of presaturation dispersal, and not of saturation dispersal. Obviously the most valuable contribution of Lidicker (1975) was his appreciation of dispersal occurring at densities lower than carrying capacity.

On the basis of my analysis I suggest that we abandon the use of saturation dispersal altogether; this term is indeed misleading. A theoretically more appropriate distinction would be between *adaptive (or ESS-type of) dispersal and non-adaptive dispersal.* Adaptive dispersal represents presaturation dispersal (occurring during periods of high specific growth rate) and ambient dispersal (occurring at low and high densities; i.e., *parts of* what Lidicker called saturation dispersal). Their dispersal tendency is independent of density and has nothing to do with a possible existence of an excess of animals in their home patch (a feature often associated with dispersal; see Tamarin 1980*a*). Non-adaptive dispersers represent those forced out of their home population by competitively superior individuals; they are the losers and include the majority of what earlier has been called saturation dispersal. These non-adaptive dispersers represent the surplus individuals. I have indicated the relative magnitude of these two types of dispersal in Fig. 5.8.

Adaptive dispersers are expected to consist of adults (being in good reproductive condition) as well as young animals all of whom are healthy. Their heritability for dispersal tendency is expected to be high. Non-adaptive dispersers, on the other hand, will be principally nonreproductive individuals of all ages (i.e. functionally subadults; see Myllymäki 1977*a*) driven out by adult aggression (cf. Lidicker 1975; C. J. Krebs 1978*b*; Gaines and McClenaghan 1980). No heritability for dispersal tendency is expected for non-adaptive dispersers. Great caution is necessary when interpreting genetic differences between dispersers and residents (see 'Genetics and dispersal', this chapter). Only during the increase phase are genetic differences appropriately related to differences in dispersal tendencies; during the peak and crash phases when extensive non-adaptive dispersal may occur (Fig. 5.8), observed genetic

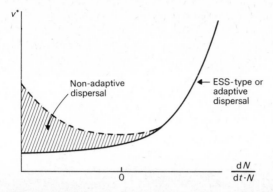

Fig. 5.8. From the reviewed models the depicted relation between the optimal dispersal rate v^* (called ESS-type or adaptive dispersal) and the specific growth rate, $dN/(dt \cdot N)$, is depicted by the solid line; see also Fig. 5.7(a). This adaptive dispersal occurs in all phases of the density cycle, but is most pronounced during the increase phase and must be distinguished conceptually from the non-adaptive dispersal rate (i.e. individuals being forced to leave their home patch due to competitive inferiority); non-adaptive dispersal is expected to be most pronounced during unfavourable or crowded conditions. C. J. Krebs (1978*b*) (see also Gaines and McClenaghan 1980) suggested similar relations. However, C. J. Krebs did not use the terms adaptive and non-adaptive dispersal; he used, somewhat misleadingly, saturation and presaturation dispersal. Nor did Krebs spell out the theoretical basis for his prediction. See main text for further discussion.

differences between dispersers and residents may primarily be due to competitive differences rather than differences in dispersal tendencies.

Superficially these are very different predictions from those of Lidicker (1975) (see also C. J. Krebs 1978*b*; Tamarin 1980*a*). In order to compare them, I have reproduced Lidicker's Fig. 5.1 (Fig. 5.9(a)). (Small alterations in Lidicker's Fig. 5.1 would make his predictions, as presented graphically, more similar to mine; (cf. Tamarin 1980*a*, his Fig. 5.1.) Essentially, the differences are as follows. Lidicker implies no correlation between v^* and $dN/(dt \cdot N)$ but a counter-clockwise rotation in the v^*-$dN/(dt \cdot N)$ plane (Fig. 5.9(b); I suggest a positive correlation between v^* and $dN/(dt \cdot N)$ but no easily detectable rotation in the v^*-$dN/(dt \cdot N)$ plane (Fig. 5.7(a)). Furthermore, Lidicker implies a positive correlation between v^* and N (Fig. 5.9(c); I predict no such correlation but a clockwise rotation in the v^*-N plane (Fig. 5.7(b)). In spite of these differences, it should be noticed that Lidicker does in fact predict a low specific rate in dense populations comparable to the low rate expected in sparse populations. The basic difficulty (or fallacy) in Lidicker's presentation is that he concentrates too much on the *number of dispersers* rather than the *rate of dispersal*; this common fallacy has lead to great confusion in the literature (see pp. 83–7 and McClenaghan and Gaines 1976).

Lidicker's concept of a dispersal sink Lidicker (1975, p. 117) defined a 'dispersal sink' as a place into which dispersers move; 'such a sink will generally be some

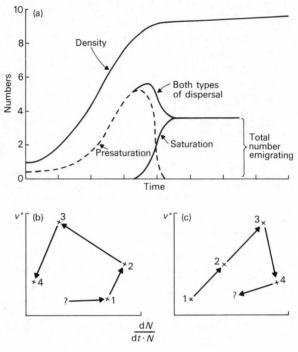

Fig. 5.9. Relation between changes in density and both saturation and presaturation types of dispersal as depicted by Lidicker (1975, his Fig. 5.1). (a) is redrawn from Lidicker (1975) as accurately as possible. From this, the predicted dispersal rate, v^* (i.e. saturation and presaturation dispersal combined), is related (b) to the specific growth rate, $dN/(dt \cdot N)$; and (c) to the density N; compare (b) and (c) with Fig. 5.7(a) and (b) and see the main text for further discussion. The numbers 1,2,3, and 4 represent consecutive stages of the population growth depicted in (a); 1 is the bottom phase and 4 is the peak phase. Note, however, that if (a) is drawn slightly differently [e.g., as done by Tamarin (1980*a*, his Fig. 5.1)], the corresponding (b) and (c) may resemble those in Fig. 5.7 more closely.

empty or unfilled suitable habitat, or perhaps marginal or even unsuitable habitat in which at least short-term survival is possible'. However, on the basis of my analysis, the term 'dispersal sink' is misleading because it represents two fundamentally different concepts; one with reference to adaptive dispersers, another with reference to non-adaptive dispersers.

For the adaptive dispersers, the sink is likely to be another rodent habitat. In this case, the sink should be viewed as a patch where dispersers establish another deme, i.e. a patch with less crowding than in its previous patch; 'refuge areas' (as in fact used by Lidicker 1975) might be a better term for this concept (e.g., Frank 1956; Southwood 1977). *For the non-adaptive dispersers, on the other hand, the sink is most likely to be equivalent to the transition habitat.*

Individuals which have left a patch because of competitive interactions are unlikely to be able to enter another rodent-suitable patch. For them the sink is appropriately viewed as a graveyard. This seems to be, for instance, Tamarin's (1977, 1980*a*) and Beacham's (1980*b*) interpretation of dispersal sink.

On this basis I suggest that we abandon the use of Lidicker's term 'dispersal sink'. The underlying ideas are better understood by applying the terms suggested by Hansson (1977*a*): *donor habitats* and *reception habitats*, both of which are rodent habitats (Fig. 5.1). A donor habitat represents the patch from which an adaptive disperser emigrates, and a reception habitat is the one into which an adaptive disperser enters. Non-adaptive dispersers also leave donor habitats but must remain in the transition habitat until their death. Alternative terms which I contend clarify some of the difficulties with the sink concept, are suggested by Myllymäki (1974, 1977*a,c*) and M. H. Smith *et al.* (1978).

Some empirical information

Preamble In studying dispersal in small-mammal populations, as in any animal population, a fundamental problem is to distinguish individuals that are dispersers or migrants from those making routine exploratory movements. Most of the data referred to in the next two subsections ('Adaptive dispersal' and 'Dispersal and cyclicity') are generated by removing all residents from a defined area. Dispersers are then operationally defined as any vole caught in these areas after the initial population has been removed. Verner (1979) has questioned the validity of considering these true dispersers. He considers the generated data a result of the 'vacuum effect' (Stickel 1949) where only individuals adjacent to the vacated area move into it by extending their home range. I assume the patterns are real and not artefacts of the vacuum effect. This is supported by some scanty data provided by Abramsky and Tracy (1980), their Fig. 4; see also C. J. Krebs 1979, p. 69). For further technical comments relevant to the problem of identifying dispersers, see Gaines and McClenaghan (1980).

Adaptive dispersal Lidicker (1975) has summarized data on small mammals demonstrating that individuals may leave their place of birth or their home range long before the carrying capacity of the patch is reached. The house mouse (*Mus musculus*) is a well-known example. Additional examples are bank voles (*Clethrionomys glareolus*), California vole (*Microtus californicus*), meadow vole (*M.pennsylvanicus*), prairie vole (*M.orchrogaster*), long-haired rat (*Rattus villosissimus*), California ground squirrel (*Spermophilus beecheyi*), cotton rat (*Sigmodon hisipidus*), and nutria (*Myocaster coypus*). In general, all cyclic microtines (cf. C. J. Krebs *et al.* 1973; Stenseth 1978*a*; P. R. Grant 1978*b*) seem to exhibit this type of behaviour during the increase phase.

Many authors have, however, reported a positive relation between, what they call, the dispersal tendency and density (e.g., Van Vleck 1968; Joule and Cameron 1975; Gaines and McClenaghan 1980, their Table 2; but see McClenaghan and Gaines 1976). The fallacy in most of these studies is that the *number of dispersers* is correlated with *density*. That this relation is positive should be no surprise; when the density (or the number of potential dispersers) increases, the number of actual dispersers will increase although the dispersal rate may be constant or even decrease with density.

In a study on the meadow vole (*M. pennsylvanicus*) C. J. Krebs *et al.* (1973) gave one of the first examples demonstrating the existence of presaturation dispersal *and* the lack of saturation dispersal (Fig. 5.10); they demonstrated

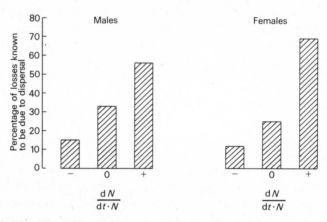

Fig. 5.10. Percentage of losses known to be due to dispersal for control populations of meadow vole (*Microtus pennsylvanicus*) during different phases of the density cycle. Dispersal is most pronounced during the increase phase. (Data taken from C. J. Krebs *et al.* (1973, their Table 2.)

that losses due to dispersal were greatest in the increase phase. C. J. Krebs *et al.* (1976) gave more detailed data on Townsend vole (*M. townsendii*); these are further analysed in Fig. 5.11. There is, as predicted, a positive relation between v^* and $dN/(dt \cdot N)$ (Fig. 5.11(b)) whereas there is no relation between v^* and the density N (Fig. 5.11(a)). This is indeed strong evidence for the existence of adaptive or presaturation dispersal *and* equally strong evidence against the existence of saturation dispersal in this case. Gaines *et al.* (1979*b*) gave yet another example for the prairie vole (*M. orchrogaster*) demonstrating exactly the same patterns (Fig. 5.12). Finally, Gaines *et al.* (1979*a*) gave similar data for the bog lemming (*Synaptomus cooperi*) exhibiting the same patterns, although a less clear relation between v^* and $dN/(dt \cdot N)$ resulted from this study (Fig. 5.13). Altogether these studies support the general predictions derived in this chapter. Further data on cricetid rodents, in general consistent with these predictions, are summarized in Table 5.1; only 8 per cent of available studies suggest a positive relation between dispersal rate and density, whereas approximately 50 per cent suggest a positive relation between dispersal rate and

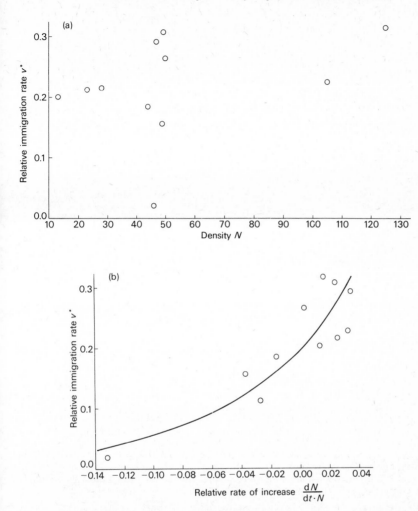

Fig. 5.11. Rate of immigration v^* into a vacated area, relative to the population size in a control area for the Townsend vole (*Microtus townsendii*). The relative rate of immigration is plotted (a) against population density N and (b) the specific rate of increase $dN/(dt \cdot N)$. The curve in (b) is fitted by eye. (Data from C. J. Krebs *et al.* (1976); their Table 2.)

specific growth rate. Due to insufficient data I have not included Beacham's results (1980*a,b*) in Table 5.1. His study suggested that the highest dispersal tendency occurred at peak density. However, the dispersers in the peak phase were predominantly smaller individuals, some of which matured before they were forced (?) to leave. Thus, these individuals may properly be classified as non-adaptive dispersers (see Fairbairn 1978*b*; Christian 1970 for a discussion relevant to this point).

Data on other small mammalian groups are even more sparse. R. J. Berry and

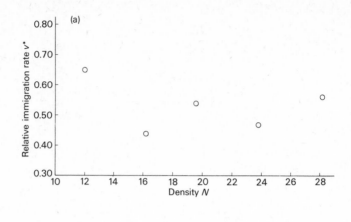

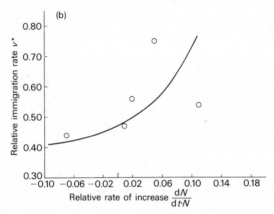

Fig. 5.12. Rate of immigration v^* into a vacated area, relative to the population size in a control area for the prairie vole (*Microtus ochrogaster*). The relative rate of immigration is plotted (a) against population density N and (b) the specific rate of increase, $dN/(dt \cdot N)$. The curve in (b) is fitted by eye. (Data from Gaines *et al.* (1979*b*); their Table 1.)

Jakobsen (1974) reported relevant data on the house mouse (*Mus musculus*) (Fig. 5.14). No relation exists between v^* and N nor between v^* and $dN/(dt \cdot N)$. Winberg and Keith (1976) and Keith and Winberg (1978) in a study on the snowshoe hare (*Lepus americanus*) indicated that a positive relation exists between v^* and N; no data on the population growth rate were available. Finally Dobson (1979) found no significant relation between v^* and N in a study of Californian ground squirrels (*Spermophilus beechey*).

Elsewhere (Stenseth 1978*a*, his Table 2) I have reviewed information consistent with the prediction that pregnant females which have had one litter are likely to disperse to another patch just before the next parturition. In that paper I have described various scanty data sets on five microtine rodent species

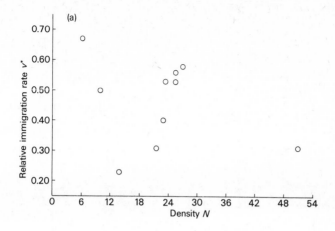

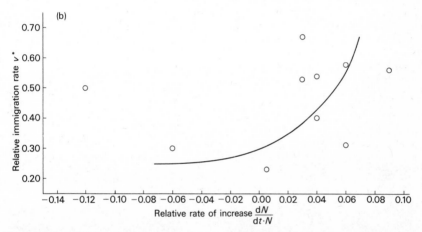

Fig. 5.13. Rate of immigration $v*$ into a vacated area, relative to the population size in a control area for the bog lemming *(Synaptomys cooperi)*. The relative rate of immigration is plotted (a) against population density N and (b) the specific rate of increase, $dN/(dt \cdot N)$. The curve in (b) is fitted by eye. (Data from Gaines *et al.* (1979a); their Table 1.)

exhibiting this type of behaviour. As predicted, this type of dispersal seems to be particularly pronounced during the increase phase.

Dispersal and cyclicity Lidicker (1975, p. 108) suspected (without any further justification) that presaturation dispersal (i.e. adaptive dispersal) might be common in cyclic species; my analysis has given theoretical support to this. The study of Rasmuson *et al.* (1977) on the field vole *(Microtus agrestis)* is important in this context. They found higher locomotor activity for animals from a cyclic population in northern Sweden than for animals from a stable population in southern Sweden (see Stenseth *et al.* 1982); these traits were genetically controlled. In Fig. 5.15, I have compiled all available data

TABLE 5.1 . *Dispersal rate (measured as recovery ratio) and percentage loss due to dispersal and their correlation to density and specific growth rate. Recovery ratio is estimated as the number of animals caught on the removal grid divided by the number of animals caught on the control grid*

Species, sex†, and location	Average dispersal rate per week	Correlation between dispersal rate and density	Correlation between dispersal rate and specific growth rate	Correlation between percent loss due to dispersal and specific growth rate	Reference
Microtus ochrogaster					
Illinois, US	u	ns	ns	u	Verner (1979)
Indiana, US ♂♂	u	u	u	+	Myers and Krebs (1971)
Indiana, US ♀♀	u	u	u	+	
Kansas, US	0.23	ns	ns	ns	Gaines *et al.* (1979b); see fig. 5.12
Microtus pennsylvanicus					
Illinois, US	u	ns	ns	u	Verner (1979)
Indiana, US ♂♂	u	u	u	+	Myers and Krebs (1971); see Fig. 5.10
Indiana, US ♀♀	u	u	u	+	
Massachusetts, US	0.06	u	u	+	Tamarin (1977)
Vermont, US	u	ns	u	u	Van Vleck (1968)
Microtus breweri					
Muskeget Isl., Mass., US	0.15	u	u	ns	Tamarin (1977)

Microtus townsendii					
Vancouver, B.C., Can.	0.12	+	+	u	C. J. Krebs *et al.* (1976, their Table 2); Krebs, Redfield, and Taitt (1978); see Fig. 5.11
Vancouver, B.C., Can.	u	u	u	+	C. J. Krebs and Boonstra (1978)
Synaptomys cooperi					
Kansas, US	0.16	ns	ns	ns	Gaines *et al.* (1979*a*); see Fig. 5.13
Peromysens maniculatus					
Vancouver, B.C., Can.	u	ns	+	u	Fairbairn (1978*a*)
Vancouver, B.C., Can.	0.48	u	+	u	
Samuel Isl., B.C., Can.	0.29	u	u	u	Sullivan (1977)
Saturna Isl., B.C., Can.	0.18	u	u	u	
Peromyscus polionotus					
South Carolina, US	u	ns	ns	u	Garten and Smith (1974)
Sigmodon hispidus					
Texas, US ♂♂	u	ns	u	u	} Joule and Cameron (1975)
♀♀	u	ns	u	u	
Reithrodontomys fulvescens					
Texas, US ♂♂	u	ns	u	u	} Joule and Cameron (1975)
♀♀	u	ns	u	u	
Summary††		8.3	37.5	66.7	

Symbol designations are as follows: +, statistically significant positive association (at 5 per cent level); ns, nonsignificant association; u, data unavailable.

† If no sex information, the results apply to both sexes combined.

†† Per cent significant positive correlations (calculated on the basis of the number of studies for which the relevant information is available).

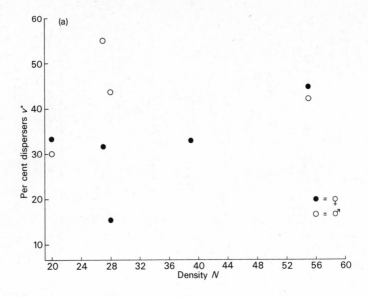

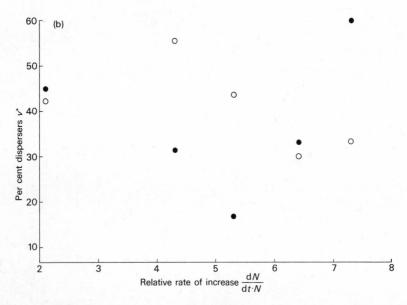

Fig. 5.14. Percentage dispersers v^* or the relative rate of dispersal for the house mouse (*Mus musculus*). Data on males and females separated. The dispersal rate is plotted (a) against population density N and (b) the specific rate of increase, $dN/(dt \cdot N)$. (Data from R. J. Berry and Jakobson (1974); their Table 3.)

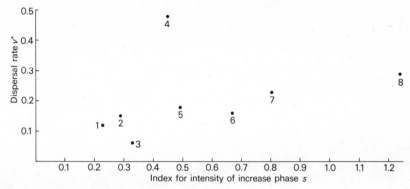

Fig. 5.15. Relation between average dispersal rate v^* for particular populations and an index for the intensity of the increase phase, s, for eight small-rodent populations; data on dispersal rates are taken from Table 5.1. The index s is estimated as the average specific rate of increase (per month) *during* the increase period; i.e. $s = (1/n)\Sigma \quad dN/dt \cdot N)_i$ where only positive $dN/(dt \cdot N)_i$ are included. Data points are: 1. Townsend vole (*Microtus townsendii*), C. J. Krebs *et al.* (1976, 1978); 2. beach vole (*M. breweri*), Tamarin (1977); 3. meadow vole (*M. pennsylvanicus*), Tamarin (1977); 4. deer mouse (*Peromyscus maniculatus*), Sullivan (1977); 5. deer mouse (*P. maniculatus*), Sullivan (1977); 6. bog lemming (*Synaptomys cooperi*), Gaines *et al.* (1979a); 7. prairie vole (*M. ochrogaster*), Gaines *et al.* (1979b); 8. deer mouse (*P. maniculatus*), Sullivan (1977). Including all points, a correlation coefficient equal to 0.335 is found. Deleting data-point 4, a correlation coefficient equal to 0.716 is found. No relation, as depicted in this plot, is found if the index for cyclicity used by Stenseth and Framstad (1980) and Stenseth *et al.* (1982) is applied. This is presumably due to the fact that no typical microtine rodent cycle (see Fig. 5.5) occurs for the populations from which the depicted information derives. In connection with these data it is important to realize that any given species might be characterized by different degrees of cyclicity in different habitats (e.g., Stenseth and Framstad 1980; Stenseth *et al.* 1982). Thus, a ranking of species according to cyclicity as done by C. J. Krebs (1979, p. 76) may be misleading.

(known to me) relevant to this issue; as can be seen a significant positive correlation between this measure of cyclicity and the population specific dispersal rate is found. The information in Fig. 5.15 is indeed important. It suggests that high dispersal rate is *not* associated with high stability as suggested by several recent theoretical studies (see the 'Introduction' to this chapter and the section below, 'Can dispersal drive the small-rodent density cycle?').

Seasonal migrations in lemmings Microtine rodents, particularly lemmings, living in arctic regions are reported to exhibit a very pronounced seasonal migration pattern (see J. D. Endler 1977 and the 'Introduction' to this chapter for terminological notes). Kalela (1961) and Kalela, Kilpeläinen, Koponen, and Tast (1971) described seasonal migration in the Norwegian lemming (*L. lemmus*). Batzli (1975, p. 248) summarized these findings. Studies of foraging patterns revealed that winter activity is concentrated under good snow cover in alpine habitats, particularly snow beds, grass-sedge heaths, and dwarf shrub heaths.

These habitats flood when the snow melts (i.e., decreased suitability), and the lemmings move into areas with better drainage and/or better cover (i.e., areas with higher suitability). Summer activities are concentrated in moist areas with fresh vegetation, including a well-developed layer of bryophytes, throughout the season.

Seasonal movement patterns of the Ob lemming (*L. obensis*) and the brown lemming (*L. trimucronatus*) resemble those of the Norwegian lemming and appear also to be responses to patterns of flooding and snow depth (cf. Batzli 1973, 1975; Batzli, White, MacLean, Pitelka, and Collier 1980). In winter the animals congregate in the lowlands, particularly polygon troughs and meadows where snow depth is greater. At melt-off these areas become flooded and lemmings are more concentrated in the uplands.

Thus, the pronounced dispersal tendencies seen in the rodent species living in these seasonally changing environments is consistent with the predictions of Andersson (1980).

Conclusion Overall much of the data support the predictions of the models. On this basis, I am confident that the theoretical framework summarized in the section on 'Evolution of dispersal', this chapter, is worth further consideration. Attempts should now be made to distinguish between adaptive and non-adaptive dispersers. The next step would then be to design specific field experiments to test the detailed predictions concerning the characteristics of these types of dispersal.

To disperse or not to disperse: what are the consequences?

Preamble

There are two effects of dispersal; (1) demographic and population dynamics; and (2) evolutionary.

As density increases, some animals may be forced to leave; being the non-adaptive dispersers, this type of dispersal should properly be classified together with other mortality factors (e.g., Lidicker 1965). Watson and Moss (1970, 1979) and Tamarin (1980*a*,*b*) would probably call this socially-induced mortality or a 'safety valve' (Lidicker 1962, 1975). Fairbairn (1977) further pointed out that these mortality factors may differ between sexes; only one of the sexes may experience non-adaptive dispersal.

Unless the total dispersal rate is too high, the non-adaptive dispersers will stabilize the population density (Maynard Smith 1968, p. 22; Hoff 1980). However, it is not *a priori* clear whether adaptive dispersers stabilize the population density or not. It is, in particular, unclear whether microtine rodents exhibit cycles *as a result of their vagility*, or *in spite of their vagility*; this is discussed in the next section.

The evolutionary effects of dispersal occur both in the donor and reception habitats. Gene-frequency changes in the donor habitat are due to adaptive as well as non-adaptive dispersers. The gene-frequency changes (including founder effects) in the reception habitat are mainly due to the adaptive dispersers. High dispersal rate may furthermore negate the effects of drift and microgeographic selection. For a thorough discussions of these aspects, see Wright (1931, 1940), Mayr (1963), Dobzhansky, Ayala, Stebbins, and Valentine (1977), M. H. Smith *et al.* (1978), 1980*a*), and Gaines and McClenaghan (1980). Lack of dispersal in a patchy environment may also have pronounced evolutionary effects; one aspect of these (the optimal sex ratio) is discussed in the section 'Nomadism', this chapter (also see Chapters 4, 7, and 8).

Can dispersal drive the small-rodent density cycle?

The problem and some observations Several authors have recently emphasized the importance of dispersal in generating the regular density cycles seen in several northern small-rodent populations (e.g., C. J. Krebs *et al.* 1973; C. J. Krebs and Myers 1974; C. J. Krebs 1978*a*; Tamarin 1978*a*; Rosenzweig and Abramsky 1980; Gliwicz 1980*a*). The data depicted in Fig. 5.15 certainly suggest a positive relation between cyclicity and dispersal tendency.

There is no reason to doubt the correctness of these data. However, the inferred importance of dispersal in generating cycles is difficult to accept on the basis of existing theory (see references in 'Introduction', this chapter). The only possible exception I am aware of, although it is difficult to understand, is the model by Abramsky and Van Dyne (1980); they seem to have found a case where dispersal begets cyclicity.

The existing models may, however, not really be relevant to the problem of whether dispersal can generate cycles. Their main shortcoming is that they really only consider the quantitative effects of dispersal. Data discussed in the section 'Genetics and dispersal', this chapter (Fig. 5.2), suggest, however, that dispersal will have important qualitative effects on both the donor and recipient populations. This is one of the important aspects in Chitty's (1960, 1967, 1977) theory as interpreted by C. J. Krebs and Myers (1974) and C. J. Krebs (1978*a*) (see also Stenseth 1981); further theoretical work is necessary before this idea can be tested seriously.

A new hypothesis for microtine population cycles

The idea Recently, Charnov and Finerty (1980) have expanded an earlier suggestion of Hamilton (1971, p. 77) concerning dispersal and aggression. The basic idea is that aggressiveness increases during periods of high dispersal because of the resulting decrease in relatedness between neighbouring individuals (see also Hamilton 1972); this high aggression may halt the population increase and even cause the population to crash. Below I further develop this idea on the basis of the results presented in the latter part of the section on the 'Evolution of dispersal', this chapter (Fig. 5.16). A habitat complex as depicted in Fig. 5.1 is essential.

Stage 1 (Fig. 5.16) Consider a population at low density just after the previous crash. Due to the low rate of adaptive dispersal just prior to and during the crash, neighbouring individuals are, at this time, likely to be relatively closely related. Furthermore, the population is unevenly distributed into what Lidicker (1973) calls 'survival pockets'; these survival pockets *may*, for instance, be favourable places where survival is above average for the habitat complex. As explained in the concluding subsection of 'Evolution of dispersal', the adaptive dispersal rate is low (but greater than zero) at this time. Thus, individuals continue to breed mainly in their patch of birth and thereby increase or maintain the high degree of relatedness; kin selection will under such conditions favour docile individuals having a high reproductive rate and interfering little or not at all with each other. See Chitty (1960, 1967), and Stenseth (1978a, his Table 3; 1981) for features characterizing such docile individuals.

Stages 2 and 3 (Fig. 5.16) As the population continues to build up and more rodent patches in the habitat complex become populated, an increased rate of adaptive dispersal is favoured by natural selection. Such a high level of dispersal causes non-related individuals to come into close contact with each other. Due to a reduction in relatedness between neighbouring individuals, kin selection will now favour more selfish and aggressive individuals. Aggressive individuals (often being large) are predicted (and observed) to have low reproductive as well as survival rates (Stenseth 1978a, his Table 1; Stenseth 1978d; see also Christian 1971). This increased aggressivity will, subsequently, lead to an increased non-adaptive dispersal rate (i.e. a further increased mortality rate) and a local halt to the population growth. Subsequently the adaptive dispersal rate will decrease and the over-all population growth in the habitat complex will cease.

The Crash (Fig. 5.16) The increase in aggression, described above, leading to an average reduction in absolute fitness, may cause an immediate crash (Crash 1). This crash may be due to intrinsic factors only; i.e. the interaction between dispersal and the resulting decreased relatedness between neighbouring individuals (Crash 1a). Alternatively, (Crash 1b), an extrinsic random disturbance of the type described by Stenseth (1977a, 1978b) may cause the crash as a result of the average low absolute fitness. Notice that the latter type will (similar to the Crash 1a type) result in regular cycles even though the disturbance is a randomly varying factor; the regularity is due to the relatively fixed length of time that necessarily must elaspe before the population can recover from its previous crash.

A qualitatively different type of crash may be understood within this framework (Stage 3 and Crash 2 in Fig. 5.16). A stable population density (with reduced adaptive dispersal rate) may conceivably occur when abiotic and

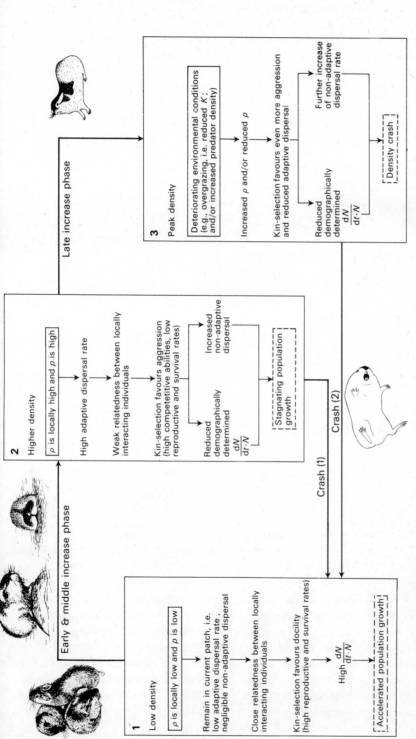

Fig. 5.16. A dispersal and kin-selection hypothesis for explaining small-rodent density cycles. Dispersal is the driving force in modifying demographic and behavioural parameters. See main text for further explanation.

meteorological conditions are favourable. However, such a general high rodent density may have a significant effect on their food supply as well as causing a buildup of predators (e.g., Hansson 1979*b*). This will cause a further increase in the rate of non-adaptive dispersal and the proper mortality rate, both of which lead to a decreased population growth rate; a crash seems unavoidable, particularly in the case of a randomly varying environment. Again, a fairly regular cycle will result. This latter type of cycle is a result of intrinsic, extrinsically random variation and deterioration of the habitat.

Dispersal together with intraspecific aggression is in all cases causing the termination of the increase phase. Thus, this theory suggests a possible answer to the question repeatedly asked by Chitty (1960, 1967, 1970, 1977); 'What prevents unlimited increase in the population?' (see C. J. Krebs and Myers 1974, p. 270).

Chitty's classification of the different types of crash The model depicted in Fig. 5.16 suggest an interpretation of Chitty's (1955) (see also C. J. Krebs and Myers 1974, pp. 284–8) classification of various types of crash. He defined three types (Fig. 5.17). Type M represents a rapid decline, in which numbers

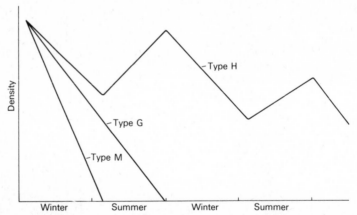

Fig. 5.17. Hypothetical diagram of three types of population decline recognized by Chitty (1955). (Redrawn from C. J. Krebs and Myers (1974); their Fig. 9.)

fall to a low during the winter and early spring after a peak year. Type G represents a more gradual decline bringing the density to a low by the end of the summer. Type H is the most gradual one; numbers fall slowly over one to two years with some recovery during the breeding season. In its extreme form, Type H may represent a stable population at high density. Of these, decline types we may interpret Type M as corresponding to a crash 1 in Fig. 5.16, and Type H as corresponding to a Crash 2 in Fig. 5.16. Type G is somewhere in between, but may most properly be viewed as a Crash 2.

Two habitat configurations corresponding to two patterns of density fluctuations? Getz (1978) concluded that habitat complexes may be classified into two categories: (1) large, relatively contiguous, stable habitats; and (2) small, isolated, patchy, and ephermeal habitats; or large and contiguous, but unstable habitats. Both may be interpreted within the framework of Fig. 5.1: Getz' first habitat configuration is one with rodent patches relatively close together; dispersing through the transition habitat is not too difficult. Such a habitat configuration was essentially assumed when discussing the model depicted in Fig. 5.16. The second configuration described by Getz, is one with very extensive transition habitat. Dispersal is indeed perilous in such a habitat complex, nevertheless, some dispersal is always adaptive. Getz assumed that a typical prairie vole (*Microtus ochrogaster*) and California vole (*M. californicus*) habitat is of the first type, whereas that of meadow vole (*M. pennsylvanicus*) is of the second type. In this connection it is indeed encouraging to observe that the prairie vole has a much higher dispersal rate than the meadow vole.

Getz (1978) summarizes several characteristics of populations living in the two types of habitats (see also Getz 1962; Christian 1970). The following features are important in our context.

Populations in habitat type 1 Reproduction is reduced at high densities, pair bonding (a sign of a high social organization?) exists, pregnant females display the Bruce effect (i.e. embryos are aborted or resorbed when strange males are present; Stehn and Richmond 1975; Kenney, Evans, and Dewsbury 1977) and sib-matings are prevented (e.g., Batzli, Getz, and Hurley 1977).

Populations in habitat type 2 There seems to be no correlation between population density and reproduction, no pair bonding (a sign of low social organization?); pregnant females display only a weak Bruce effect and no pronounced inhibition of sib-matings (e.g., Batzli *et al.* 1977).

Cyclic and irruptive density changes On this basis it seems justified to hypothesize essentially two different types of density fluctuations – the *cyclic* and the *irruptive* types. The cyclic type occurs in habitat type 1 and in populations in which reproduction is reduced as a result of exposure to strangers. The irruptive type occurs in habitat type 2 and in populations in which reproduction is accelerated as a result of social contacts. In the first case, animals react by increasing their competitive abilities; in the other case, animals react by increasing their own reproductive output.

I hypothesize that the characteristics described by Getz (1978) may be extended along a continuum in both directions. Species or populations may then be placed along this continuum of relative scale. Thus, I presume, for example that, most *Microtus* species (including *Microtus ochrogaster* and *M. pennsylvanicus*, both of which were discussed by Getz 1978; see also Getz and Carter 1980), and *Lemmus* spp. live in a habitat type 1 compared with *Myopus schisticolor* and *Dicrostonyx* spp. (at least the northernmost *Dicrostonyx-*

populations) (see Stenseth 1978*b*). The *Microtus-Lemmus* species may, on a relative scale, be assumed (!) to react to dispersers by inhibition of reproduction; the *Dicrostonyx-Myopus* species to react to dispersers by accelerating reproduction. The first group may be said to be cyclic, the second group to be irruptive (see, e.g., Pitelka 1973, his Fig. 3; Mysterud 1961; see also the section on 'Nomadism, site tenacity, and optimal sex ratios', this chapter).

These suggestions are indeed based on speculations and somewhat circular reasoning; however, as they are suggested on the basis of empirical findings, and because a consistent pattern emerges, I believe these ideas to be worth further developing and testing. For further discussion of similar ideas, see Tamarin (1980*b*).

Some observations Any given population in a particular habitat complex may go through the sequence 1–2–1 and 1–2–3–1 in a randomly alternating manner; thus, according to the theory depicted in Fig. 5.16, cycles are predicted to have a fairly inconsistent pattern. In particular, the decline phase is expected to be highly variable. This is also what is seen in most microtine cycles studies for a sufficient length of time (e.g., Pitelka 1973; Myrberget 1973; C. J. Krebs and Myers 1974; Batzli 1981; Batzli *et al*. 1980).

Given that density cycles occur, natural selection will favour different strategies in different phases of the cycle. Stenseth (1978*a*) found that the extremes of the demographic strategies favoured by natural selection, correspond to the two types (docile and aggressive) hypothesized by Chitty (1960, 1967, 1970); ample empirical evidence is consistent with the hypothesized existence of these demographic types (C. J. Krebs and Myers 1974; Stenseth 1978*a*; C. J. Krebs 1978*a*). Together with my earlier analysis (Stenseth 1978*a*), the above discussion clearly demonstrates that the observed polymorphism may be a consequence of the density cycles rather than the cause of the density cycles, as assumed in Chitty's theory. Genetics may, according to Fig. 5.16, be said to drive the cycle, but in a very different manner from that assumed in Chitty's theory.

The experiments manipulating the sex ratio of local populations carried out by Redfield, Taitt, and Krebs (1978*a,b*) may easily be interpreted as consistent with the model depicted in Fig. 5.16. For instance, Redfield *et al*. (1978*a*) removed males of the creeping vole (*Microtus oregoni*) from one grid and transferred them to a grid from which all females were removed; these females were transferred back to the male-removal grid. Both sexes were removed from a third grid. Essentially, these removal experiments produced (directly or indirectly) a high dispersal rate. Social disruption and increased reproductive output were concluded to occur in these experimental populations. This is consistent with the presumption that the creeping vole is an irruptive species around Vancouver. On the basis of the observed difficulties in finding creeping vole populations around Vancouver (D. Chitty, personal communication), this species may be assumed to live in a habitat type 2. Similar conclusions were drawn from a comparable experiment on the Townsend vole (*M. townsendii*) (Redfield *et al*. 1978*b*), a species that also may be assumed to be of the irruptive type (C. J. Krebs *et al*. 1976; C. J. Krebs 1979). However, it is also encouraging to notice that Petrusewicz (1957, 1963) observed a significant increase in the specific growth rate after introduction of strangers in a house mouse (*Mus musculus*) population.

Comparing dispersal and the level of wounding or other measurements of aggression should provide important information relevant to the hypothesis. Unfortunately, those (e.g., Rose and Gaines 1976; Rose 1979; Fairbairn 1978*b*; Beacham 1979*a*, 1980*b*) who have studied these aspects seem to have studied irrelevant quantities with respect to the present hypothesis. Rather than only measuring the attributes (e.g. aggression) of dispersers, they ought to have measured the behavioural effects of dispersal.

The final piece of information suggesting the validity of this hypothesis is the fact that C. J. Krebs and DeLong (1965) could not demonstrate a numerical response by adding food or fertilizing the field. Furthermore, other studies have indicated that there is no obvious food shortage during the vole peaks (Golley 1960; Chitty, Pimentel, and Krebs 1968); note in particular that *Microtus* spp. often consume less than 3 per cent of available food (e.g., Batzli 1975; N. R. French, Grant, Grodzinski, and Swift 1976; Ryszkowski, Goszczynski, and Truszkowski 1973; P. R. Grant 1978*b*). Finally, it is interesting to observe that Cole and Batzli (1978) and Getz, Verner, Cole, Hofmann, and Avalos (1979) found the highest peak densities in good-quality habitats, but were unable to prevent a decline by adding high-quality food.

The Krebs or fence effect One of the striking experimental responses of vole populations is the unusually high densities produced by fencing natural populations (C. J. Krebs, Keller, and Tamarin 1969; Boonstra and Krebs 1977; Gipps and Jewell 1979). This was called the *Krebs effect* by MacArthur (1972). Presumably, this is due to frustrated dispersal (Lidicker 1975); i.e., dispersal is strongly motivated but is prevented by some barriers. Presumably, within a fence, frustrated adaptive dispersers remain active in the population whereas frustrated non-adaptive dispersers die *in situ*. In these experiments a decline is, if it ever occurs, eventually brought about by starvation.

On the basis of the model depicted in Fig. 5.16, these experiments are easily understood; *immigrants*, i.e. strangers, are prevented from entering the population. Thus, the degree of relatedness remains high, and unlimited increase is not prevented before the population runs out of resources.

The emphasis is on the *lack of immigration*. C. J. Krebs (1978*b*), Tamarin (1980*a*), and Beacham (1980*b*) seem to place more emphasis on the lack of a *dispersal sink* in these fenced populations. On the basis of the previous discussion, I predict that adding a sink to a fenced population actually may stabilize the population at a high level by allowing the non-adaptive dispersers to move into the sink. However, I contend that this effect is rather peripheral to the effort in trying to understand why cycles occur.

The importance of immigration (as opposed to emigration and the existence of a dispersal sink) is, however, emphasized by the experiment carried out by Gaines *et al.* (1979*b*) and Abramsky and Tracy (1979). See Rosenzweig and Abramsky (1980) for a discussion of this point.

Nomadism, site tenacity, and optimal sex ratios

Hamilton (1967) pointed out that R. A. Fisher's (1930) argument for a 1:1 sex ratio at birth being optimal depends on the assumption of population-wide random mating; that is, the argument depends on the occurrence of extensive dispersal (G. C. Williams 1979; Bulmer and Taylor 1980*a*; see Chapter 7). Hamilton demonstrated that a female-biased sex ratio was optimal in a model in which mating occurred within small local subgroups before dispersal of mated females occurred. These arguments were extended by Stenseth (1978*b*) and Maynard Smith and Stenseth (1978) with reference to the wood lemming (*Myopus schisticolor*), an irruptive species; in this species a Y-linked mutant, Y^*, causes individuals having the genotype XY^* to be females; furthermore, the Y^* causes a meiotic drive so that XY^* females only produce X-eggs (cf. Fredga, Gropp, Winking, and Frank 1976, 1977; see also Kalela and Oksala 1966). The analysis carried out by Maynard Smith and Stenseth (1978) suggests that inbreeding (e.g., the lack of dispersal) may make the observed female-biased sex ratio in the wood lemming evolutionarily stable; however, in this case a rather extensive degree of inbreeding is necessary. These theoretical results have been confirmed by much more realistic models (Bulmer and Taylor 1980*b*; Thorsrud and Stenseth 1980).

The collared lemming (*Dicrostonyx groenlandicus*), another irruptive species, but a relatively unrelated species to the wood lemming (Kowalski 1977) is also observed to have a female-biased sex ratio (J. F. Hasler and Banks 1975; Stenseth 1978*b*; Gileva and Chebotar 1979). The much more vagile Norwegian lemming (*Lemmus lemmus*), a close relative of the wood lemming, lives in a more continuous habitat complex; thus, it is consistent with the framework described on pp. 93–8 to observe an approximately 1:1 sex ratio in the Norwegian lemming (Stenseth 1978*b*). On this basis it seems reasonable to suggest that the wood lemming and the collared lemming both have a female-biased sex ratio because of their similar ecologies and not, as suggested by Bengtsson (1977), because of the occurrence of a unique mutation in a common remote ancestor to these lemming species.

Conclusion and summary

In the earlier sections of this chapter, I have reviewed models and data relating to the evolution of dispersal in small rodents. Both on the basis of theory and data, it is concluded that an increased dispersal rate with increasing positive specific growth rate is adaptive; presaturation dispersal is the most important component of what is called adaptive dispersal. Dispersal occurring at saturation densities should, in most cases, be called non-adaptive dispersal to indicate its pathological nature. Further empirical studies on these different forms of dispersal are of great importance for the improvement of our understanding of small-rodent biology. The genetic basis for these dispersal types should also be investigated; only adaptive dispersal is expected to have a genetic basis.

The problems of whether or not dispersal may drive the density cycle commonly seen in microtine rodents have been discussed. A new hypothesis, originally suggested by Charnov and Finerty (1980), for microtine cycles has been further developed. The basic idea is that aggressiveness increases during periods of high dispersal because of a resulting decrease in relatedness between interacting individuals; thus, dispersal may conceivably be seen to drive the cycle. The development of mathematical models for this general idea should be undertaken.

Finally optimal sex ratios under various regimes of inbreeding (i.e. low dispersal rates) are discussed. When outbreeding is prevented (i.e., when incest is an optimal strategy), the optimal sex ratio is predicted to be female-biased. Here, empirical studies are necessary to test various theoretical models. In particular, studies on whether or not siblings or other close relatives breed with each other are of great importance with respect to general evolutionary biology and to density cycles in microtine rodents.

We are now at a stage where interesting patterns of dispersal and cyclisity are emerging; dispersal may indeed be one of the essential factors causing the regular density cycles in microtine rodents. But the underlying mechanisms may be very different from those commonly assumed.

Although some theoretical studies are still needed, it is now far more important to gather new data on various types of dispersal. We know when to expect dispersal; we also have some clear ideas about the consequences of various dispersal regimes. The testing of these expectations is paramount before further progress is possible.

6
Intraspecific differences in movement
IAN R. SWINGLAND

Movement enables a species to minimize the possibility of extinction through over-exploitation of resources and to extend its range. But to move may expose an individual to an increased probability of death when it could stay where it is more safely. Why move? Many aspects of moving in terms of time and energy budgets (Pyke, Chapter 2 and Mace *et al.*, Chapter 3), population structure and dynamics (Stenseth, Chapter 5; Rogers, Chapter 9, and Taylor and Taylor, Chapter 10), and genetics (Shields, Chapter 8), are discussed elsewhere in this book. This chapter is concerned with species in which some individuals are sedentary and some migrate.

Several movement models are described to illustrate ways in which intraspecific movement can be examined with the help of the evolutionary stable-strategy concept (Maynard Smith and Price 1973). Two case histories are examined in the light of the models, one an insect and the other a large vertebrate. A review then collates evidence within the animal kingdom of such movement differences and whether such differences are environmentally or genetically controlled. The genetic consequences of intraspecific differences in movement are discussed together with some of the life-history features which are associated with this tendency.

The most striking difference between the two behavioural forms of *Schistocerca gregaria* (a migratory locust) is in their different movement patterns. One form (*gregaria*) is intensely gregarious and migratory, while the other form is relatively solitary (thus *solitaria*) and migrates at night rather than during the day (Rainey 1976, 1978). In the migratory locusts (*Locusta, Nomadacris, Schistocerca*) this kind of phase polymorphism (Kennedy 1961a; Uvarov 1966, 1977; Albrecht 1967) is under genetic influence as well as subject to the effects of density on morphology, reproduction, and migratory behaviour. Kennedy (1961b) thought that this polymorphism was an adaptation to fluctuations between a habitat where resources are abundant and concentrated (i.e. *solitaria*) and one where resources are erratic and dispersed (i.e. *gregaria*). (*Solitaria* has longer wings and a smaller body than *gregaria*.)

In migratory locusts the differences in movement are not related to differences in age or sex but there are many cases in other species where the manner of movement is conditional on sex or age or both. Many raptors, like other widespread bird species in the northern hemisphere, are completely migratory in the north of their breeding range and completely sedentary in the south, while in the range in between some individuals leave and others stay behind (partial migration).

Frequently, partial migration involves a greater proportion of juveniles than adults and more of one sex than the other. In some falcons the males stay behind in large numbers and in accipiters, the females. In one accipiter, the goshawk (*Accipiter gentilis*), this sex difference is more marked in the juveniles than in the adults (Newton 1979). The proportions fluctuate annually in relation to food supply. For instance, when there were relatively large numbers of migrant goshawks captured in Wisconsin, the majority were adults and, when relatively few individuals migrated, they were predominantly juveniles. In 'poor' food years (those in which large number migrated), the migration occurred earlier than normal. What is noteworthy is that in the poorest of years (1972–73) females outnumbered males for the first time in 25 years (Mueller, Berger, and Allez 1977). This behaviour is paralleled by some seed-eating birds (Newton 1979) such as the chaffinch (*Fringilla coelebs*). In this species it is the juveniles and females which migrate in those areas where partial migration occurs, while the adults and males are sedentary.

Where the proportion of migrants fluctuates, individuals are probably responding to food conditions at the time and it is by this means that annual variations, particularly in relation to sex and age, can be explained. Food shortage would therefore be both the ultimate and proximate factor in influencing which individual migrates (Newton 1979). Newton says, when discussing the role of inheritance in partial migration, that there are two possibilities; either some individuals are programmed to migrate and others are not, or that the same individuals are programmed to migrate in certain conditions and not in others. The two possibilites are not mutually exclusive but the latter best accounts for the observed variations in raptor migration. This is a conditional strategy, an individual movement pattern which is conditional on environmental factors or on phenotypic charactors which can either be qualitative or quantitative. In the case of raptors and seed-eaters the behaviour is dependent on environmental factors and on the sex and/or age of the individual.

The evolution by natural selection of movement patterns would be impossible unless there were genetically-determined differences between the movement patterns of individuals. Such genetically-determined alternatives are called strategies. But which alternative, or strategy, will persist? Maynard Smith and others (Maynard Smith 1972, 1974*b*, 1976*b*; Maynard Smith and Price 1973; Maynard Smith and Parker 1976; Dawkins 1980) have developed the idea of evolutionary stable strategies (ESS) to resolve such questions. A strategy is evolutionarily stable against a list of alternative strategies if a majority of the population employs that strategy and if individuals exhibiting those alternatives do not do better. In order for two movement strategies to persist in a population the average pay-off to a migrator must be exactly equal to that of a nonmigrator. However, it seems extremely unlikely that in most cases the net costs and benefits of two strategies would, just by change, be exactly equal. There are therefore only three possibilities: *either* the population is not stable and selection is increasing the proportion of the more successful strategist, *or* the two movement patterns are not really strategies at all (i.e. there is only one single strategy in the population), *or* there is frequency-dependent selection maintaining the two strategies.

Movement models

1. *Conditional strategy – a single pure strategy.* The relative merits of migrating and non-migrating may not be the same for all individuals in the population. Competitively superior individuals, for instance male and/or adult finches, may do better by staying, whereas juveniles may do better by accepting the cost of migrating in return for a habitat in which there is less competition. In this case we would not necessarily expect the juveniles to have the same pay-off as the adults; each is doing what is best for it, but the juveniles may well be 'making the best of a bad job'. There is really only *one strategy*, 'if adult stay put, if juveniles migrate', but it involves a decision based, in this case, on the age of the individual and results in *two behaviours*. This has been called a conditional strategy because of the conditional nature of the decision rule.

2. *Mixed ESS.* A mixed ESS can either be mixed within an individual (an individual mixed ESS) or within a population (a genetic polymorphism). To distinguish these two possibilities is difficult. If we take a simple case, say of 'when to leave?' during emigration, it will illustrate the problem.

A population could be genetically monomorphic and each individual be programmed to leave at a time such that a mixture of frequencies of the various strategies follow some specific distribution whose mean is adjusted so that the success of leaving early is on average equal to that of leaving late, This is an inidividual mixed ESS.

Or, alternatively, it could be that the population is genetically polymorphic for emigration. In this case each individual pursues a pure strategy, so the population contains a mixture of pure strategists. This is a genetic polymorphism.

These two explanations of a mixed ESS are mathematically identical but biologically very different. It is probable that in most cases the individual mixed ESS is the more likely, as it is difficult to imagine a genetic process in which a specific distribution of leaving times is maintained in successive generations.

A mixed ESS is most probably maintained by frequency-dependent selection (Maynard Smith 1979; Clarke 1979). Heterozygous advantage and neutrality are two other possible mechanisms. Each of these two is supported by an argument which is not only lacking unequivocal evidence but such evidence is mainly drawn from theoretical studies or from experiments on organisms removed from their natural environment. The evidence for frequency-dependent selection is derived from these sources as well as from studies on actual or potential selective agents (Clarke 1979). Frequency-dependent selection is capable of maintaining genetic polymorphism (R. A. Fisher 1930; Haldane 1932) and it is important in promoting and supporting genetic diversity (Clarke 1975).

A balanced genetic polymorphism will result if the fitness of a genotype is related to its frequency (R. A. Fisher 1930). If the fitness of a genotype increases as its frequency in a population declines, there will be an increased proportion of the alleles composing the genotype in the next generation; conversely should the fitness decline as the frequency increases, the trend will be reversed.

Frequency-dependent selection is a kind of balancing act favouring rare genotypes *and phenotypes* (Kojima 1971; Ayala and Campbell 1974). In populations where some adults migrate and others do not, and where food resources are limited (see section on giant tortoises) the benefit to a migrant would decrease if more individuals started migrating and a concomitant advantage would accrue to each of the remaining nonmigrants. The migrant's reproductive output would be impaired while that of the nonmigrant would be increased. Ultimately, the critical proportion of migrants to nonmigrants would be restored.

3. *Learnt movement patterns.* Of the three kinds of movement models proposed, learnt movement patterns are at one end of a spectrum of genetic control and mixed ESS are at the other. Conditional strategies lie somewhere in the middle, for although they are conditional on some characteristic, they are still genetically determined. Even though a learnt movement pattern is acquired during the lifetime of an individual, there is an implied genetic predisposition to learn at some particular stage in the life history (e.g. during adolescence, imprinting in ducklings).

An animal can learn a movement pattern either from other individuals or from its own experience. Where it learns from others, such as its parents (e.g. Canada geese, *Branta canadensis*, Raveling 1976), or at random from unrelated individuals (e.g. Starlings, *Sturnus vulgaris*, Perdeck 1958, 1964), then the process is analogous to genetic inheritance. However, this poses the question 'so why learn, why not just inherit?' It is probably more economic to learn than to inherit as it takes up less genetic space. This analogy between movement patterns learnt from others and genetic selection implies that two movement types in a population ought to be maintained in the same way as genetic selection, namely by frequency-dependent selection. For this reason it is difficult to see how an analogy can be sought for a conditional strategy.

Individuals may learn by their own experience (e.g. lesser black-backed gulls, *Larus fuscus*, Baker 1980; Canada geese, Raveling 1976; also see R. R. Baker 1978). In this type of learnt movement pattern the frequency distribution of the different patterns is created anew in each generation. Selection, acting as an optimizing agent, is replaced by a judgement on the part of each individual on the basis of experience. Although something akin to frequency-dependent selection could be acting, it is unlikely as the final distribution of individuals must exist at the time that each individual is making a judgement about which areas to use, and it would also present a problem for intermediate stages in migration. It is, however, easier to see the analogy with a conditional strategy. An animal could make judgements about whether an area is suitable for itself as a juvenile or as an adult, all at the same time, then uses that area at the appropriate stage in its life-cycle. The pay-off from using an area as a juvenile is not necessarily the same as from using the same area as an adult.

A mixed ESS or a conditional strategy

A difficulty arises in distinguishing between a mixed ESS and a pure conditional strategy. For example, why do the Aldabran giant tortoises show individual differences in movement and why are these differences correlated to differences in carapace shape (see 'Giant tortoises'). The population may be at a mixed ESS *or* some tortoises are long and narrow and others short and wide, and it would not pay a short and wide individual to migrate. In the latter case all tortoises have the same pure conditional strategy, 'if short and wide do not migrate, if long and narrow migrate'.

Maynard Smith (1979) suggested that for a strategy to be interpreted as a mixed ESS, two criteria have to be met. Firstly, for a true mixed strategist, the pay offs, when adopting different policies, should be the same; and secondly, if any component of the supposed mixed ESS became more frequent, individuals exhibiting that component would have a lower pay-off (i.e. lifetime reproductive successes would be negatively frequency-dependent).

When reviewing particular field studies involving intraspecific differences in movement, can we say whether a pure conditional strategy, an individual mixed ESS, a genetic polymorphism, or a learnt movement pattern exists? Such few studies that have been done in the field are inconclusive to a greater or lesser degree. To collect data enabling distinctions to be made requires time; for instance to distinguish between a individual mixed ESS and a genetic poly-morphism, it would be necessary to know whether an individual always did the same thing or whether it performed both kinds of behaviour.

Two case histories

To illustrate some of the movement models which have been described, two case histories are mentioned in detail: dung flies and the giant tortoises of Aldabra.

Dung flies Male dung flies (*Scatophaga stercoraria*) wait for females which visit cowpats to mate and lay their eggs (G. A. Parker 1978). The rate at which they arrive on a cowpat falls off as the dropping grows stale. A male is faced with a problem: if he is the only male, he would maximize his mating success by leaving each cowpat as soon as it begins to get stale providing the time it took to find anther fresh pat did no lose him more successful matings than he would gain by staying. But he is not the only male. How long should he stay at a particular cowpat? If all males moved from one fresh cowpat to another, they would soon become overcrowded and competition for females would be intense. Perhaps then, a more appropriate male strategy should be to stay by a pat even though it becomes stale because although fewer females would arrive at least he would be faced with decreasing competition from other males. With either strategy the pay-off, namely successful matings, would drop off with time. Selection will favour those strategies that maximize the rate of resource uptake (i.e. successful matings) and if the time spent moving between cowpats is significant, those that act to minimize time costs for given gains will be selected. In this case there is a mixture of strategies of different staying times. The frequency of these various strategies follows a negative exponential distribution whose mean is adjusted so that the mating success of an early leaver is on average, equal to the mating success of a late leaver.

In the case of *Scatophaga* and *Sepsis cynipsea*, another dung fly (see G. A. Parker 1978), the mechanism by which the mixed ESS of emigration thresholds is maintained is not known. Parker considered that the most likely possibility is that the population is genetically monomorphic and that individuals are following a mixed strategy.

The only alternative is a genetic polymorphism; that the population is genetically polymorphic and individuals are pursuing a pure strategy. G. A Parker (1978) thought this possibility was unlikely (see section on 'mixed ESS').

Giant tortoises Individual giant tortoises (*Geochelone gigantea*) on Aldabra Atoll show distinct and definite movement patterns (Swingland and Lessells 1979). There is a marked seasonal shift of individuals to the coast during, or just before, the rainy season. Not all individuals migrate to the coast but the ones that do differ with respect to sex ratio and body size from those that remain inland.

The major gain from migrating is access to coastal food at the beginning of the rains which influences reproductive output (Swingland 1977; Swingland and Coe 1978). The major cost of migrating is the lack of shade, which tortoises need during the middle of the day to prevent overheating and subsequent death (Swingland and Frazier 1979). This lack of shade also restricts the foraging distance from the few isolated clumps of trees and bushes on the coast (Swingland 1977); more die on the coast than inland.

Those females which migrate to the coast have a significantly larger number of pre-ovulatory follicles and/or corpora lutea than nonmigrants and it was argued that such a difference will persist until the offspring are recruited into the breeding population (Swingland and Lessells 1979). Migrant females (particularly small ones) have a higher probability of death than nonmigrant females so that lifetime reproductive success in terms of surviving offspring may differ little between the two movement types. The pay-off to migratory females will be frequency-dependent; the more individuals which go to the coast the more the benefit (food) will decrease, while the cost (risk of death from over-exposure to the Sun) must increase as there is only a limited amount of shade available.

Males migrate in proportionately larger numbers than females; but we do not know whether males improve their reproductive achievement by migrating or not. The cost of migrating is lower for a male than a female; disproportionately more females than males die, suggesting competition for shade (Swingland and Lessells 1979).

In a recent analysis of the shape of the carapace it has been shown that migrants are longer and narrower than nonmigrants (Swingland, Parker, and North 1981). Although carapace shape in the Galapagos giant tortoises (*Geochelone elephantopus*) is inheritable, quite distinct carapace modifications can be exerted depending on the environment in which the individual grows (Fritts, unpublished manuscript). Nevertheless, although carapace shape and the environmental characteristics of the different islands appear to be related, young tortoises from differnt islands reared under identical conditions still show morphological differences suggesting there is genetic control. Such pronounced and different carapace shapes are not found between one island and another on Aldabra Atoll but they were found between the Indian Ocean islands before man extinguished nearly all the tortoise populations (Arnold 1979).

In short we cannot say whether differences in carapace shape are genetic and correlated with movement pattern, or environmental and the result of differences in movement pattern. What is definitely implied is that there are individual differences and that tortoises tend to do the same thing from year to year, i.e. migrate or not (see also Swingland and Lessells 1979). Thus, if it is a mixed ESS, it is far more likely a genetic polymorphism than an individual mixed ESS.

Further support for this possible existence of a mixed ESS comes from the circumstantial evidence suggesting that lifetime reproductive successes of migrants versus nonmigrants are negatively frequency-dependent (Swingland and Coe 1978; Swingland and Lessells 1979).

The other possibilities (conditional strategy, learnt movement pattern) cannot be dismissed. We do not know whether the pay-offs to migrants and non-migrants are exactly equal; if they are not, then it would tend to support a conditional strategy rather than genetic polymorphism. Similarly, we have no evidence denying the existance of learnt movement patterns. Young tortoises do appear to move haphazardly; sometimes staying inland, sometimes on the coast. They may be sampling the different habitats at different times, perhaps following more adult individuals. R. R. Baker (1978) put forward the hypothesis of familiar-area maps in which individuals acquire a knowledge of the area in which they are and move accordingly. However little, if any, incontrovertible information exists to support either Baker's hypothesis or learnt movement patterns in giant tortoises. More work will have to be done to show whether the intraspecific differences in movement are an example of genetic polymorphism, learnt movement patterns or a conditional strategy.

A review

A bean aphid (*Aphis fabae*) may not move more than a centimetre from its birthplace until it dies, while another genetically identical sibling may migrate a thousand kilometres before parturition occurs (B. Johnson 1958; L. R. Taylor 1957b, 1975). On a smaller scale the males of the digger bee (*Centris pallida*) either cruises around looking for spots where females are about to emerge from the earth and dig down and mate underground *or* wait in trees or shrubs some distance from the female emergence sites and attempt to intercept flying females who have been missed by the 'cruisers' (Alcock, Jones, and Buchmann 1977). Alcock (1980) has also illustrated intraspecific movement differences in other Hymenoptera. These and many other differences in the way invertebrates move are frequently associated with morphological and physiological characters (Lees 1975; Dry and Taylor 1970; Shaw 1970, 1973) such as flight polymorphism (Brinkhurst 1959; Lees 1966, 1967; Vepsalainen 1974; Young 1965a,b). In aphids (*Aphidoidea*), overheating (Dry and Taylor 1970), predator attack (Roitberg, Myers and Frazer 1979) and the age of the host (R. A. J. Taylor 1979a) provoke migratoriness; and in the field crickets (*Gryllus*, Gryllidae spp.) temperature changes (Ghouri and McFarlane 1958), density (Fuzeau-Braesch 1961; Saeki 1966a), photo-period (Alexander 1968; Masaki and Oyama 1963; Mathad and McFarlane 1968; Saeki 1966b; Tanaka *et al.* 1976), and diet (McFarlane 1962) have all be shown to affect the proportion of long-winged individuals in laboratory populations.

Attempts to define a genetic basis for flight behaviour polymorphisms in insects have produced ambiguous results. Single-locus genetic models have been proposed to explain observations on wing polymorphisms in water striders (Gerridae, Hemiptera) (Poisson 1924; Ekblom 1941; Brinkhurst 1959). Vepsalainen's (1973, 1974a,b, 1978) and Jarvinen and Vepsalainen's work (1976) on water striders had identified both genetic and phenotypic intraspecific polymorphism in these hemipterans. Populations of the different species show an array of wing-development strategies varying from winglessness (*Gerris najas*) to fully-winged (*G. rufoscutellus*). Genetic polymorphism is illustrated in *G. lacustris* in Finland. This species lives in rivers and lakes and experiences cool summers. It has a dimorphic winter generation and a short-winged summer generation; the same species living elsewhere in ditches, rivers, and pools is seasonally dimorphic having a long-winged winter generation (diapause) and a short-winged summer generation. Even further south *G. lacustris* has a genetically dimorphic winter and summer generation. *G. lateralis, G. asper, G. najas* are all genetically dimorphic. The single-locus genetic model has also been invoked in the field cricket, *Gryllus desertus* (Sellier 1954). Experiments on the genetic basis of intraspecific variation in movement in the flour beetle (*Tribolium castaneum*) showed that dispersal is probably determined by the genotype at a single, sex-linked locus (Ritte and Lavie 1977). The allele for dispersal is dominant. Males dispersed more readily than females of the similar genotype. Those two workers also showed that 'dispersers' had a predictably shorter developmental time and higher fecundity than non-dispersers (Lavie and Ritte 1977). According to the authors these experiments suggest heterozygous advantage as the maintenance mechanism for the movement polymorphism (contrasting Clarke 1979). They also mention the influence of environmental conditions on these movement patterns (Lavie and Ritte 1980).

In contrast to these examples, where the intraspecific movement differences appear to be under strong genetic control, there are example of environmentally-induced dimorphism (for instance, in some populations of the water strider, *Gerris lacustris,* mentioned earlier) rather than genetically-based polymorphism.

Genetic variation in threshold response to environmental factors is the most likely maintenance mechanism for intraspecific differences in movement (Harrison 1979); this threshold is specific to individuals (Southwood 1961). When particular factors in the environment become unfavourable they act as stimuli to influence ontogeny and cause some individuals to develop into migratory forms (Kennedy 1956; C. G. Johnson 1963), incorporating a genetic basis into a more sophisticated environmental cueing system (Taylor 1979*a*). These characteristics are most likely determined by multiple loci rather than single loci. Individuals may be sensitive to particular environmental factors only during a limited period of their life cycle. The relative threshold within genotypes, when different environmental variables will begin to affect an individual phenotypically, are probably the result of modifiers selected under specific environmental regimes. In the gerrids the frequency of long-wings will be determined by genotype with the homozygous dominant short-wing requiring higher temperatures to develop than the heterozygote, which needs in turn, higher temperatures than the homozygous recessive long-wing. The phenotypically-dimorphic populations of *G. lacustris* (for example) have lower temperature thresholds than genetically dimorphic ones. Vepsalainen (1978) explained the dominance of the short-wing allele by arguing that as habitats become isolated, the cost of growing long wings increases producing a dominant allele for short wings. Additionally increasing day length stimulates the development of short wings rather than long wings, while high temperatures induce some individuals to develop long wings even when photoperiod is increased. C. G. Johnson (1969) reviewed density effects on insect migration; in particular he highlighted the fact that, although different environmental factors can affect flight, variances between individuals are still high (C. G. Johnson 1976).

The interplay between environmental factors and genotype in controlling movement patterns in fishes has also been investigated but the results have often been unclear (Northcote 1969). Campbell (1972) in his study of the brown river trout and sea trout (both of which are the same species *Salmo trutta*) showed that although they differ markedly in their individual movement patterns they come from a single geographic interbreeding population. The proportions of the progeny of any mating which move to the sea and those that remain in the River Tweed in Scotland may vary from year to year but what influences the proportions is unknown. Electrophoretic analysis of blood serum proteins did not reveal consistent differences between the two forms.

Another example of intraspecific differences in movement involves the *leiurus* and *trachurus* forms of the three-spined stickle-back (*Gasterosteus aculeatus*) (Hagen 1967; Bell 1976; Wootton 1976). Apart from the genetically-based differences in seasonal habitat preferences, morphologically-different individuals are also genetically distinct (Hagen and McPhail 1970; Hagen 1973; Bell 1976). The different forms are maintained by selection (specifically predation) rather than gene flow (Moodie 1972; Hagen and Gilbertson 1973; Bell 1976). Positive assortative mating is estimated at 62–65 per cent within morphs (Hay and McPhail 1975). The migratory *trachurus* form is phenotypically uniform and stable; probably caused by the selection in more uniform environments and the greater probability of gene flow prohibiting drift. The more sedentary *leiurus* form, on the other hand, has a complex mixture of phenotypic patterns presumably induced by the many different habitats and predators. Wootton (1976) remains in doubt as to the functional significance of migration and joins Bell (1976) in indicating that the behavioural polymorphism allows the exploitation of a range of environments. This does not help in explaining the evolution of population divergence on the basis of individual selection.

Among reptiles, the marine turtles are probably the best known migrants. Information on the green turtle (*Chelonia mydas*), the ridley (*Lepidochelys olivacea*), the leatherback (*Dermochelys coriacea*), and the hawksbill (*Eretomochelys imbricata*) show a general pattern of intraspecific differences in movement. Turtles feeding in the same area may divide to breed on completely different beaches while those breeding on the same beach can divide and feed in widely separated feeding ranges (Hirth 1971). There are no data on whether this is genetically, environmentally, or culturally controlled.

About a thousand pairs of eiderduck (*Somateria mollissima*) breed in the Forvie Estuary in Scotland; two-thirds are migratory and the rest sedentary. The migrants leave the breeding ground in summer and return the following spring. By the time they return all the sedentary birds are already paired while the migrants pair during the migration. The nonmigrant birds nest in the estuary and the migrants near the sea. The gene frequency (from analysis of the egg albumen protein) is the same from year to year within the two colonies but consistently different between them (H. Milne and Robertson 1965). It does seem in this instance that the two colonies are entirely separate populations.

There are many other examples of bird species showing both migratory and sedentary behaviour (field sparrows (*Spizella pusilla*) and dickcissels (*Spiza americana*) Fretwell 1972, 1978; juncos (*junco hyemalis*) Ketterson and Nolan 1976; bluegrouse (*Dendragapus obscurus*) Zwickel *et al.* 1977; and other species (Alerstam and Enckell 1979). Leap-frogging (where the migratory population overfly the non-migratory population) has been documented in *Passerella iliaca* (Swarth 1920; Lincoln 1952) and in other species (Salomonsen 1955).

Harris (1970) cross-fostered young of the more sedentary herring gull (*Larus argentatus*) with the migratory lesser black-backed gull (*Larus fuscus*) and a significant proportion of the herring gull chicks, when adult, migrated (although not so far as their foster parents) — suggesting environmental and genetic influences. When lesser black-backed young were cross-fostered by the herring gulls, the former still migrated — implying genetic influences are dominant to environmental ones.

In some mammals, like the wildebeeste (*Connochaetes taurinus*) sedentary and migratory individuals are found within a population. In the Ngorongoro Crater, where grass is usually available throughout the year, the wildebeeste population is mainly sedentary but in dry years some Ngorongoro individuals will migrate to the surrounding Serengeti Plains; there the population is migratory (Talbot and Talbot 1963). Sedentariness is only found in seasonally stable areas (Estes 1969, 1976; Pennycuick 1975). Lidicker (1973) used the terms 'saturation' and 'pre-saturation' dispersal when referrring to small mammal populations. Saturation dispersal results from a 'overspill' of individuals in a population nearing carrying capacity whereas pre-saturation dispersal implies regulation below carrying capacity (see Stenseth, Chapter 5). He (Lidicker 1962, 1975) ascribed the following individual selective agents in favouring pre-saturation dispersal as the principle mechanism

1. More mating contacts;
2. Increased recombination and therefore more genetically diverse young (cf. G. C Williams 1975);
3. Avoidance by the individual of the problems of population increase;
4. Increased efficiency of resource use by dispersers.

Chitty (1967) suggested that the genetic composition of the population would change profoundly during population fluctuations and that aggressive behaviour between individuals would influence demography.

Chitty's and Lidicker's hypotheses have since been taken up and examined on innumerable species by countless workers (e.g. Myers and Krebs 1971; C. J. Krebs *et al.* 1973, 1976; Hillborn and Krebs 1976; Lidicker and Anderson 1962; C. J. Krebs 1966; Lidicker 1973; DeLong 1967; P. K . Anderson 1970; Kozakiewicz 1976; Kalela 1961; Kalela *et al.* 1971; Curry-Lindahl 1962; Clough 1965, 1968; Fairbairn 1977; Storer, Evans, and Palmer 1944; Stickel 1968; J. A. King 1955; Joule and Cameron 1975).

Nevertheless, the principle question remains: are there any genetic differences between migrants and residents within a species? The information available so far is sparse and unsatisfactory. The heterozygote Tf^c/Tf^e at the tranferrin locus is over-represented in dispersing female voles during the population increase phase (Myers and Krebs 1971; C. J. Krebs *et al.* 1973) and these females have a slight reproductive advantage over other Tf genotypes. Hillborn (1975) found that presumptive siblings in four species of *Microtus* dispersed suggesting a major heritable component of dispersal behaviour. Rasmuson *et al.* (1977) supported this suggestion by showing the heritability of high levels of dispersal activity in a northern cycling population of *Microtus agrestis* and low levels in a southern non-cycling population; hybrids between the two populations showed intermediate behaviour. Berry (1977) clearly describes an island house-mouse population consisting of residents on the peripheral cliff slopes and summer vagrants which move out from these areas to the central part of the island. One of the two movement-types is selected against in winter. Moreover, the movement-types are associated with different skeletal structures and different genotypes.

An unusual example of a nongenetic polymorphism and differential movement concerns the interaction between the dimorphic achenes of the bristly ox-tongue (*Picris echiodes*) and the natural small-mammal dispersers (e.g. *Clethrionomys glareolus, Microtus agrestis, Apodemus sylvaticus*). The inner achenes are wind dispersed but the outer achenes are dispersed by mice and voles. This is an example of a somatic seed polymorphism and is not the result of genetic segregation. In contrast, *Aegilops speltoides* has two kinds of dispersal units, borne on different individuals, and is an example of a genetic polymorphism (Sorenson 1978).

Habitat selection, genetic consequences, and life histories

The movement behaviour of individuals can be determined by the range of habitats available. A species can occupy a breadth of envionments either because the species has a number of divergent individuals each a habitat specialist, or because individuals are habitat generalists (e.g., Roughgarden 1972). If different sections of the population or species are each adapted to their specialized habitat, then polymorphism will result in populations or species with broad habitat use (Van Valen 1965). This niche-variation hypothesis states that species with broader ecological niches should be more variable than those with narrow niches because of the action of disruptive selection. In this pattern of selection the individuals closest to the mean (for some qualitative or quantitative characteristic, e.g. behaviour or size) are discarded and the extremes are mated together. (Of course, this results in considerable phenotypic diversity.) To test this niche-variation hypothesis, individuals from different areas would have to be compared to show habitat use was wider in some areas and then evidence would have to been found to show that a character important in exploiting that habitat was more variable in areas where the habitat use was broader. Studies have produced evidence for the hypothesis e.g. (P. R. Grant, Grant, Smith, Abbott, and Abbott 1976) and against (e.g., Soule and Stewart 1970).

Related to this hypothesis are several genetic models of disruptive selection induced by the distribution of a population between different habitats (or niches). These models demonstrate that if the fitness of different genotypes differ between habitats then polymorphism can be established with different equilibrium gene frequencies in the different habitats (e.g. Levins and MacArthur 1966; Bulmer 1972). Even in the presence of high gene-flow between the different habitats such differing gene frequencies can be maintained.

Maynard Smith (1966) has argued that genes which differ in fitness from habitat to habitat will tend to become associated with genes affecting habitat loyalty because an animal which selects a habitat in which it is fitter will leave more offspring than one which does not. But apart from some work on a polychaete worm (*Spirobis borealis*) by MacKay and Doyle (1978), which showed that disruptive selection can produce a response in terms of habitat selection, little if any experimental evidence exists to support this argument.

The density of a population can influence the movement patterns present and, apart from the frequency of genes affecting habitat use, it can influence the difference in fitness between two habitats. A good environment can become crowded and a new arrival may 'do better' by opting to settle in a less good and probably less crowded environment (Fretwell and Lucas 1970). The presence of other animals may lower fitness in the good habitat to the level of fitness in the less good, less crowded one because of competition.

In a crowded population of long-lived adults, vacant territories due to recent deaths will be rare and scattered. Dispersal (e.g. itinerant adolescents) could expose youngsters to a large number of potential vacancies even at the expense of encountering young competitors who are also looking for openings (c.f. learnt movement pattterns). It has been shown theoretically that, where the availability of the most preferred habitat is limited by competition and other suitable habitats are rare, a young animal (in this case marine plankton) may go through an exploratory phase before selecting a habitat provided that the average increase in fitness (from information gathered during that phase) outweighs its chance of death (Doyle 1975). As the exploratory phase proceeds, the plankton become less choosy (Knight-Jones 1953). Where an animal has limited time in which to find a suitable habitat after which time it dies, it becomes worth opting for a less good site if searching ability is low or if the difference between suitable and less suitable sites is small (Levins 1968).

It is sometimes asserted that intraspecific differences in movement are more common in *r*-selected than *K*-selected species. This is because (it is argued) *r*-selected animals are more vulnerable to environmental fluctuations, and therefore to extinction, because they are smaller than *K*-selected animals.

The truism that small animals are *r*-selected and large animals are *K*-selected has been superceded by Southwood *et al.* (1974) and Stubbs (1977) who have shown that most large animals are *K*-selected but only some insect species have life histories which are *r*-selected; some are *K*-selected.

In fluctuating populations, selection favours a high reproductive rate which reduces the ability of the population to damp environmental fluctuations which increases fluctuations in the population and so on (Horn 1978). This form of self-reinforcement exists for both extremes of *K*-selection and *r*-selection; animals may show extreme characteristics more often than an intermediate set of characters (Horn 1978). Intraspecific differences in movement may be cited as an extreme characteristic.

Are intraspecific differences in movement patterns more common in unpredictable or fluctuating environments? It is believed by some that environments with high temporal heterogeneity will select for individuals with broad ecological niches (e.g., Levins 1968) and that populations of such individuals might be more likely to show intraspecific differences in movement. The corollary is that species in temporally homogeneous environments should not show dichotomy in movement patterns because they will have narrower niches. It is argued that animals in the equable tropics face more constant environments than temperate zone animals and they therefore should have narrower niches than their temperate counterparts (e.g. Sheppard, Klopter, and Oelke 1968). However there are no conclusive results; animals in the tropics do not necessarily face more constant environments than temperate species and there are numerous examples of intraspecific differences in movement patterns in the tropics (e.g. R. R. Baker 1978).

7
Mating systems and the evolutionary consequences of dispersal

PAUL J. GREENWOOD

For some time now zoologists have attempted to explain much of the variability in the dispersions and social organizations of species in terms of two main factors, the spatial and temporal variations in their food supplies and the risks of predation (e.g. Crook 1965; Crook and Gartlan 1966; Lack 1968; Schoener 1968; Clutton-Brock and Harvey 1977a; see also Chapter 3). A parallel development has also resulted in a reappraisal of mating systems in a similar vein (e.g. Orians 1969; Bradbury and Vehrencamp 1976, 1977; Emlen and Oring 1977; Wittenberger 1979). Whereas in the past mating systems were usually classified on the basis of the nature of the pair bond (e.g. monogamous, polygamous, promiscuous), nowadays the classification is based more on the underlying ecological, behavioural, and phylogenetic factors which determine the evolution of particular mating systems. This has meant that evolutionary questions about spacing patterns, group sizes and structures, and the mating systems of animals can now be tackled within one framework.

The mating systems of animals are inextricably linked to their spatial and group dynamics and therefore, ultimately, to their population structures. As such, the extent to which animals move from area to area or group to group will have profound consequences for species' social organizations. It is usually assumed that the widespread dispersal of individuals will result in the disruption of co-adapted genotypes and in a reduced likelihood of related individuals living in close proximity. Limited dispersal, on the other hand, will favour the genetic differentiation of populations, an increased probability of inbreeding, and the evolution of co-operative behaviour between relatives (Wright 1943, 1946; Hamilton 1964, 1971, 1972; see also Chapter 8). However, despite a long-standing theoretical interest there have been few attempts to synthesize studies of mating systems with those on dispersal and philopatry.

It is the purpose of this chapter to draw together some of the advances which have been made in the understanding of the ecological and evolutionary factors which influence mating systems with the somewhat more disparate body of information on dispersal. In particular I will argue that patterns of dispersal are closely related to the type of mating systems and that differences between mating systems have important implications for the population and social structures of species.

The chapter will concentrate mainly on two types of mating system both of which commonly occur in higher vertebrates. These are resource-defence and mate-defence mating systems. Resource defence is where members of one sex, usually the limited sex, control access to a resource which is crucial to the acquisition of mates. Resource defence need not necessarily be directed towards the protection of food supplies in the classical sense of the all-purpose territories of some birds but can include other features such as nest sites and roosting areas. Mate defence is where individuals of one sex primarily defend members of the other sex rather than the resources associated with their dispersion (see J. L. Brown 1964; Bradbury and Vehrencamp 1977; Emlen and Oring 1977). Both resource-defence and mate-defence mating systems include examples of polygamous as well as monogamous bonds. The first part of the chapter will briefly consider the association between mating systems and dispersal and the factors which give rise to dispersal, in particular sex differences in movement. The remainder of the chapter concerns the implications of limited and, in many cases, differential dispersal for the evolution of co-operation, disruption, and sex ratios. Examples will mainly be taken from higher vertebrates but where important principles are illustrated by other groups of animals these will be mentioned.

Mating systems and dispersal

Many species of animals undergo large-scale dispersal at one or more stages in their life history (see Chapters 6 and 8). Others are much more conservative with offspring breeding in localities close to their birth place and adults re-occupying a previous site of reproduction. In many instances it may be a group rather than a site to which individuals are faithful though frequently the two are concomitant. Animals subject to large-scale movement can be referred to as dispersive; more sedentary species as philopatric. Whilst there are predictable life-history differences between extremes of the two types, splitting the continuum at some point is at the moment an arbritary decision (see Chapter 8). I will be dealing with species at the philopatric end of the distribution and in this respect the vast majority of higher vertebrates probably come within this category.

In most philopatric species there are marked differences in dispersal between juveniles and adults. Natal dispersal, the movement between birth and breeding site, is almost invariably more extensive than breeding dispersal, the movement between successive breeding localities. I will not consider in depth the evolution of age differences in dispersal (for a comprehensive review see R. R. Baker 1978); some of the consequences, however, are dealt with in later sections. Instead I will concentrate on the evolution and implications of differential dispersal of the sexes which commonly occurs in species which, on the whole, are faithful to both birth and breeding site.

Sex differences in dispersal

In the vast majority of birds, females, both as juveniles and adults, are more likely to disperse than males. Two such examples of sex-biased natal dispersal are illustrated in Fig. 7.1 and 7.2. The great tit is a monogamous territorial

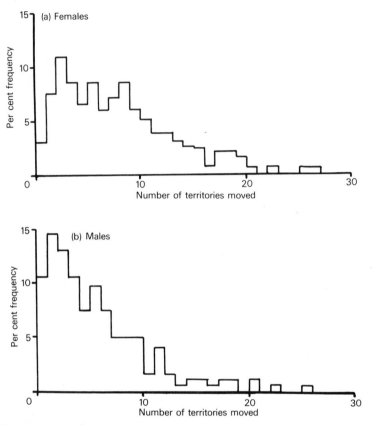

Fig. 7.1. Frequency distribution of natal dispersal in the great tit for (a) females and (b) males. (From P. J. Greenwood *et al.* (1979*a*).)

passerine which breeds mainly in deciduous woodland. On average, males move 4.6 territories from birth to first breeding site, females 7.1 territories (P. J. Greenwood *et al.* 1979*a*). Similarly, as adults, females are more likely to leave a previous breeding site than males. Breeding dispersal is however over much shorter distances than natal dispersal; the majority of adults return to within one territory's width of their previous nesting site (Harvey, Greenwood, and Perrins 1979). This pattern is one which is repeated in many other bird species (P. J. Greenwood 1980).

Differential dispersal is taken a stage further in group breeding birds such as

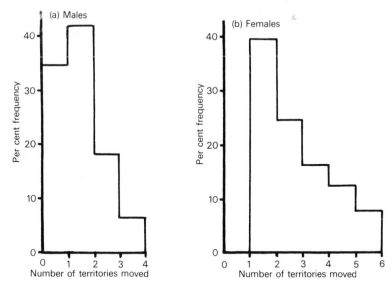

Fig. 7.2. Frequency distribution of natal dispersal in the communally breeding Florida scrub jay. (From Woolfenden and Fitzpatrick (1978).)

the Florida scrub jay (Fig. 7.2). Males are again the sedentary sex and for the first three or four years may act as non-breeding helpers on their natal territory. A large proportion of males eventually inherit their natal territory or one adjacent to it. Females tend to leave at a younger age and move further from their territory of birth to breed (Stallcup and Woolfenden 1978; Woolfenden and Fitzpatrick 1977, 1978).

Amongst mammals the usual pattern of differential dispersal is reversed; males are much more likely than females to leave their natal and breeding area or group. The sex difference in natal dispersal of the prairie deer mouse is illustrated in Fig. 7.3. A higher proportion of females stay within their home range of birth than males. In group-living species with more highly-structured social organizations the sex difference can be more marked, analogous to the pattern which prevails in some communal birds. For instance, in a long-term study of the olive baboon, fifty individuals have been recorded as transferring from one troop to another; all but two were males (Packer 1979).

Within both birds and mammals there are exceptions to the prevalent patterns of differential dispersal in their groups. An over-all summary of sex differences in dispersal is shown in Table 7.1. The bird species with male-biased dispersal are all from one family, the Anatidae, whereas the mammals with female-biased dispersal are a much more hetergeneous group both in terms of ecology and taxonomy. They range from the comparatively asocial pika to the more complex organization of the chimpanzee and, reproductively, from the effective monogamy of the African wild dog to the extreme polygyny of the white-lined bat.

The types of mating system which occur predominantly in the two taxa show

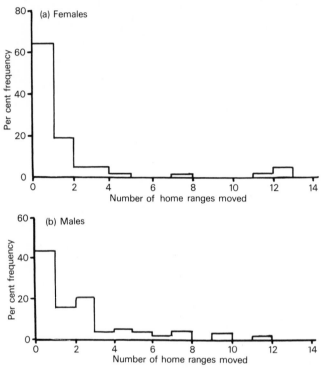

Fig. 7.3. Frequency distribution of natal dispersal in the prairie deermouse. (From Dice and Howard (1951).)

TABLE 7.1. *The numbers of species of birds and mammals with female-biased, male-biased, and no sex difference in natal and breeding dispersal. (From P. J. Greenwood 1980)*

	Predominant dispersing sex					
	Natal			Breeding		
	Male	Female	Both	Male	Female	Both
Bird species	3	21	6	3	25	1
Mammal species	45	5	15	21	2	2

a broad correlation with the over-all difference between them in sex-biased dispersal. Most bird species are monogamous and have a resource-defence mating system where males compete for resources in order to attract females. I have argued elsewhere that males will be better able to acquire a territory or breeding site at home than elsewhere – analagous to the home advantage enjoyed by football teams – whereas females without the constraint of establishing a territory choose between the available resources of different males (P. J. Greenwood 1980). The outcome of this difference in selective forces acting on males and females will be the appearance of a female bias in dispersal. Most mammals have a mate-defence mating system where males are primarily concerned with gaining access to and defending females and not the resources associated with their dispersion. Such a system may well have evolved partly as a result of the asymmetry in parental investment between the sexes. Whatever the ultimate cause of the mating system, the outcome is one of greater dispersal of the limited sex and philopatry of the limiting sex.

Amongst the exceptions in the two groups, the Anatidae have a pattern of dispersal which may at first sight appear atypical of monogamous species. Their behaviour, however, is much more akin to species with a mate-defence mating system in which male-biased dispersal is predicted. In the long-tailed duck, a male closely guards the female until egg-laying and then deserts (Alison 1975). Even in the lesser snow goose where pair bonds are more long-term and paternal investment greater, Mineau and Cooke (1979) suggest that the male's aggressive behaviour during the breeding season is not a form of territory defence but a means of protecting the female from rape. Outside the breeding season and following pair formation, it is the male which returns to the natal colony of the female to breed, irrespective of whether he is breeding for the first time or re-pairing with a new female (Cooke, MacInnes, and Prevett 1975).

Explanations for the mammalian exceptions are less obvious and whilst some species can be included within the proposed framework (e.g. white-lined bat with a resource-defence mating system (Bradbury and Vehrencamp 1977)) it is probable that others cannot (see P. J. Greenwood 1980). One possibility is that female-biased dispersal in the African wild hunting dog is the product of a mating system which involves defence of males by females, analagous to male dispersal when males defend females. Frame, Malcolm, Frame, and van Lawick (1979) point out that in the African wild dog it is females which are competing for access to groups of males and not vice versa. Why a mating system with male philopatry and female dispersal should have evolved in the first place is not clear. It is also likely that there are species where it is females which are the primary resource defenders in which case dispersal should be biased in favour of males. Some of the polyandrous bird species (e.g. jacana) may have breeding systems of this kind.

There is a broad association between the type of mating system and the direction of the sex bias in dispersal. However, a unitary hypothesis is unlikely

122   *The ecology of animal movement*

to account for the patterns across all species although a number of major inconsistencies in earlier explanations have been eliminated, particularly those based on the distribution of monogamy and polygamy in birds and mammals (Table 7.2). Two major factors which may give rise to sex differences in dispersal will now be considered before discussing some of the evolutionary consequences.

TABLE 7.2. *Mating systems and dispersal in birds and mammals. A summary of the main features associated with resource-defence and mate-defence mating systems. (Modified from P. J. Greenwood 1980)*

Resource defence	Mate defence
High male investment in resources, in presence or absence of mate(s)	Low male investment in resources, particularly in absence of mate(s)
Low female investment in resources	High female investment in resources
Intermale competition for resources	Intermale competition for mates
Mainly monogamous	Mainly polygamous
Male philopatry	Female philopatry
Greater female dispersal: (i) Reproductive enhancement – female choice of male resources, both natal and breeding dispersal	Greater male dispersal: (i) Reproductive enhancement – increase access to females, both natal and breeding dispersal
(ii) Inbreeding avoidance – mainly natal dispersal	(ii) Inbreeding avoidance – mainly natal dispersal

Reproductive enhancement Sex differences in reproductive costs and benefits are a major source of differential dispersal. The relative advantages which the two sexes gain from either dispersal or philopatry are probably the substantive reasons underlying the observed sex differences in dispersal. With resource-defence mating systems, the resource defender may be severely constrained in establishing and maintaining an initial site and in the potential for vacating one for another (see P. J. Greenwood 1980). Where males are the resource defenders, this sex difference in site attachment may in fact predispose them towards the evolution of traits atypical of males in other circumstances. There is an indication that site defence and paternal care are causally related both in fish and birds (see Ridley 1978). The tendency of females to move from one defended area to another to select a site is presumably a means of maximizing reproductive success.

When species have a mate-defence mating system the relative investment in resources tends to shift in favour of the limiting sex; in most mammals this is the female. A number of ecological and physiological factors may account for that shift. One suggestion is that particular spatial and temporal patterns of resource distribution, mediated in some cases through the stability and size of female groups, favours the evolution of mate-defence mating systems (Bradbury and Vehrencamp 1977). In addition, the asymmetry in reproductive physiology of male and female mammals probably makes dispersal extremely costly for females encumbered with offspring. Both factors will favour polygamy and intermale competition for mates. In the Anatidae, the high initial investment of females in eggs, the production of precocial young, and the difficulties of defending transitory food resources similarly predispose them towards a mate-defence system exemplified by female philopatry, intermale competition for mates, mate defence of females, and male dispersal.

Inbreeding avoidance Close inbreeding in species which normally outbreed is usually assumed to be harmful, although there is still a shortage of data from natural populations (see P. J. Greenwood *et al.* 1978; also Packer 1979; Ralls, Brugger, and Ballou 1979). Given the observed patterns of dispersal in higher vertebrates, one means of avoiding inbreeding in mammals is through male natal dispersal and, in birds, through female natal dispersal. Once individuals are established at a site or within a group, many of them remain there for the rest of their reproductive life. Thus, inbreeding avoidance is mainly achieved by the movement of young individuals away from their natal area. It is less likely that sex differences in breeding dispersal are important in avoiding inbreeding.

One major unresolved question is whether dispersal functions in an evolutionary sense as an inbreeding avoidance mechanism. In many species, the absence of close inbreeding may simply be an effect rather than a cause of dispersal. Any movement from the site of origin is likely to reduce the probability of encountering a relative as a potential mate. To illustrate the problem of disentangling cause from effect, it is worthwhile looking at the timing of dispersal in one philopatric species, the communally breeding Florida scrub jay studied by Woolfenden and Fitzpatrick (1978). Young females start breeding at an earlier age than males and to do so leave their natal group. Inbreeding avoidance as a cause of dispersal is implicated in the statement by Woolfenden and Fitzpatrick (1978) that scrub jays within a family do not mate with each other. Nevertheless, the impetus for female dispersal may come from the opportunity to breed elsewhere at a younger age than on the natal territory. One is left with the speculation that, in the absence of that selective force, females would eventually leave to avoid inbreeding but at an older age.

There are few data on the mechanisms which enable animals to avoid inbreeding. When inbreeding depression results from the mating of close relatives, those individuals which disperse will be at an advantage since the chances of encountering a relative as potential mate will be reduced. Most incestuous pairs of great tits have territories close to the site of birth (P. J. Greenwood, Harvery, and Perrins 1978). Under these conditions inbreeding avoidance could evolve as a demographic trait with no additional requirement that recognition of relatives need occur. However, more subtle patterns of dispersal could arise if animals were able to distinguish between close kin and non-relatives, or between familiar and unfamiliar individuals. As such, dispersal as an inbreeding-avoidance mechanism would be a consequence of an underlying behavioural interaction.

There is a limited amount of experimental data on quail which indicates that close kin are avoided as potential mates through a process of learning by familiarity at a particular stage in their early development rather than by an innate means of recognition (Bateson 1978; see also Chapter 8). There is also some circumstantial evidence from natural populations that there may be a behavioural basis to inbreeding avoidance which results in dispersal and that the process is mediated through the recognition of individuals as familiar or unfamiliar animals. Female olive baboons do not present to males born into their own troop but they may solicit males from neighbouring ones (Packer 1979). Similarly, chimpanzees alter their ranging patterns in a way which appears to minimize the possibility of close relatives mating together (Pusey 1979). The acorn woodpecker provides an example of conditional dispersal where movement between groups is dependent upon the presence or absence of a close relative as a potential mate. The acorn woodpecker is a communally breeding species where young males leave their natal territory if their mother is still in residence but lacking a mate; females depart when there is also a vacancy for a breeding bird but their father is still alive (Koenig and Pitelka 1979). On a more anecdotal level, Sherman (1980) comments that the movements of male Belding's ground squirrels are predictable only in that they do not visit their natal area to compete for mates. Natal areas tend to be occupied by close relatives. The obvious conclusion from much of this work is that such behaviours have evolved as a means of avoiding close inbreeding, though for most species the actual costs of inbreeding and the means of recognition are unknown.

Higher vertebrates may be atypical when compared to other taxonomic groups in the over-all lack of species which are known to habitually inbreed although Stenseth (1978b; Chapter 5) has predicted that cyclic inbreeding will be much commoner than previously supposed in species of rodents. Even so, close inbreeding appears to be much more widespread in other groups (Hamilton 1967; see also Chapter 8). Once this mode of reproduction is well established, reproductive costs will be minimal and selection to avoid such matings unlikely.

The consequences of philopatry and dispersal

Kin associations

Co-operative behaviours or other aid-giving traits are more likely to occur between kin than non-kin as a gene with such an effect is more likely to be shared by relatives (Hamilton 1964). Individuals may increase their inclusive fitness if relatives rather than non-relatives benefit from these acts. Philopatry to an area or group will increase the probability that related animals are in close proximity. Limited dispersal may indeed be an important prerequisite for the evolution of kin-selected traits. One probable example is the association of aposematic coloration with distastefulness in a number of insect larvae. The larvae of these species frequently occur in groups and it is presumed that members of these aggregations are siblings. Distastefulness may not confer direct advantages to an individual sampled by a predator but could benefit close relatives within the group. A similar advantage has been proposed to explain the association of close kin amongst tadpoles of the American toad. Here there is evidence that tadpoles preferentially choose to group with related individuals (Waldman and Adler 1979). This does not imply an innate means of kin recognition. It could simply be a learnt response to some cue such as an olfactory response to being present in the same batch of eggs as another individual. However, a recent study of the Cascades frog has shown that naive individuals prefer to associate with siblings rather than nonsiblings. This suggests that in one species at least some unspecified innate component is the basis of the discrimination rather than one which involves familiarity during rearing (Blaustein and O'Hara 1981).

For many examples of this nature it is often difficult to determine the evolutionary pathway which has resulted in groups associating along kin lines. Any tendency to aggregate, for example at a rich food source, in response to predators or simply as a passive byproduct of parental patterns of reproduction may result in sibling associations which are merely demographic consequences of limited dispersal and local dispersion. There are many advantages to philopatry and grouping which do not depend upon the presence of kin (see Chapter 8). However, once this type of population structure has evolved it will enable kin-selected traits to spread more easily and rapidly. A gene for, say, distastefulness could then act to reinforce any initial tendency for aggregation and philopatry and a whole series of life-history features may co-evolve in consort.

The examples described above involve characters which are not sex-limited. In species with more complex social systems, co-operation or kin associations have been frequently reported more commonly in one sex.

Sex differences in kin association Classic examples of sex differences in kin association occur among the eusocial insects. In the social Hymenoptera, non-reproductive female workers which are offspring of the queen assist in the rearing of reproductive siblings. Female workers can exhibit two forms of philopatry, environmentally to a locality such as a hive and genetically to a group of individuals. Relatedness within groups is maintained during swarming and transfer from one site to another and may be important in the establishment of new colonies. Such a female-dominated kinship system is mainly a product of the unusual reproductive habits of eusocial insects and the limited and brief role of males. In other species, however, it is more likely that a sex bias in dispersal will produce demographic patterns of kin association analogous to those of the eusocial insects. To what extent then do population structures and social organizations of higher vertebrates reflect the observed sex differences in dispersal?

In philopatric species some individuals will inevitably breed near to close relatives. In the great tit, for example, one-quarter of first-year males nest within one territory's width of their birth place (Greenwood *et al*. 1979*b*). Of those which are breeding in the same year as their father, 8 per cent are in an adjacent territory. Even so, there is little evidence of a direct reproductive advantage from this arrangement. It may simply be an incidental consequence of philopatry. Nevertheless, the possibility of kin-mediated advantages occurring in relatively asocial birds needs to be explored in more detail.

Familial proximity has reached a degree of sex-biased co-operation in some communal birds. These species tend to be characterized by low adult mortality, long juvenile periods, stable populations, and a limited availability of territories. In the Florida scrub jay, young male offspring provide more assistance to their parents than females both in the provision of food for nestlings and in territorial defence. Females leave their natal group when 2–3 years old; males assist for longer and, if they survive, eventually inherit their territory of birth or one adjacent to it (Woolfenden 1975; Woolfenden and Fitzpatrick 1978). Females as the dispersers may contribute less as helpers to minimize their costs prior to leaving. It is also possible that young females can avoid potential exploitation by their parents (Vehrencamp 1979) since their eventual reproduction does not depend upon territorial inheritance from them.

In mammals, high maternal investment has probably predisposed them to the greater incidence of co-operative behaviours between females than seems to occur among males in birds. Patterns of dispersion which are effectively extensions of the family unit are widespread amongst asocial species as well as those with more complex social organizations. Even so, co-operative behaviour is not an automatic feature of social groups nor of kin proximity. Whether such traits evolve will depend in part on the prevailing ecological conditions. As with birds, a vast range of grouping patterns are the result of resource distribution or predator pressure promoting cohesion irrespective of the presence or absence of relatives. But, since dispersion and philopatry will limit the potential source of individuals for groups or social units, family cohesiveness may well be a primary condition in the evolution of co-operation leading progressively to the selective recruitment of close kin.

When species do not live in groups, it is often the case in mammals that females occupying adjacent home ranges are close relatives, if dispersal is limited. This is the usual arrangement in many of the Sciuridae. In the detailed study of Belding's ground squirrel with its matrilocal population a whole series of behaviours have been reported for females which are in line with the observed pattern of dispersion. These include affiliative rather than agonistic behaviours which may be especially important when females are establishing nest burrows, sharing of parts of the home ranges allowing access to food and shelter, co-operative defence against conspecifics, which reduces the chances of intraspecific infanticide (see below), and alarm-calling when predators are present. The degree of co-operation between females decreases with decreasing genetic relationship (Sherman 1977, 1979, 1980).

In group-living species limited dispersal has also favoured the evolution of co-operative acts between close kin, such as the communal suckling of lion cubs (Bertram 1975, 1976). Each lioness appears to gain in personal fitness from the synchronization of litters and co-operative rearing so that kin proximity need not have been obligatory for such behaviour to occur but may have facilitated its evolution. Since the females within a pride are related and there is no costly prereproductive dispersal the over-all gain, in terms of inclusive fitness, is that much greater.

A sex bias in co-operative rearing occurs in the opposite direction in another social carnivore, the African wild hunting dog. Males are the philopatric sex in this species and it is the nonreproductive males which assist the breeding pair in the provision of food for the young (Frame *et al.* 1979), an arrangement comparable to communal birds with a similar sex difference in natal dispersal.

Group splitting Whilst the examples from group-living species described above have dealt with the patterns within groups, it is now clear from studies of primates that kin associations may be important factors during group fission. As group size increases the average degree of relatedness between members of the group will tend to decrease whilst competition for resources will increase. Clutton-Brock and Harvey (1976) suggested that under these circumstances the effect of aggression directed by dominant animals towards unrelated sub-ordinates will culminate in an eventual split. Since mother and daughter occupy adjacent positions in dominance hierarchies, they predicted that group fission will occur along matrilines with the lowest-ranking ones splitting away first. There are observations on the rhesus monkey which fit their predictions. Groups do split along matrilines and do so when the average degree of relatedness falls below the level of first cousins. Fission restores the degree of relatedness to a higher level in the new groups (Chepko-Sade and Olivier 1979). Although this one example fits the demographic picture outlined by Clutton-Brock and Harvey, the mechanisms underlying the split and the advantages to be gained from leaving need to be investigated in more detail. In addition, the models will also need to take account of other factors. It has been suggested that in both the Japanese monkey and the olive baboon, group splitting is prompted by changes in the male dominance hierarchy (Furuya 1969; Nash 1976).

Dispersal and Kin Association Under most conditions the dispersal of an individual away from its natal area or group will decrease the probability that relatives are in its neighbourhood. Some species do however have a pattern of dispersal whereby individuals can still derive benefits from kin association. The usual means of maintaining a degree of co-operation is for related animals to disperse together. In lions, pride males, the dispersing sex, are on average more closely related to each other than the females which are philopatric. Brothers co-operate in taking over prides. Once in possession a large group of lions has a high reproductive success. They can hold a larger number of lionesses for longer periods than either single or smaller groups of males (Bygott, Bertram, and Hanby 1979). Similarly, in the Hanuman langur, groups of related males may act together in order to seize a harem of females from the resident male. Unlike the lion, only one of the male intruders will remain with the female group; his prior assistants will be expelled (Hrdy 1977).

Dispersal and disruption

One consequence of leaving a group or locality is that dispersing individuals will encounter nonrelatives whose reproductive interests differ markedly from their own. In many species of mammals males will frequently attempt to move from one group to another whilst those already established in a group may try to repel them. Incoming males are potential competitors for mates and often have difficulties gaining entry to a group. This type of conflict is well documented in species of group-living primates (e.g. olive baboon, Packer 1979).

   When an animal is successful in moving to a new area or group, a further conflict of interest can arise between a newly arrived individual and residents. Particularly vulnerable are females with dependent offspring; their young may be killed by the new arrival. When such infanticide is committed by a male, the behaviour is interpreted as a means whereby he can sire his own offspring following the early resumption of oestrous in the adult females (Hrdy 1977). This type of infanticide has now been reported in a wide range of mammals (see Hrdy 1977; Sherman 1980). In other species the conflict is resolved in a different way although the outcome is similar to that brought about by male infanticide. Female prairie voles will abort if exposed to an unfamiliar male (Stehn and Richmond 1975) – an effect well known from studies of laboratory mice (Bruce 1959) – whilst post-parturition female collared lemmings will kill their own young in similar circumstances (Mallory and Brooks 1978). It is assumed that the male would eventually kill the young so that it is in the female's interest to terminate the investment.

   One study (Sherman 1980) has looked in more detail at the incidence of infanticide in relation to the site of origin of the perpetrator. In Belding's ground squirrel, unweaned young are frequently killed by conspecifics, usually adult females or young males. Killing by a female often occurs after she herself has lost her own young to predators. She will leave her own home range and move to a less vulnerable site, attempt to kill any young there, and settle close by. These females do not kill the young of relatives or neighbours. Sherman (1980) suggests that the main reason for the infanticide by females is to reduce the amount of competition for safe breeding sites. Yearling males, on the other hand, probably kill for food but, like females, do not attack close relatives.

The emphasis so far in this section has been on the success that dispersers may have in their competitive encounters with unrelated residents. In many species countermeasures have evolved to minimize the threat from intruders. In Belding's ground squirrel, close relatives co-operate in chasing away potential infanticidal immigrants. As a result, those females with more close relatives as neighbours are less likely to lose their offspring (Sherman 1980). This type of rejection system is taken a stage further in some invertebrates where there may be specific individuals responsible for the vetting of potential immigrants. For example, in a social sweat bee there are guards at the entrance to the colony burrow. Greenberg (1979) found that the proportion of arrivals which gained access to the colony was positively related to the average coefficient of relationship between the guard and the intruder. Discrimination between individuals was based on an olfactory cue with a genetical component in its variation. Those with an odour familiar to the guard, in other words close relatives, were allowed access; others were rejected.

Even when immigrants do succeed in establishing themselves in a group it may still be possible for residents to thwart their reproductive interests. One recently reported example concerns the function of infant-carrying by male chacma baboons. As with many primates, infants are at risk from immigrant males (see above). It appears that the frequency of infanticide is low in this species because of the protection afforded to unweaned offspring by resident males who are the putative fathers (Busse and Hamilton 1981).

Sex ratios

The philopatry of individuals to their natal area has profound consequences not only for the evolution of co-operative behaviour between kin but also for the equilibrium sex ratio. In a population which mates at random, Fisher's sex ratio theory predicts equal investment in males and females (R. A. Fisher 1930). But when species show limited dispersal in one or both sexes then conditions for the production of a 1:1 sex ratio may be broken. It was Hamilton (1967) who first proposed that in species with limited dispersal where brothers are competing with each other for mates the evolutionarily stable sex ratio should be biased in favour of females. More recently P. D. Taylor (1981) has explored a whole range of possible sibling interactions and their possible effects on the equilibrium sex ratio. To what extent then are the social organizations and patterns of dispersal of species likely to give rise to deviations from a 1:1 sex ratio.

When species live in small mating populations with limited dispersal between groups then individuals of both sexes will be closely related and inbreeding inevitable. Hamilton (1967) lists a number of invertebrates with this population structure where females maximize their reproductive success by producing an excess of daughters and sufficient sons to fertilize them (however, see Colwell 1981). Hamilton's theory of local mate competition has since received quantitative support from a study of a parasitic wasp (Warren 1980). However, as P. D. Taylor (1981) points out, there may be two factors favouring a female bias in the sex ratio. First, competition between male siblings for mates will be reduced and, second, an excess of daughters will increase the reproductive success of sons. Whilst the incidence of local mate competition in conjunction with sibmating appears fairly widespread in invertebrates, it has been rarely reported in other groups. One possible example is the wood lemming, a cyclic microtine, which also produces a female-biased sex ratio. Stenseth (1978*b*) proposes that sibmating is common in this species particularly during phases of low population density (see also Chapter 5).

Although studies of local mate competition have concentrated on species which habitually inbreed, inbreeding *per se* has no effect on the equilibrium sex ratio (Colwell 1981; P. D. Taylor 1981). As such, we might expect sex ratio distortions in birds and mammals as a consequence of their sex differences in philopatry. P. D. Taylor (1981) considers a number of possible intrasex interactions which may affect the sex ratio – competition or co-operation between brothers in relation to competition or co-operation between sisters. Unfortunately, his model deals exclusively with interactions within the same generation and not between generations. Even so, there are a number of species for which the model may be appropriate.

In the thick-tailed bushbaby studies by Clark (1978) the sex ratio of offspring is biased in favour of males. Typical of most mammals, males disperse and females remain in their natal area. Clarke suggests that the female siblings are competing with each other, and probably also their mother, for food and space. Under these circumstances the sex ratio bias is in the predicted direction. It is also possible to have the reverse situation where the sedentary sex is co-operating rather than competing. In terms of Taylor's model, if, for example, male helpers in communal birds are enhancing the potential reproductive success of their sisters then this would favour a sex ratio in favour of males. By considering the interaction between generations males more obviously are likely to be aiding their parents. In doing so they may effectively become the cheaper sex to produce due to their post-fledging contribution (Trivers and Hare 1976). Both in terms of Fisher's theory and Taylor's additional sibling interaction a male bias in the sex ratio would be predicted. Finally, the African wild hunting dog provides an example where two very different sex-limited sibling interactions may both contribute to the male bias in the sex ratio of pups. First, subordinate males assist their parents or brothers in the rearing of young through the provision of food from co-operative hunting. Second, sisters often compete for access to male groups in order to become the one breeding female in the pack (Frame *et al.* 1979).

Summary

Ecological factors such as the distribution of food are major determinants of mating systems. In turn, different types of mating system produce predictable patterns of dispersal. In birds and mammals, a female bias in dispersal is usually associated with a resource-defence mating system; a male bias with a mate defence-mating system. Differential dispersal has important effects on the evolution of kin associations and co-operative behaviour between members of the philopatric sex whilst those animals which disperse are frequently involved in aggressive encounters with non-relatives. The occurrence of sex-biased kin groups also produces conditions which influence the equilibrium sex ratio.

8
Optimal inbreeding and the evolution of philopatry

W.M. SHIELDS

Introduction

Animal dispersal has been defined as the movement of an individual away from its site of origin to a new area or succession of areas. Dispersal is expected to result in settlement, and if a propagule survives, breeding activity should eventually be centred in the new area (Howard 1960). Two distinct but related questions about the adaptive value of such movements are

1. What determines whether an animal disperses or not;
2. What selective pressures affect dispersal patterns, especially, how far propagules are likely to move?

Question 1 has been addressed in this volume (see Chapters 4 and 5) and elsewhere (e.g., Howard 1960; Lidicker 1962, 1975; Mayr 1963; Murray 1967; Cohen 1967; Gadgil 1971; Van Valen 1971; Hamilton and May 1977; Bengtsson 1978; R. R. Baker 1978). Here, I will assume that dispersal is adaptive and focus primarily on question 2.

Dispersal patterns, and especially distances, will have important consequences in determining population size and genetic structure (e.g., Dobzhansky and Wright 1943). Genetic population structure, in turn, is a primary force in determining the balance of random and selective factors which will control the evolutionary process (for reviews, see Wright 1977, 1978). Dispersal patterns in nature should be at least partially determined by the selective value of their genetic consequences.

In addition to strictly genetic factors, other more ecological pressures are likely to be important in adaptively shaping dispersal strategies. For example, average dispersal distances will often reflect the average distances separating needed resources. The availability of predator refuges, food, or suitable nest sites should influence a propagule's decisions about where and when to settle, and therefore the distance it actually moves before settling.

As a first approximation, dispersal patterns may be considered adaptive responses to different combinations of independent genetic and ecological selective pressures. The primary focus of discussion will be on what these pressures are, the relative importance of each type, and how they might interact to produce the dispersal patterns observed in nature.

Dispersal: philopatry and vagrancy

Whatever its adaptive nature, dispersal is characteristic of life. Whether powered by external forces (passive) or their own locomotive power (active), dispersive propagules are produced by all organisms at some stage in their life cycle. There are usually three species-characteristic patterns of dispersal. When one stage of the life cycle (usually breeding adult) shows site tenacity (i.e., a relatively stable attachment to a specific place), propagule dispersal is usually philopatric or vagrant (Fig. 8.1). When no stage shows site attachment, then individuals can be considered spatially nomadic, as they wander more or less

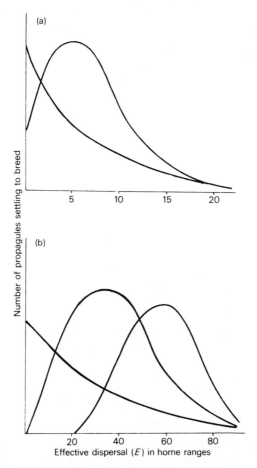

Fig. 8.1. Hypothetical distributions of propagule dispersal distances reflecting the actual patterns observed in many organisms in nature: (a) philopatric dispersal with median effective dispersal (E = median dispersal/average home-range size) less than 10 units; (b) vagrant dispersal with $E > 10$ home range units.

widely throughout their lives. Philopatry is observed in most vertebrates and many invertebrates (for reviews, see J. D. Endler 1977; R. R. Baker 1978; Shields 1982, as well as in many plants (for review, see D. A. Levin and Kerster 1974). Vagrancy is more frequent in benthic marine invertebrates, marine fishes, and in different plant species (for review, see Shields 1982. Nomadism is widely distributed phylogenetically, being found in every major animal group, but is rarer than either philopatry or vagrancy (see above references).

The word philopatry was originally used to describe an absence of dispersal, that is, if propagules remained at their birthplace (e.g., Mayr 1963). I define it more broadly as relatively localized dispersal, with propagules either remaining at their birthplace, or as near to it as local ecological conditions will permit (see below). Since the important genetic consequences of dispersal depend on local density as well as on absolute distances moved, spacing between breeding individuals is an important contributor to genetic structure (Shields 1982; for a similar conclusion reached independently, see P. J. Greenwood *et al.* 1979*a*). I will define effective dispersal E, then, as the absolute distance a propagule moves before settling, divided by the average distance separating settled individuals (e.g., home-range diameters). Since the distribution of dispersal distances is usually highly skewed towards the origin (Fig. 8.1(a)), mean dispersal may not be the best measure of central tendency. Operationally, then, I define as philopatric any dispersal distribution with a *median* effective dispersal less than ten units in magnitude. In philopatric species, absolute dispersal distances will be the same order of magnitude as the distances separating sedentary and site-tenacious breeders. Thus, an average propagule in a philopatric species will move fewer than 10 home ranges from its birthplace (Fig. 8.1(a); Table 8.1).

Vagrancy, in turn, is characterized by median effective dispersal greater than ten units in magnitude. While precise distance estimates are usually unavailable for vagrant species, owing to the difficulties of following propagules for great distances, the implication is that the average vagrant propagule moves much further than the average philopatric propagule. This can hold regardless of the shape of the dispersal distribution (Fig. 8.1(b)). Vagrancy is characteristic of the pelagic larvae of many marine invertebrates which travel hundreds or even thousands of kilometres before settling (for reviews, see Thorson 1950; Scheltema 1971).

As defined, philopatry and vagrancy have opposing genetic consequences. A philopatric species will usually interbreed in smaller and more isolated demes than are typical of vagrants. If all the propagules produced by a population remain at or near their birthplace, then the genetic relatedness of potential mates will be higher than if dispersal were less localized. By limiting effective population size (N_e, see Kimura and Crow 1963), philopatry results in inbreeding. As the intensity of philopatry increases, it will be mirrored by increased levels of inbreeding (Jacquard 1975). Increased tendencies to vagrancy, in contrast, are expected to produce larger demes and therefore increased levels of

TABLE 8.1. *Effective dispersal (median absolute dispersal/mean home-range diameter) for selected vertebrate species*

Species††	Absolute†			Effective‡		
	M	F	Avg.	M	F	Avg.
Amphibians						
Plethodon glutinosus	17.5	14.3	——	4.3	3.5	——
Reptiles						
Chelonia mydas	——	1400.0	——	——	7.0	——
Uta stansburiana	17.7	12.8	——	0.8	1.0	——
Sceloporus olivaceus	40.8	31.3	——	1.4	1.7	——
Birds						
Diomedea immutabilis	19.0	24.0	——	4.0	5.1	——
Muscicapa hypoleuca	——	1000.0	——	——	2.5	——
Parus major (1)	475.0	775.0	——	3.0	5.0	——
Parus major (2)	558.0	879.0	——	4.4	6.9	——
Parus caeruleus	——	——	700.0	——	——	9.3
Sitta europaea	——	——	900.0	——	——	5.9
Sylvia atricapilla	——	——	240.0	——	——	3.5
Passer montanus	——	——	300.0	——	——	3.0
Zonotrichia leucophrys	375.0	450.0	——	2.6	3.1	——
Melospiza melodia (1)	——	——	265.0	——	——	3.7
Melospiza melodia (2)	——	——	225.0	——	——	1.6
Mammals						
Peromyscus maniculatus	103.0	57.0	——	1.6	1.0	——
Tamias striatus	45.1	24.9	——	2.0	1.2	——
Spermophilus beldingi	449.7	47.0	——	2.6	1.3	——
S. richardsonii	——	——	50.0	——	——	0.4

† Median (or when not available mean) dispersal distance in metres.
‡ Effective dispersal (absolute dispersal/average home-range diameter) assuming circular home ranges.
†† Data from, *P. glutinosus*, Wells and Wells 1976; *C. mydas*, Carr and Carr 1972; *U. stansburiana*, Tinkle 1967; *S. olivaceus*, Blair 1960; *D. immutabilis*, H. I. Fisher 1976; *M. hypoleuca*, Berndt and Sternberg 1968; *P. major* (1), Bulmer 1973; *P. major* (2), P. J. Greenwood *et al.* 1979*a*; Perrins 1956; *P. caeruleus* and *S. europaea*, Berndt and Sternberg 1968; *S. atricapilla*, Bairlein 1978; *P. montanus*, Balat 1976; *Z. leucophrys*, M. C. Baker and Mewaldt 1978; *M. melodia* (1), Nice 1964; (2), Halliburton and Mewaldt 1976; *P. maniculatus*, Dice and Howard 1951; *T. striatus*, L. Elliot 1978; *S. beldingi*, Sherman 1977; *S. richardsonii*, Michener and Michener 1977.

outbreeding. We can consider dispersal magnitude a continuum ranging from intense philopatry to extreme vagrancy. This dispersal continuum helps control effective population size resulting in a concomitant breeding-system continuum. This can range from intense inbreeding ($N_e < 100$) to a potential outbreeding maximum of species-wide panmixia.

In this framework, relatively intense philopatry occurs in a taxonomically varied group of organisms. Philopatric species differ widely in trophic status, in intrinsic capacities to disperse (mobility), in whether dispersal is passive or active, and in the latter case in whether the species migrates or not (for reviews, see Mayr 1963; Endler 1977; R. R. Baker 1978). Is there a common denominator underlying this diversity which might implicate a primary function for philopatry? I suggest that philopatry's general genetic consequence of increasing inbreeding intensity by limiting population size is its primary function. More explicity, I suggest that the different combinations of morphological, physiological, and behavioural characters, which on a proximate level result in philopatry, often serve the same ultimate fuction of maintaining optimal levels of inbreeding.

The hypothesis, then, is that philopatry evolved and is currently maintained by natural selection, because it promotes inbreeding. Given the dogma on the disadvantages of inbreeding (e.g., Lerner 1954; Crow and Kimura 1970; Wright 1977; Maynard Smith 1978; but see Wright 1932; Carson 1967; Allard 1975), the obvious question is, does inbreeding possess sufficient advantages to have ever favoured the evolution of philopatry?

Optimal levels of inbreeding

A definition of inbreeding

Inbreeding has been defined explicitly and used implicitly in a number of fundamentally different ways (Jacquard 1975). The resulting ambiguity has produced confusion about the potential consequences of inbreeding and thus its potential selective value. A vast and traditional literature on inbreeding depression, perhaps in concert with a hidden legacy of human incest taboos (Carson 1967), has accentuated the *potential* disadvantages of inbreeding. This bias has apparently overshadowed an equally old, but sparser and less visible, literature on potential advantages of inbreeding (for review, see Shields 1982).

Intuitively inbreeding is thought to occur if related individuals mate. Relatedness is often viewed from two perspectives

1. Relatives are expected to share identical genetic material at some portion of their loci;
2. They share such alleles because they have been transmitted in replicated form from shared ancestors (Wright 1922).

If two individuals do share alleles identical by descent from common ancestors, they are considered relatives. Should they mate, they would be inbreeding. In this framework all intraspecific matings must be somewhat inbred. Given a continuous limit on population size, then, if one goes back far enough, any pair of contemporaries will share ancestors. This conclusion is not novel and many have remarked that such mild inbreeding must occur in every species (e.g., Darlington 1958; Falconer 1960; Kimura and Ohta 1971). Yet the point is often considered trivial, and so inbreeding has been arbitrarily and often implicitly redefined to encompass specific regions of what is actually an inbreeding continuum.

The more ancestors two individuals share, and the fewer unique individuals in their combined pedigrees, the more related they would be. The more related they are, the more intense the resulting inbreeding would be should they mate. This natural inbreeding continuum operates at highest intensity with selfing (incest in biparental species), reaching a practical minimum at species-wide panmixia (Table 8.2; Wright 1922). Thus, inbreeding intensity is inversely proportional to the size of a species-typical deme, itself defined as a random mating population. On the basis of what I feel are the important genetic consequences, I define global outbreeding as random mating over many generations in large demes $(N_e>10\ 000)$. While specific matings within such an outbred group are likely to entail inbreeding (e.g., occasional full-sib mating is expected), this 'nonrandom' component of the total inbreeding (Allen 1965; Crow and Kimura 1970) is relatively unimportant for this discussion. More important is the species-typical 'random' component of the inbreeding which is controlled by effective population size $(N_e$, Kimura and Crow 1963). Global inbreeding, and thus an inbred species, will be characterized by consistent random mating in small demes $(N_e<1000)$. This inbred portion of the continuum can then be subdivided into mild $(100<N_e<1000)$, intense $(2<N_e<100)$, or extreme inbreeding, this last being associated with incest $(N_e=2)$ or obligate selfing $(N_e=1)$.

TABLE 8.2. *Pedigree characteristics for various levels of inbreeding*

Breeding pattern	Possible	Ancestors in common (per cent)	Number of unique ancestors
Selfing	Yes	100	t†
Incest‡	Yes	100	$2t$
Mixed	Yes	$0<x<100$	$(2^{t+1}-2)<x<2t$
Perfect outbreeding	No	0	$(2^{t+1}-2)$

† t is the number of generations prior to the mates in question.
‡ Incest includes continuous parent–offspring or full-sibling matings.

The consequences of inbreeding

As defined, inbreeding is expected to generate genetic consequences different from those associated with wider outbreeding. These include

1. A primary genotypic (intralocus) effect of reducing individual heterozygosity;
2. A primary genomic (interlocus) effect of maintaining associations of alleles at different loci.

The latter consequence would increase the possibility, that when genomes are considered at *all* loci, members of inbred demes would be more similar than members of more outbred demes. The degree of: 1. heterozygote decay (or its complementary fixation) and 2. genetic homogenization, expected in each generation will vary as positive functions of the inbreeding intensity in that generation (for reviews, see Carson 1967; Jacquard 1975; Allard 1975; Wright 1977).

In a single generation, heterozygote decay results from the production of higher frequencies of homozygous loci in zygotes than had characterized their parents (segregational load). In the absence of selection, mutation, and migration, this fixation (drift) is expected to be cumulative, eventually resulting in complete homozygosis (measured by F as a fixation index; Fig. 8.2). In the presence of any form of heterotic selection (for a discussion of various kinds of heterotic selection, see Berger 1976), the net loss of heterozygosity can be reduced or eliminated. Inbreeding will reduce heterozygosity during zygote production, but selective factors will then increase the proportional survival of heterozygotes from the zygote stage = parenthood. This will raise the frequency of heterozygous genotypes in subsequent pools of parents (Wright 1977).

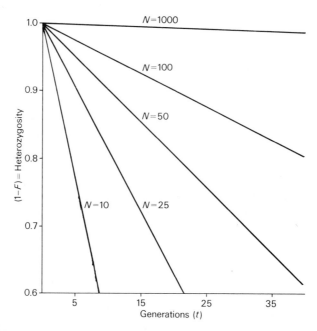

Fig. 8.2. Theoretical relation between heterozygosity (or its inverse F, the inbreeding coefficient) and effective population size (N), as a function of the number of generations of isolation in demes of particular size. As a measure of heterozygosity, the values illustrated assume no mutation, migration, or selection is occurring. As a measure of correlation, F is independent of such factors.

Regardless of selection, consistent inbreeding will *always* increase the correlation between uniting gametes (also measured by F; Fig. 8.2). Here, F estimates the proportion of alleles in a genome that have shared a common selective history (for at least 2 generations). Thus, with inbreeding, alleles A_1 and A_2 are more likely to find themselves consistently associated with alleles B_1 and B_2 at a second locus, rather than with B_4 or B_7 (Shields 1982). Inbreeding increases the probability that allelic associations across loci will be faithfully transmitted from parents to progeny. This genomic consequence can be viewed as 'organizing the entire populational genotype into a sort of giant supergene' (Allard 1975, p. 124). It results in increases in the genetic similarity of parents and their offspring relative to that expected under wider outbreeding. In the end, continuous inbreeding necessarily increases the genetic homogeneity (and not necessarily the homozygosity) of a local deme (for a comprehensive review, see Shields 1982).

The selective value of inbreeding

Whatever its intensity, inbreeding is likely to entail both costs and benefits. Cost can be measured in the inbreeding depression resulting from increased levels of genotypic segregation. If phenotypes are, in any measure, the result of intralocus dominance interactions, including overdominance, segregation will entail a genetic load consisting of two independent components. Since recessiveness is usually associated with deleterious effects (Crow and Kimura 1970), the increased segregation of homozygous recessives associated with inbreeding is expected to increase the dominance load relative to wider outbreeding. This is simply the unmasking of deleterious recessives commonly associated with inbreeding. Similarly, if any form of heterotic selection is operating, then the increased fixation associated with inbreeding is expected to result in an additional overdominant load. For both dominance and overdominance components, inbreeding depression is usually measured in the number of homozygous progeny *lost*, relative to the increased number of heterozygotes which would not have been lost under less intense inbreeding (for reviews see Falconer 1960; Wright 1977).

Inbreeding's less publicized benefits can be assessed in terms of the *outbreeding* depression associated with the recombinational load of wider outbreeding. If phenotypes are, in any measure, the result of interlocus epistatic interactions, then outbreeding will entail a recombinational load (G. C. Williams 1975; Maynard Smith 1978). By disrupting harmoniously interacting combinations of alleles carried at different loci and randomly recombining them into novel associations in progeny, outbreeding will reduce fitness relative to a more conservative asexuality. By preserving coadapted genomes, inbreeding can mimic asexuality's conservatism. Indeed, if mutation rates are high enough, inbreeding can be *more* conservative than asexuality. Unlike asexuality, which suffers a mutation-accumulating ratchet, inbreeding can edit novel mutations by recombining them into waste zygotes, simultaneously producing complementary zygotes with reduced mutational loads (Shields 1982). Finally, inbreeding will reduce the genetic 'cost of meiosis' (Williams 1979) as relatedness between mates, and thus parents and progeny, is increased (Shields 1982). Thus inbreeding's benefits result from its conservatism, from a capacity to faithfully transmit parental genomes which have proved themselves.

The possibility that extreme outbreeding, even within species, might entail such costs, and thus increased inbreeding benefits, was initially discussed by Wright (1932) and Mather (1943). Later experimental work supported their views (e.g., Table 8.3; Wigan 1944; Dobzhansky 1970; for a review see J. D. Endler 1977). More recent theoretical work (e.g. Shields 1982), as well as direct (e.g., Ryman, Allendorf, and Stahl 1979) and indirect (e.g., M. C. Baker and Mewaldt 1978; Bateson 1978) empirical evidence, has suggested that even mild *inbreeding* ($N_e < 1000$) can entail significant *outbreeding* depression (Fig. 8.3).

TABLE 8.3. *Comparison of viability, fecundity, and longevity of intra- and interpopulation hybrids of* Drosophila pseudoobscura *and* D. willistoni. *Intrapopulation value of measured fitness components is standardized at 1.00. (After Wallace's (1968) summary of Vetukhiv's data.)*

Species	F_1	F_2
Viability		
D. pseudoobscura	1.18	0.83
D. willistoni	1.14	0.90
Fecundity		
D. pseudoobscura	1.27	0.94
Longevity		
D. pseudoobscura (16°)	1.25	0.94
D. pseudoobscura (25°)	1.13	0.78–0.95

Extreme inbreeding can entail a cost owing to its genotypic consequences. Reduced levels of inbreeding, however, may entail balancing costs, owing to the independent recombinational load and meiotic loss associated with its genomic consequences (Shields 1982). The conclusion, often overlooked (e.g., Ralls *et al.* 1979), yet often reiterated (e.g., Wright 1932; Wigan 1944; Alexander 1977; Bateson 1978; Price and Waser 1979; R. H. Smith 1979; Shields 1982, is that there may be a genetically optimal balance between too close inbreeding and overwide outbreeding.

Optimal inbreeding

If we assume that inbreeding's primary function is the transmission of successful parental genomes, then maximum fidelity should yield maximum benefit. Maximum fidelity might be expected to occur at maximum inbreeding inten-

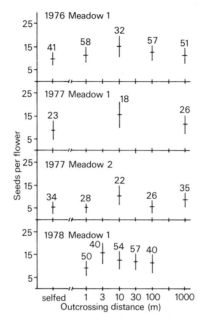

Fig. 8.3. The number of seeds set per flower in the montaine perennial *Delphinium nelsoni* as a function of the geographical and presumably genetic distance separating parent plants. Significant incest (selfing) and outbreeding (distances >100 m) depressions are indicated. (From Price and Waser (1979).)

sities. Yet extreme inbreeding is also associated with maximum costs measured in segregational load. One ideal might be to minimize the cost/benefit ratio (Fig. 8.4). Since the genomic consequences of incest can be generated at reduced levels of inbreeding (Jacquard 1975), maximum benefits can be maintained with lesser inbreeding intensities with the added bonus of reduced segregational loads. Factors influential in determining optimal levels of inbreeding in qualitatively predictable, but quantitatively unknown ways include the relative importance of dominance (favouring outbreeding) and epistatic (favouring inbreeding) interactions in determining fitness and the absolute and relative rates of occurrence of both favourable and deleterious mutations that operate through each type of interaction.

Based on genetic consequences alone, the optimal level of inbreeding will rarely include incest. Driving the optimum away from incest are the costs of segregation. Random mating in larger populations increases the probability that deleterious mutations carried by potential mates will have arisen independently. Compared with incest, this will reduce segregational load and may increase transmission fidelity (Shields 1982). With wider outbreeding, deleterious mutations are also more likely, at least initially, to be carried in

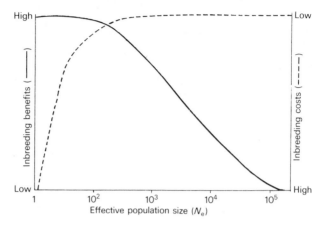

Fig. 8.4. A hypothetical representation of the costs (- - - -) and benefits (------) associated
with different inbreeding intensities. The minimum cost/benefit ratio for particular conditions will reflect an optimal inbreeding intensity for those conditions. Primary determinants of the shape and position of the curves include the relative importance of epistasis
versus dominance interactions in determining fitness. As the proportional contribution
of epistasis increases, inbreeding's benefits increase with increasing inbreeding. As the importance of dominance increases, the disadvantages of inbreeding increase with increasing
inbreeding (see text for more discussion).

heterozygous condition, before being edited via recombination. If such
mutations recur with appreciable frequency, the developmental adjustments we
call dominance can evolve, permitting production of progeny with unaltered
parental phenotypes (R. A. Fisher 1931; Manning 1976b). Since selection acts
on phenotypes, this phenotypic masking of mutation will increase inbreeding's
adaptive conservatism.

   Any adaptive tendency to reduce inbreeding will ultimately be limited by
the increasing likelihood that as mates become less related, increased recombinational load will balance reductions in segregational load. With decreased relatedness, not just recent mutations but also many favoured ancestral alleles
will have arisen in mates independently. If these interact epistatically, interbreeding will disrupt previously coadapted associations. This will generate
a recombinational load and entail a meiotic loss. These outbreeding costs will
then drive the optimum back towards more intense inbreeding.

   Relative fecundity is expected to affect the shape of the cost curves and
thus the inbreeding optimum. Increases in the number of progeny produced in
the range of low fecundity ($<10^4$ offspring/female/lifetime) should reduce the
impact of any segregational load. By producing excess zygotes, trade-offs between immediate genetic and inevitable ecological deaths associated with higher
fecundity, can be made. Such selection regimes result in higher permissible
inbreeding intensities (B. Wallace 1968).

Based solely on these genetic considerations and as a first approximation, most low-fecundity species could minimize the cost/benefit ratio of sex by avoiding incest, while inbreeding relatively intensely in small demes ($2<N_e<$ 1000). With much higher fecundity ($>10^5$ progeny), selection is expected to favour increased variance in families of progeny (G. C. Williams 1975). Only then would the optimum shift to the more progressive outbreeding end of the continuum ($N_e>10$ 000), for it is only such high-fecundity organisms that can afford the gamble of disruptive outbreeding (G. C. Williams 1975; Maynard Smith 1978).

While the details of the optimal inbreeding model are by no means proven, sufficient evidence consistent with its broader implications has accumulated that I will assume its truth (for reviews, see Wright 1977, 1978; Shields 1982). I do so in order to use it as a genetic base for a more detailed analysis of animal dispersal. The model provides a theoretical base to explore how independent genetic and ecological pressures interact in shaping animal dispersal.

Philopatry: origin and adaptive value

Historical antecedents

Much has been written about the proximate control of philopatric dispersal, but often with little regard for its ultimate consequences (e.g., C. G. Johnson 1966; Hilden 1965; Bekoff 1977). More numerous are the important studies exploring the consequences of philopatry without specifically addressing the question of function (e.g., Miller 1947; Ehrlich and Raven 1969; Levin and Kerster 1974; Endler 1977; Wright 1977). Still remaining is a substantial body of work which does deal directly with philopatry's origin and potential functions.

Supporters of models of dispersal as biotic adaptation appear to view philopatry as a result of opposing selection pressures that ultimately benefit populations or species. Either tacitly (e.g., Howard 1960; Johnston 1961; Halliburton and Mewaldt 1976) or explicitly (e.g., Wynne-Edwards 1962; Mayr 1963; Van Valen 1971), all view dispersal distributions as compromises between individual selection reducing and group selection promoting dispersal. The group benefit of increased dispersal is thought to result from the consequences of increased gene flow (e.g., Howard 1960; Lidicker 1962; Mayr 1963; E. O. Wilson 1975), pioneering new or currently empty habitats or niches (e.g. Johnston 1961; Mayr 1963; Udvardy 1969), or countering the effects of local population extinction (e.g. Wynne-Edwards 1962; Gadgil 1971; Van Valen 1971). Supporters of models of individual adaptation view any group-level consequences as fortuitous effects rather than as adaptive functions (e.g. Lidicker 1962, 1975; Hamilton and May 1977; and especially Murray 1967). Given the constraints on the efficacy of group selection (G. C. Williams 1966*a*; Maynard

Smith 1976*a*; but, see D. S. Wilson 1980), I will concentrate on models of individual advantage (for a review of all the models, see Shields 1982).

Models of chance

In sessile or sedentary organisms, philopatry is often considered a chance result of an inability to move longer distances. Individuals are either not mobile enough (e.g., snails) or disperse passively relying on outside agencies which necessarily limit dispersal (e.g., pollen, seeds, many insects). Both arguments are blinded by proximate factors. Both mobility and mode of dispersal are subject to selection and so could have evolved differently. A conclusion that in the case of low-mobility species philopatry is adaptation, rather than chance, is supported by species with similar mobilities or dispersal syndromes which nonetheless do disperse vagrantly. The few individuals in a philopatric species that move considerably farther than average also imply that increased magnitudes could become the population mean should this be beneficial.

The philopatry of very mobile and migratory organisms cannot be attributed solely to chance. A common alternative is that 'homing' serves to bring the sexes together after separation during non-breeding seasons (e.g., Orr 1970; D. A. Hasler, Scholz, and Horral 1978). This argument is usually applied when the spatial distribution of breeding sites is limited relative to the distribution of nonreproductive individuals (e.g., island-breeding pinnipeds, sea turtles, or birds, Orr 1970; Carr and Carr 1972). If a species' breeding range includes more than one nesting colony, getting the sexes together does not require *birth-site* philopatry. Individuals that returned to a general breeding area and then chose breeding sites at random from *all* those available would still find mates. They would also increase the level of 'beneficial' outbreeding. Despite considerable experience with entire breeding regions, most return to their birth colonies, often settling within metres of their birthplace (e.g., Austin 1949; Kenyon 1960; Carr and Carr 1972; H. I. Fisher 1976). Neither 'they cannot do anything else' nor 'to bring the sexes together' unequivocally answers the question, why philopatry?

Models of somatic benefit

A second set of models explains philopatry, or at least adult-site tenacity, as a result of somatic benefits gained directly by propagules. The imply that non-dispersing organisms are likely to benefit because

1. Movement *per se* increases the risk of being preyed upon, of succumbing to energetic stress, or both;
2. Individuals familiar with an area are likely to be more efficient at finding and using resources, escaping predation, and at defending resources from conspecifics (e.g., Burt 1940; Blair 1953; Lack 1954, 1958; Leggett 1977; Southwood 1977; L. R. Taylor and Taylor 1977; Hamilton and May 1977; Gauthreaux 1978; Bengtsston 1978; R. R. Baker 1978).

These models are also less than satisfactory because (1) and (2) do not apply to passive dispersers, (1) does not readily apply to migrant individuals, since these could reduce the risk of movement by settling closer to their wintering areas, with the result that they actually disperse further than their philopatric fellows, and (2) applies only to adult site-tenancy, as juveniles which disperse even short distances are likely to enter unfamiliar areas (but, see R. R. Baker 1978). Why do such juveniles move 4 or 5 home ranges and not 10 or 100, if they have already left the familiar? In the case of a mobile bird, the difference is but a few-minutes flight. While each of the factors discussed here could contribute to the total selective value of philopatry in particular cases, none is general or inclusive enough to be totally satisfying.

Econgenetic models

Ghiselin (1974) eloquently advances a 'phenotypic' hypothesis to explain philopatry,

Seals can hardly be expected to know, through abstract reasoning, when to seek a better place for giving birth. Rather they ought to go where they were born, where everyone else goes, or both. That mother succeeded in a given place is clear and sufficient evidence that daughter may.

He appears to argue that if one's parents were successful in an area, a reasonable assumption since one is alive and ready to reproduce, then one is likely to be successful in the same area (for similar views, see Murray 1967; Harden-Jones 1968; Alcock 1975; Forester 1977; D. A. Hasler *et al.* 1978).

The problem is that evidence that is clear and sufficient to the investigator may be neither to the seal. The hypothesis should be able to explain why seals that wander more widely are *less* successful than those which follow Ghiselin's dictum. As stated, the 'explanation' neither explains nor predicts any disadvantages for wider dispersal. It isn't really an argument, but rather a circularity. Why return home – because mother was successful. Why was mother successful – because she returned home. I believe that the power of this argument stems from an underlying, and perhaps unconscious, reliance on the ecogenetic hypothesis presented in greater detail by others. Clearly the 'phenotypic' hypothesis relies on an assumption that progeny bear some minimal genetic resemblance to their parents.

The ecogenetic hypothesis, *sensu strictu*, suggests that philopatry both permits and maintains local adaptations, thereby benefitting individuals and populations (e.g., R. A. Fisher 1930; Mayr 1942, 1963; Dobzhansky 1970; Blair 1953; Ford 1964; E. O. Wilson 1975; Immelmann 1975; D. A. Hasler *et al.* 1978). R. A. Fisher (1930, pp. 140–2) has even provided an intuitive model of how individual selection could favour philopatry.

... the instincts governing the movements of migration, or the means adopted for dispersal or fixation, will influence the frequency with which the descendants of an organism, originating in one region, find themselves surrounded by the environment prevailing in another. The constant elimination in each extreme region of the genes which diffuse to it from the other, must involve incidentally the elimination of those types of individuals which are most apt to diffuse. If it is admitted that an aquatic organism adapted to a low level of salinity will acquire, under Natural Selection, instincts of migration, or means of dispersal, which minimize its chances of being carried out to sea, it will be seen that selection of the same nature must act gradually and progressively to minimize the diffusion of germ plasm between regions requiring different specialized aptitudes.

His argument relies on two assumptions

1. That the ecological environment is spatially heterogenous, and specifically that the scaling of that heterogeneity is small enough that, propagules moving more than 10–12 home ranges (Table 8.1) will often encounter selectively different conditions;
2. That these different regions favour genetically different and 'incompatible' selective peaks (Wright 1956).

Without ever stating it explicitly, all the ecogenetic arguments rely on philopatry's genetic consequence of increased inbreeding. It is inbreeding which protects a local gene complex by preventing a genome's dilution by 'unrelated' alleles from selectively different areas.

Empirical evidence from a wide variety of organisms is consistent with the ecogenetic hypothesis. For example, it is obvious that wide dispersal will be maladaptive in plant species living in soils on heavy metal gradients. Free interbreeding between genetically resistant individuals and those beyond the influence of contamination often results in progeny poorly adapted to either condition (for review, see Antonovics, Bradshaw, and Turner 1971). Similarly, there are two populations of sockeye salmon (*Oncorhynchus nerka*), one breeding on the inlet, the other on the outlet side of a lake (Brannon 1967). Developing young of the downstream population show a positive rheotaxis, swimming upstream. Under identical conditions members of the upstream population display a reversed taxis swimming downstream. Under natural conditions, this behavioural difference would be adaptive, bringing both groups' fry to the lake for further development. Experimental crosses of adults from the two populations simulated straying or reduced philopatry. The resulting broods showed segregation for the taxis. In each brood, some fry swam upstream, some downstream, some alternated between both. In nature, such an outbred family would probably lose over half its members owing to the disruption of this single trait (Brannon 1967). Under these or similar conditions, philopatry will obviously be adaptive for individuals in R. A. Fisher's (1930) sense.

In his monograph on heterogeneous environments, Levins (1968) presents a similar but expanded model. He predicts that species perceiving environments as coarse-grained (consisting of large spatially separable patches with different selective properties) should remain in their natal patches and adapt to local conditions through inbreeding. Symmetrically, he suggests that organisms perceiving environments as fine-grained (consisting of less discrete patches) should outbreed, as they adapt to a wider range of circumstances. His analysis extends and echoes Ford's (1964) conclusion that low levels of dispersal will permit independent adaptation to local environmental differences, while wider dispersal will only permit adaptation to average conditions over a wider range of environments.

If these models are true then

1. Dispersal distances should reflect the spatial scale of environmental grain for particular species;
2. Philopatry implies small patches differing from nearby patches in some selectively important property.

For these predictions to hold, the environment of the great tit, for example, should show significant ecological differences every 2–3 km, since their median dispersal is less than 1 km (Table 8.1).

While such small-scale ecological differences are demonstrable for metal-stressed plants or the sockeye salmon discussed above and may even apply to the great tit, they are less than obvious for many other philopatric species. An exception is provided by the Laysan albatross (*Diomedea immutabilis*) which lives in an apparently homogeneous environment (H. I. Fisher 1975, 1976). The Laysan albatross is a large (2–3 kg), long-lived ($\simeq$25 yrs), mobile, and migratory seabird. During the nonbreeding season, individuals are pelagic and range widely over the entire North Pacific. They breed in loose colonies in sandy, sparsely-vegetated areas on 11 islands in the leeward Hawaiian archipelago (Palmer 1962). They begin breeding at 8–9 years and almost invariably settle on their natal island (only 2 of 107 breeders banded in the nest were known to have changed islands though searches were made). In a specific study of site tenacity, H. I. Fisher (1976) reported that first-time male breeders nested an average of 19 m, and females 24 m, from their birth sites.

Why philopatry in the Laysan albatross? Since they can home precisely, there must be some difference among nest sites perceptible to the birds. Yet, is it likely that sparsley vegetated beach sand in the interior of a single mid-Pacific island is selectively heterogeneous on a scale of decametres? After all, the Laysan is a large, mobile, homeostatic vertebrate. It is forced to endure greater variation in ecological conditions during its life than is offered on a single breeding island. It is exactly the sort of creature that Levins (1968) suggests will perceive its environment as fine-grained. Its only use of the island is as an egg and chick respository, and it creates the microenvironment itself. There are no natural predators on the islands, and food is usually gathered far offshore (200–800 km), even during the breeding season. When this apparent lack of heterogeneity is compared with the intensity of philopatry the species displays, the ecogenetic hypothesis suffers.

A single exception should not overturn an otherwise useful hypothesis, but the albatross is one of many. The same coincidence of philopatry and homogeneous environments occurs in many other island-nesters (e.g., Austin 1949; Richdale 1957; Warham 1964; Coulson 1971; Stirling 1975; Carr and Carr 1972). In terrestrial species, the apparent lack of dispersal between populations of house mice in single barns belies the hypothesis (Selander and Yang 1969; Selander 1970). The hypothesis is endangered further by philopatric snails (Selander and Hudson 1976) and butterflies (Ehrlich, White, Singer, McKechnie, and Gilbert 1975) living in apparently homogeneous fields. It is certainly improbable as an explanation of the reproductively isolated populations of brown trout observed in a single lake (Ryman *et al*. 1979).

The insufficiency of the strict ecogenetic hypothesis as a universal explanation of philopatry generates a dilemma. Are we to accept it for coarse-grained species and look elsewhere for an independent answer for philopatric but fine-grained species? Or do we go one step further and find, if possible, a single common denominator that can explain both cases? If outbreeding were advantageous for fine-grained species, as Levins (1968) has suggested, the Laysan albatross could easily outbreed. Since they are mobile and are observed to visit many nesting islands as pre-breeding juveniles (Van Ryzin and Fisher 1976), they could return to the region and choose nest sites at random from all those which are both familiar and available. This would easily generate species-wide panmixia. At the very least, they might choose sites at random on their natal island. Of course, they do neither, being intensely philopatric. I would suggest that it is the resulting inbreeding *per se* that is adaptive. Based on the arguments of optimal inbreeding developed above, I suggest that inbreeding can be an adaptive response to spatial heterogeneity in the *genetic* environment independent of ecological scale.

A genetic model

Reproductive isolation is an essential element in current thinking about species and speciation. Central to the concept are the two major classes of pre- and post-mating isolation. The hybridization of nonconspecifics that are genetically different can result in the disruption of the 'unity of the genotype' (Mayr 1963) and lead to a 'swarm of genetic endowments of low fitness' Dobzhansky *et al*. 1977). Should this occur, it is usually defined as post-mating reproductive isolation. This hybrid dysgenesis, resulting from the disruption of coadapted genomes is an integral part of all the currently accepted models of speciation (e.g., Mayr 1963; Dobzhansky 1970; V. Grant 1971; White 1978). While the idea is not new, I believe that it contains a number of hidden implications.

Primary among these is that given sufficient genetic differentiation, interbreeding *will* be maladaptive. Since degree of relatedness is a function of genetic similarity, one can safely conclude that outbreeding, at least at the extreme level of interspecific hybridization, will be maladaptive. Given post-mating isolation, biological characters which prevent wasteful hybridization are expected to be favoured by natural selection. This is the rationale behind the 'Wallace effect' (V. Grant 1963), which implies that pre-mating isolating mechanisms like behavioural discrimination, can evolve as adaptive responses to prior post-mating factors (for reviews, see Dobzhansky 1970; White 1978).

There is no compelling reason for believing that the logic of the Wallace effect, or the concept of post-mating isolation, need be limited to species and speciation. Post-mating isolation is qualitatively, if not quantitatively, equivalent to intraspecific outbreeding depression in its probable causes and consequences. A selection coefficient (s) of 1.0 (100 per cent loss) is not necessary to affect the evolutionary process. Selection coefficients of 0.001 (0.1 per cent reduction in fitness), though difficult to measure empirically, can act effectively though at much slower rates (R. A. Fisher 1930; Haldane 1932). The question is: what level of genetic differentiation (unrelatedness) is likely to induce enough outbreeding depression to favour increased inbreeding? After reviewing pertinent evidence, I have concluded elsewhere (Shields 1982) that quantitatively minor differences (especially in the regulatory genome, see Oliver 1979) theoretically can and in many cases actually have produced the required depression (for other reviews, see Thoday 1972; Endler 1977). If this is generally true, then selection can be expected to favour characters which promote intraspecific inbreeding at the appropriate intensities.

In a panmictic group, each individual is expected to carry a random sample of all the alleles available in the group's gene pool. Since diploid organisms are limited to two alleles per locus, polymorphic loci will produce ancestral genomes (G), which can be represented as different subsamples of the total pool (e.g., G', G'', ...G^N). Assume that G' contains a relatively rare allele (A) at one locus and a novel mutation $(B{\Rightarrow}b)$ at a second. If the Ab association is positively epistatic, then G' may produce Ab offspring possessing higher fitness than those carrying alternative combinations (e.g., $AB,\!-\!B$). G', then, is expected to have greater reproductive success than either G'' or G^N. If G' and its Ab carrier offspring outbreed (i.e., mate with individuals outside the Ab family), b will be dispersed throughout the group with little chance of being reassociated with the rare A. If the novel b combinations produced $(-b)$ are less adapted than Ab, then outbreeding will result in a fitness depression.

If Ab individuals bred among themselves (i.e., parent—offspring or full sibmating), the probability of transmitting the Ab combination would increase. The resulting inbred offspring would possess greater fitness potential than their outbred competitors. As a result, selection would concurrently favour any character that was causally associated with the increased inbreeding.

If genes interact epistatically, a single novel mutation can, if it is coadapted with a low frequency variant or complex of variants at other loci, induce enough outbreeding depression that selection could favour relatively intense inbreeding. The substrate variant must be rare for, if it were common, outbreeding would not disrupt the novel association and little fitness depression would be expected. For carriers of novel combinations, inbreeding's advantage will increase with the intensity of positive epistasis. If the novel mutation interacts poorly with alternative genetic backgrounds available in the group (i.e., negative epistasis), outbreeding's costs will increase for carriers and extend to unrelated individuals should they mate with carriers.

Assuming the epistasis and intraspecific outbreeding depression are commonplace (for review, see Endler 1977), intense inbreeding will often have been adaptive. If the members of a particular family have a genetically-based tendency to philopatry, they are expected to inbreed more intensely. Philopatry's causal association with inbreeding, then, will increase fitness relative to vagrancy's association with increased outbreeding depression. Given time and the recurrence of novel epistatic combinations, the genetic substrate controlling philpatry is expected to increase in frequency, ultimately fixing in the wider population. Favoured combinations arising after philopatry is the rule will be assimilated more quickly and efficiently (fewer maladapted progeny) and will reinforce the entire process. When a major portion of every individual's genome consists of coadapted complexes of interacting alleles (assumed to be the case at present for most organisms), it would not be necessary for novel combinations to arise to maintain philopatry. Disruption of extant combinations would be expected to produce enough outbreeding depression that the advantages of inbreeding and philopatry would be maintained.

In this context philopatry can be considered a pre-mating isolating mechanism, that evolves in response to low-level post-mating isolation (i.e., intraspecific outbreeding depression). It evolves through a Wallace effect, but in miniature, since it occurs within species. Nonetheless, its causes and consequences are qualitatively equivalent to those associated with the adaptive evolution of pre-mating mechanisms separating species. Its advantage over vagrancy is that it limits population size and increases the level of inbreeding, thereby protecting coadapted genomes. From this perspective, and at least for mobile animals, the evolution of philopatry might be considered equivalent to the evolution of allopatry. Only then does the population subdivision and isolation by distance characteristic of so many organisms, in spite of their enormous capacities for dispersal and increased panmixia, finally become explicable (Ehrlich and Raven 1969).

Philopatry, then, can be considered an adaptation to heterogeneity in the *genetic* environment. In sexual species, sufficient conditions for its evolution and maintenance would be genetic variation in dispersal tendencies, and conspecifics separated in space that carry different and incompatible coadapted genomes. Two processes could produce such spatial heterogeneity in the genetic environment. The most obvious, and perhaps most common in nature, is external ecological heterogeneity. Selective differences are expected to change with distance as a result of differences in the external environment. This can lead to the assimilation of different alleles or allele combinations in geographically separate areas. This is equivalent to R. A. Fisher's (1930) ecogenetic hypothesis as outlined above. The second process offers a tentative explanation for philopatric species in ecologically homogeneous environments (e.g., Laysan albatross). Ecological heterogeneity is not a necessary conditon for either the initiation or maintenance of genetic differentiation. Given a large, ecologically homogeneous patch, we would still not expect a *single* favourable mutation to arise independently in two or more individuals in any finite population. More likely, two different mutations, or sets of mutations, each favoured, but each generating that favour through different proximate mechanisms, would be assimilated by two individuals and hence their families. Should the resulting genetic complexes be incompatible, selection would favour inbreeding and philopatry in spite of ecological homogeneity. This second process, then, is the direct result of selection's opportunism and the random and independent nature of mutation.

The necessary conditions of the ecogenetic model have been demonstrated in both laboratory and field (R. A. Fisher 1930; for review, see Shields 1982). The potential for random differentiation producing conditions favourable to philopatry has been explored in less detail. A series of laboratory experiments on the development of adaptation in novel environments has indicated that the process can occur (e.g. King 1955; W. W. Johnson 1974; Palenzona, Mochi, and Boschieri 1974; Palenzona, Allicchip, and Rocchetta 1975).

In his classic studies, J. C. King (1955) performed a series of crosses to explore the genetic basis of DDT resistance in *Drosophila melanogaster* flies. He varied the selection intensities via pesticide dosage, and also varied the life stage (larva or adult) subjected to the poison. All treated lines showed increased resistance to DDT after selection. All lines were started from the same base stock, and replicates were subjected to identical selective conditions. Nonetheless, when lines (including replicates) were crossed, F_2 hybrids showed disruptions of the resistance. King concluded that, '....different lines stemming from the same original stock can build different systems (genetic) within a dozen generations'. The hybrid breakdown indicated that the systems were incompatible as well as different. In the end inbreeding and philopatry are expected to be associated with spatially heterogeneous genetic environments regardless of the factors which actually control the process of differentiation.

This genetic model suggests that increased inbreeding is the *primary* function of most instances of philopatry. It can explain philopatry in coarse- and fine-grained species, in mobile and sessile organisms, in active and passive dispersers, in sedentary or migratory animals, and even in plants. It is the most general and universal explanation of philopatry currently available. Yet it is by no means an exclusive explanation. Either during or after the evolution of philopatry, secondary functions (e.g., site-familiarity, the costs of movement) are likely to add to the advantages of philopatry via their somatic benefits. Under changed conditions it is even conceivable that these could assume the role of philopatry's primary function in particular cases. Unless there is significant contrary evidence, however, the assumption that philopatry serves to increase inbreeding intensity remains the simplest hypothesis. As such, we can integrate it with the notion of optimal inbreeding and additional ecological considerations in order to develop testable predictions.

Synthesis: predictions and evidence

If the optimal inbreeding arguments are correct, most long-lived (generation> 1 year), low-fecundity organisms should inbreed. On the basis of genetic consequences alone, they should inbreed intensely, mating randomly in small, semi-isolated demes while avoiding incest ($2 < N_e < 1000$). High-fecundity organisms are expected to outbreed by mating randomly (with respect to genome) in larger demes ($N_e > 10\ 000$). The genetic model of the evolution of dispersal patterns, suggests that philopatry is *one* proximate mechanism promoting inbreeding. By fine-tuning the philopatry behaviourally, it may be the most efficient mechanism promoting optimal levels of inbreeding (see below). Vagrancy, then, could be considered a primary strategy that functions to promote increased outbreeding.

Based solely on genetic consequences we can integrate these arguments and test both hypotheses simultaneously. *Philopatry which is effective in producing relatively intense inbreeding should occur in, and only in, low-fecundity organisms. Wider outbreeding resulting from vagrant dispersal should be limited to high-fecundity organisms.*

Dispersal systems are subject to ecological pressures as well as the constraints imposed by their genetic consequences. It may be necessary to uncouple genetic requirements from dispersal requirements before the latter are free to respond to special ecological conditions. For example, organisms in ephemeral habitats, or in habitat patches which are rare in space or time, may be forced to disperse propagules widely. To meet both genetic and ecological challenges, low-fecundity organisms would be forced to inbreed through mechanisms other than philopatry. Similarly, if circumstances forced philopatry on a high-fecundity species, it would still be expected to outbreed despite the barriers to vagrancy.

Tests

With lifetime fecundities spanning seven orders of magnitude ($10^2 - 10^9$ offspring/female/lifetime) and the presence of both major dispersal strategies, the marine invertebrates offer an ideal test of these predictions. Since wide vagrancy is associated with long-lived pelagic larvae (Thorson 1950) and philopatry with non-pelagic larvae and the occurrence of parental care (e.g., Thorson 1950; Morrison 1963; Gee and Williams 1965; Menge 1975; Hartnoll 1976; Schopf and Gooch 1977; Sanders 1977; Hansen 1978; Cook and Cook 1978), an association between dispersal and fecundity immediately emerges. Thorson (1950) reviewed data from 53 benthic marine invertebrates, representing all of the major marine phyla. He demonstrated a strong association between annual fecundity and dispersal strategy (Fig. 8.5).

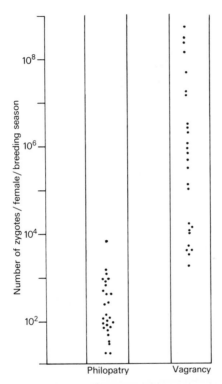

Fig. 8.5. Association between annual fecundity and alternative dispersal strategies in a sample of benthic marine invertebrates. Philopatric species are usually viviparous displaying extensive parental care, or lay large eggs and show more primitive forms of brood protection. Vagrants produce relatively long-lived pelagic larvae. (Modified from Thorson (1950).)

Many have attempted to explain the high fecundities observed in vagrant species as a consequence of the dangers of their larvae's pelagic life (e.g., Thorson 1950; Vance 1973; Shine 1978). Even if true, this argument implies nothing about the association between fecundity and mating system. There is no ecological reason why a high-fecundity species should not self-fertilize. Many are hermaphoroditic, yet most of these appear to be exclusive cross-fertilizers. In concert with their vagrancy, this is likely to ensure relatively wide outbreeding (e.g. Gee and Williams 1965; Menge 1975). Like many plant species, some of these high-fecundity animals even display a rudimentary self-sterility (e.g., Longwell and Stiles 1973, and references therein).

In contrast, self-fertility and confirmed selfing appear to occur more commonly in philopatric and low-fecundity species (e.g., Gee and Williams 1965; Hartnoll 1976). Recently, Cronin and Forward (1979) reported on an estuarine crab, *Rhithropanoplus harrisii*, which may be philopatric despite dispersing pelagically. Normally, one can assume that pelagic larvae will disperse vagrantly as they are passively carried by local currents. In this species, however, behavioural adjustments apparently result in, 'retention of these planktonic larvae in estuaries near the parent populations'. If this retention is localized, then the crab would be best described as philopatric. If true, then the crab should also possess low fecundity. In response to this prediction, Cronin (personal communication) informed me that his best guess of *R. harrisii's* lifetime fecundity was 10 000 progeny (range: 2000–25 000).

On the basis of these broad comparisons, as well as in the specific case of *R. harrisii*, the association between fecundity and dispersal pattern in marine invertebrates is consistent with the predictions. An analysis of plant dispersal reveals a similar association of low *lifetime* fecundity with inbreeding via selfing or philopatry, and of high fecundity with outbreeding via wider dispersal (Shields 1982; Hamrick, Linhart, and Mitton 1979). Thus, in both major classes of passive dispersers, the evidence fails to falsify the hypotheses. More detailed analyses of data gathered or presented in the framework of the predicted associations will be necessary before we can verify or alter these initial conclusions.

Many of the marine fishes, like their invertebrate counterparts, are very fecund ($10^6 - 10^9$ offspring/lifetime). These high-fecundity species normally produce pelagic eggs, free-swimming larvae, or both. Their pelagic propagules are then subject to passive dispersal in water currents. Vagrancy and wide outbreeding are expected to result. Adults of such groups as herring, *Clupea* spp., and cod, *Gadus* spp. also tend to show less adult site-tenacity than other vertebrates. They show additional behaviour, such as mass spawning in large groups, which would also appear to favour increased outbreeding (for review, see Harden-Jones 1968). In contrast, most fresh-water, low-fecundity fishes (e.g., salmon, Salmonidae and sunfish, Centrachidae) lay demersal eggs, pair bond, are often territorial, and show intense birth-site philopatry. All of these characters are expected to promote inbreeding (Harden-Jones 1968). Thus, in all the groups with comparative data available (i.e., plants, marine invertebrates, and fishes), fecundity and inbreeding intensity appear to be negatively correlated.

Since the remaining vertebrate classes and most of the insects are low-fecundity organisms, I would predict that *all* should be inbreeding. While this extreme view does not permit comparison, it does imply that philopatry, or some alternative mechanism promoting inbreeding, should be the rule. The vast majority of low-fecundity insects and vertebrates are philopatric (Endler 1977; R. R. Baker 1978; Shields 1982). Biparental reproduction combined with philopatry can result in optimal levels of inbreeding $(2 < N_e < 1000)$ while incest with its attendant costs may be avoided. When deme sizes are estimated from natural populations, they are small enough to generate the inbreeding intensities predicted here (Table 8.4). P. J. Greenwood and his colleagues (Greenwood and Harvey 1976; Greenwood *et al.* 1979*a*; Chapter 7, this volume) may be correct in postulating that sexual differences in effective dispersal in many vertebrates may have evolved to prevent 'too close' inbreeding. Since *both* sexes remain philopatric, and N_e therefore remains small (Table 8.4), the sexual asymmetry certainly did not evolve to prevent inbreeding *per se*, an implication P. J. Greenwood appears to have accepted (Greenwood *et al.*

TABLE 8.4. *Empirical estimates of genetically effective population sizes* (N_e) *in a variety of plant and animal species*†

Species	N_e	Source
Plants		
Phlox pilosa	75–282	Levin and Kerster (1974)
Liatris aspera	30–191	Levin and Kerster (1974)
L. cylandracea	30–200	Levin and Kerster (1974)
Lithospermum caroliniense	2–6	Levin and Kerster (1974)
Animals: invertebrates		
Cepea nemoralis	<2800	Wright (1978)
C. nemoralis	190–6500	Greenwood (1974)
Drosophila pseudoobscura	25 000	Wright (1978)
D. subobscura	10 000	Begon (1976)
D. subobscura ‡	≏400	Begon (1978)
Animals: vertebrates		
Rana pipiens	46–112	Merrell (1970)
Sceloporus olivaceus	<225	Kerster (1964)
Uta stansburiana	17	Tinkle (1965)
Melospiza melodia	65–215	Miller (1947)
Chamaea fasciata	500	Miller (1947)
Peromyscus maniculatus	10–75	Rasmussen (1964)
P. maniculatus	80–120	Wright (1978)
P. polianotus	240–360	Wright (1978)
Mus musculus	<12	Levin, Petras, and Rasmussen (1969)
M. musculus	≤10	Defries and McClearn (1970)

† Almost all estimates are high because they fail to account for all factors controlling effective population size.
‡ N_e from Begon (1976) based solely on effective dispersal; from Begon (1978) based on same populations but sex ratio, variance in reproductive success, and fluctuations in local density taken into account.

1979*b*; Greenwood 1980), though it may have escaped the attention of others (e.g., Packer 1979; Koenig and Pitelka 1979).

What of the colonizing or nomadic species of insects and invertebrates? Smaller insects with poor flying abilities disperse more passively than actively. In colonizing species, the ephemeral nature of their native habitat often makes philopatry impossible. In biparental species, the simplest method of ensuring inbreeding under such ecological conditions is to mate prior to dispersal. This is taken to its logical extreme in the species which mate prior to birth within their own mother's body (Hamilton 1967), or more often in their singly parasitized host (e.g., *Telenomus fariai*, Dreyfus and Breuer 1944). Either strategy results in continuous full-sib mating. Similar adaptations promoting intense inbreeding prior to dispersal have been demonstrated in such colonizing arthropods as mites (e.g., Cooper 1939; R. Mitchell 1970), many free-living and especially parasitic hymenoptera (Cowan 1979; Greenberg 1979; and for review, Hamilton 1967), and many other insects (for reviews, see Hamilton 1967; C. G. Johnson 1969).

Given ephemeral environments, a mobile vertebrate could also employ the passive strategy of mating prior to dispersal, with impregnated females taking the role of propagule. The only example of this potential strategy I am aware of is the wild turkey (*Meleagris gallopavo*). It is at least possible that fertilized females may be the primary turkey propagule (W. F. Porter, personal communication). I suspect that if the pattern does occur, it will occur rarely since it entails unnecessarily intense inbreeding for an active disperser. Since vertebrates are usually mobile, a nomadic colonizing strategy is available to them that will permit optimal levels of inbreeding. Travelling in social groups, consisting of the same number of individuals with the same average relatedness as is characteristic of stationary demes, will meet both the colonizing and inbreeding needs of mobile individuals. I would define faithfulness to such mobile demes as *philogamy*. Philogamy should be characteristic of such colonizing birds as crossbills (*Loxia* spp.) or Franklin's gulls (*Larus pipixcan*, Bent 1968; Burger 1972), as well as the nomadic mammals (e.g., wildebeest or zebra, Leuthold 1977; or even man). There is some evidence that nomadic vertebrates normally travel in the expected groups (e.g., Mayr 1963, and references above). There is, with one exception, little evidence regarding the relatedness of group members. In humans, such groups usually are clans or tribal units consisting of individuals with the predicted degree of relatedness. Further study of the genetic structure of nomadic vertebrates on a wider taxonomic front would permit more sensitive tests of the philogamy hypothesis.

Broad comparisons are invariably disturbed by exceptions, and this analysis is certainly not immune. For example, many of the ducks and geese (Anatidea) mate on their wintering grounds with males following their mates to the females' natal area (for reviews, see Mayr 1942, 1963). Females display birth-site philopatry, while marking indicates that many males settle far from their birthplace (e.g., Cooke *et al.* 1975; Rockwell and Cooke 1977). Mayr (1942, 1963) has suggested that this fowl dispersal strategy may have evolved to ensure wider outbreeding and increased gene flow. If true, it would stand at odds with my predictions, so I offer an alternative explanation. Waterfowl may mate on the wintering grounds because of the short duration of benign conditions on their northern breeding grounds (Newton 1977). If during the evolution of winter mating, each deme within a species overwintered in their own separate and traditional area (winter philopatry), then winter mating would still permit inbreeding. If ecological conditions changed and different demes were forced to share wintering grounds, the likelihood of interdemic outbreeding would increase.

Outbreeding, then, could be the chance and maladaptive result of the destruction of these species normal wintering areas. It would be a nonequilibrium condition rather than an adaptive strategy. This view is consistent with the destruction of wild coastal marshes in the past century. The tendency for members of different stocks of wildfowl to winter in different areas where habitat has not been altered is also consistent with this interpretation (e.g., Bent 1962; H. Milne and Robertson 1965). While this is an *ad hoc* and therefore not a very satisfying explanation of wildfowl dispersal, I do find some support for my hypothesis in this exceptional group. The entire pattern they display offers a rather simple mechanism which could ensure relatively wide outbreeding in many migratory bird species. Why do more avian species not mate on the wintering grounds? The rarity of this or alternative mechanisms which could produce wider outbreeding via dispersal strategies is indirect evidence that inbreeding enjoys some favour.

General discussion and conclusions

With some exceptions (e.g., waterfowl), the predicted negative correlation between fecundity and inbreeding intensity is observed in a variety of taxa subject to diverse ecological conditions. This robustness suggests that the causal connections implied by the genetic models may be real. The equally robust associations of philopatry with low fecundity and vagrancy with high fecundity are consistent with the hypothesis that increased inbreeding may be a primary function of philopatry. Thus genetic effects may not only be important consequences of dispersal but also may act as evolutionary causes in the origin and maintenance of specific dispersal strategies.

These models are consistent with the occurrence of phenotypic traits which appear to function in the prevention of *incest* (and not inbreeding *per se*, as is often implied). These include the classical mechanisms of physiological or mechanical self-sterility in plants (for review, see V. Grant 1975), kin recognition and incest avoidance in birds (Koenig and Pitelka 1979), mice (Hill 1974), and especially primates (e.g., Sade 1968; Itani 1972; Packer 1979).

The models are also consistent with the *incest* depression observed in many species when avoidance mechanisms fail (e.g., Greenwood and Harvey 1977; Price and Waser 1979; Packer 1979).

In contrast to more traditional views, these models admit to the benefits and potentially wide occurrence of intense inbreeding in similar kinds of organisms. They are consistent with the occurrence of mechanisms promoting selfing or less extreme inbreeding in plants (Grant 1975). They are also consistent with the otherwise problematic occurrence of regular systems of intense inbreeding in many vertebrates (e.g., frequent full-sib mating in mice, Dice and Howard 1951; lizards, Tinkle 1967; wolves and other canids, Mech 1970; or frequent father–daughter matings in deer, R. H. Smith 1979; primates, Itani 1972).

These models may contribute to our understanding of the varied dialect systems in which relatedness, and therefore potential genetic compatability, may be signalled during courtship and used in mate choice. Intraspecific dialects are characterized by spatially mosaic distributions of signals in many sensory systems. Owing to developmental differences, such signals can, at least theoretically, reliably indicate various levels of relatedness (e.g., deme or sibship membership; for reviews, see E. O. Wilson 1975; W. J. Smith 1977). Mate preferences based on such cues could result in wider outbreeding (e.g., through a rare-male effect or other negative assortment, Ehrman 1970; Treisman 1978; Jenkins 1977; Halliday 1978), or in increased inbreeding through positive assortment (e.g. Nottebohm 1969; Shields 1982).

In nature, such systems appear to result in increased inbreeding more often than not (e.g. acoustic dialects: birds, M. C. Baker and Mewaldt 1978; frogs, Capranica, Frishkopf, and Nevo 1973; olfactory dialects: bees, Greenberg 1979; fishes, Tilzey 1977; Nordeng 1977; voles, Godfrey 1958; visual dialects: birds, Bateson 1978; baboons, Abegglon 1976; Crook 1970). Such systems may be fine-tuned to promote optimal levels of inbreeding by reducing the likelihood of both incest and overwide outbreeding (e.g. Bateson 1978). This might be analagous to the fine-tuning of dispersal, with sexual biases in dispersal within the constraints of philopatry, also promoting optimal levels of inbreeding. In any case, detailed investigations of the correlations between relatedness and the behavioural cues used in mate selection, could provide important tests of the hypotheses developed here, as well as greater general understanding of mating systems.

Perhaps the most important implication of these models, if they are true, is that low-fecundity species are likely to be spatially subdivided into demes that might better be described as families. At the inbreeding intensities predicted, demes would become genetically homogeneous. Deme members would be more closely related than immediate pedigrees indicated, while differing genetically from members of other demes (Jacquard 1975). Relatedness between demes, then, would be expected to decline with increasing geographical distance (Hamilton 1975). Such a population structure would facilitate selection at levels of organization higher than the individual. For example, on D. S. Wilson's (1980) selection continuum, interdeme group selection might be genetically equivalent to the theoretically less onerous family selection discussed by others (Griffing 1976; Wade 1979). This might make the occurrence of true group adaptation less improbable. The proximity of close relatives would also generate favourable conditions for kin selection and high levels of social co-operation, if not phenotypic altruism (e.g. Hamilton 1964, 1975; Sherman 1977; P. J. Greenwood *et al.* 1979*a*; Chapter 7; but see Shields 1980). The decreased relatedness between members of different demes might then be expected to generate increased conflict between them (e.g. interdeme xeno-phobia, see Hamilton 1975).

In an earlier discussion of the evolutionary role of inbreeding, Carson (1967, p. 283) suggested, 'Even now, despite the comprehensive genetic explanation of inbreeding depression and the demonstration of the apparent absence of harmful effects in 'normally' inbreeding plants and animals, a tendency remains for the importance of inbreeding to be underrated. Perhaps this is partly influenced by human attitudes toward both inbreeding and outbreeding'. I applaud Carson's judgement, but suggest that even he may have underestimated the importance of inbreeding. I believe that he did misjudge the numbers and kinds of organisms to which that importance applied. I am also convinced that until inbreeding, whatever its guise and whatever its intensity, is brought into the mainstream of evolutionary theory, a comprehensive understanding of the evolutionary process will elude us.

9
Pattern and process in large-scale animal movement

DAVID ROGERS

> The Owl and the Pussy-cat went to sea
>   In a beautiful pea-green boat.
> They took some honey, and plenty of money,
>   Wrapped up in a five-pound note. [Edward Lear]

This chapter is mainly concerned with insects, but since the class is so numerous and so varied it provides a wide range of examples of migration strategies that seem to apply throughout the animal kingdom: hence the quote at the top of this page.

Immediately we are faced with two questions that are raised by the story of the owl and the pussy-cat: from what were they escaping, and whither were they bound? The poem leaves both questions suitably unresolved and, to date, invertebrate ecologists in their own field have done little better.

Suggested hypotheses for insect migration

All current views of insect migration stress its evolutionary importance but there has, nevertheless, been a difference of emphasis on the abiotic and biotic components which is best exemplified by the question: are insect migrants colonizers or refugees (Dingle 1972)?

The first group of ideas, advanced in its present form by Southwood (1962, 1977) stresses the abiotic component and suggests that movement is a necessary consequence of the exploitation of temporary habitats. Southwood pointed out that all habitats are temporary on a certain time scale, and that those of the insects are temporary on the time scale of the insects' own life spans (i.e. H/τ of the order of unity, where H is the period of habitat suitability and τ is the organism's generation time). To survive, such creatures *have* to move. The theory of $r-K$ selection (Macarthur and Wilson 1967) arose out of a similar appreciation of the ephemeral nature of habitats. In these ideas, therefore, the habitat is the driving force for movement, and the animal's own density is only occasionally important (Southwood 1962): migrants are thus refugees from deteriorating conditions.

L. R. Taylor and R. A. J. Taylor (1977, 1978), whilst acknowledging the abiotic component during the past and present evolution of the habit of migration, suggest that the animals come to use their own density as a cue to current and future favourability. These ideas, with earlier origins in Lidicker (1962), make movement a response to the increasingly crowded nature of inhabited regions that results in movements away from high population-density areas. There is a compensating movement towards areas of low population density, while at intermediate density there is no net movement into or out of the area. Thus the animal itself, or rather its abundance, is the proximate driving force for movement, and the emphasis is on the colonizing role of migration.

Whereas Southwood's ideas simply categorize nature, those of Taylor and Taylor, because they can be related to easily measured quantities such as population density, have greater descriptive and predictive power. Not only can density–distance and variance–mean relationships be interpreted, but also the model suggests a new mechanism of population regulation by density-dependent movement (coupled with density-independent mortality of the migrants) rather than the classical ideas of density-dependent mortality of a relatively static population (see Tamarin 1978*b* for a history of such ideas).

As both Southwood (1977) and R. A. J. Taylor and L. R. Taylor (1979) emphasize, these two groups of ideas about insect migration are two major variations on the same evolutionary theme, two points on a continuum. Where the H/τ ratio is of the order of unity, it is likely that migration will be obligatory at some life-cycle stage or time. The Taylors stress the same temporary nature of the habitat but on a rather longer time scale (H/τ greater than one) implying that, for the individual insect, facultative migration is the more suitable option. In fact both facultative and obligatory migration can be found in the insects and, to summarize the examples in R. R. Baker (1978), it appears that insect migration is usually obligatory if survival depends on the rate of habitat formation elsewhere (e.g., temporary habitats, habitats in successional communities, some seasonally suitable habitats) and facultative if it depends on the quality of the present environment (e.g., the condition of the host plant, abundance of conspecifics, or of parasites and predators).

The division into abiotic and biotic stimuli, which respectively emphasize the refugee and colonizing aspects of animal migration, points to the answer to our initial question: migrants are both colonizers *and* refugees. In short, they are survivors.

Migration: the great debate

The debate over the function of migration, especially that of insects, has been enriched by a variety of workers. But the division between the behavioural and ecological approach, identified in this volume by Taylor and Taylor (Chapter 10), often confuses the 'how?' and the 'why?' of the process. Table 9.1 is an attempt to bring some order into the discussion and identifies key European workers who have been concerned with the initiation, duration,

Table 9.1 *Major European workers associated with the three aspects of migration (take-off, travel, and arrival) at the behavioural and ecological levels*

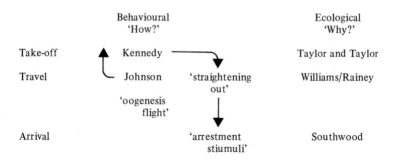

	Behavioural 'How?'		Ecological 'Why?'
Take-off	Kennedy		Taylor and Taylor
Travel	Johnson	'straightening out'	Williams/Rainey
	'oogenesis flight'		
Arrival		'arrestment stiumuli'	Southwood

and termination of migration of insects at the behavioural and physiological levels.

It should be clear that the behavioural level makes sense only in terms of the ecological one. This characteristic is typical of the hierarchy of descriptions and explanations that can be applied to any biological phenomenon. If we classify fields of biology on a simple ascending scale, for example, macromolecules, genetics, physiology, behaviour, ecology, evolution, we find that the *description* of an event occurring at any one level (e.g. behavioural) can be in

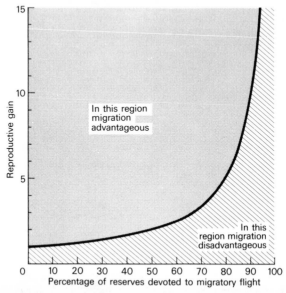

Fig. 9.1. The necessary factor by which migrant eggs must have a higher survival rate than resident eggs for migration to be advantageous (the 'necessary reproductive gain') related to the percentage of the total reserves that the individual devotes to flight.

terms of events occurring at lower levels (physiology, genetics, etc.), but the *explanation* for the event must be in terms of higher levels, ultimately that of evolution. This logical asymmetry, which has occasionally made behaviour and ecology incompatible bedfellows, must be acknowledged if any real progress is to be made. There is no convincing behavioural explanation of why an aphid differs from a fruit fly in its tendency to migrate. There is, unfortunately, a bewildering variety of ecological explanations for the difference.

It is here, when we abandon our behavioural water-wings, that we are likely to be drowned in a sea of ecological facts and ideas. In this sea all phenomena can find some explanation or another, as compatible as the crab and its hermit shell. What we need to do is to look for general rules of migration, and these rules must be based on what migration achieves (displacement, a phenomenon with ecological implications) rather than how it occurs (a phenomenom with behavioural implications).

In a summary of forty papers on the movement of highly mobile insects Kennedy and Way (1979) re-affirm Kennedy's (1975) earlier behavioural characterization of migration as a temporary syndrome of persistent, straightened-out locomotion accompanied by and indeed dependent on some inhibition of competing 'vegetative' responses. This behaviour tends to maximize the distance covered from the insects' point of origin (Southwood 1978), achieved either by steady and consistent compass orientation, maintained regardless of wind direction (C. G. Johnson 1969) or by use of air movements, spectacularly exploited by locusts (Rainey 1979) and many others (Rabb and Kennedy 1979). At present we assume that the goal for each type of movement, the maximization of offspring production, is the same, but this is more a statement of faith than a proven fact. Because the scale over which migrations take place is so large, there are few studies to rival those of the anti-locust teams (Rainey 1978) and the Rothamsted Survey (R. L. Taylor 1979), but there is an increasing awareness that such studies are required.

Much of this chapter begs the question as to what is a migrant. Should migration be distinguished from other, so-called 'trivial' movements (Southwood 1962), as it has been classically (C. G. Johnson 1969), or is any, even infinitesimally small, movement away from the place where an animal was born a form of migration (Taylor and Taylor, Chapter 10)? The latter approach belittles some impressive examples of migration, many characterized by rather specialized pre-migratory preparation, or the production of migratory morphs, but it does emphasize the continuous nature of the necessity to move, in Steinbeck's (1962) phrase, ... 'any place away from here'.

How much migration?

At some scale every environment is heterogeneous in its ability to provide the needs of the animal within it. At smaller scales environments are more homogeneous, and animals tend to live in the suitable parts. These often repeated ideas are encapsulated, in a rather corrupt but entertaining form, in Don Marquis' poem:

a louse i
used to know
told me that
millionaires and
bums tasted
about alike
to him. [*the life and times of archy and mehitabel*].

The problem from the insect's point of view is to find the suitable parts. It is obvious that this quest will involve energy expenditure but not so obvious that despite this expense there may nevertheless be a wide range of conditions under which migration can be a distinct advantage.

Consider a newly emerged adult insect with limited food reserves in a habitat which is satisfactory though not ideal for reproduction. The insect's choice is to remain and lay all its eggs, or migrate and use some of the energy for flight that would otherwise be devoted to egg production. If 50 per cent of the metabolic reserves are devoted to flight, the eggs that are eventually laid should have more than twice the survival prospects of eggs laid by any resident insect for migration to be advantageous. The general relationship in Fig. 9.1. (a reciprocal curve) shows that above the 50 per cent level there have to be considerably greater advantages in migration for it to be better strategy than residency, but below 50 per cent this is not the case.

The hidden assumption of Fig. 9.1. is that the migrants and the residents have a similar probability of survival to reproduction. If we now incorporate the probability of migrant survival relative to resident survival on a scale of 0–100 per cent (100 per cent indicating that migrants and residents survive equally well) we have the three-dimensional graph shown in Fig. 9.2(a). Obviously the smaller the probability of survival, the greater must be the potential reproductive gains of any particular level of individual investment in migration for migration to be advantageous relative to residency.

If the different reproductive gains (the dashed contours on Fig. 9.2(a) are projected on to the base of this figure, a series of indifference curves is produced (Fig. 9.2(b)) relating survival rate, individual investment, and reproductive gain. The evolutionary 'game' that the migrants play is how to reduce the necessary reproductive gains to a minimum.

It is clear from Fig. 9.2(b) that at very low levels of individual investment, a small increase in investment need be accompanied by only a small increase in survival for the investment to be advantageous. As more and more investment is made, however, it must be accompanied by ever increasing levels of survival for the trend to continue. Obviously, at some stage the marginal increase in survival rate for a unit increase in investment becomes insufficient, and the level

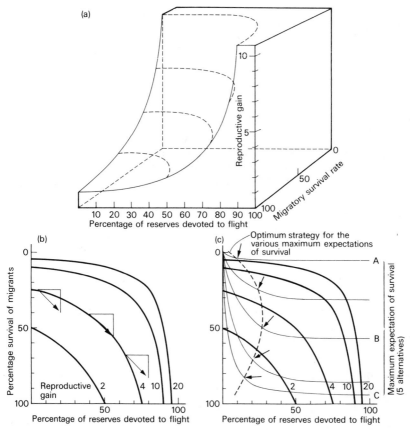

Fig. 9.2(a) Fig. 9.1 extended to cover a range of survival rates of the migrants, from 0 to 100 per cent that of the residents. (b) Contours from (a) (i.e. necessary reproductive gains) projected on to the horizontal plane, relating the percentage survival of migrants and the percentage of the total reserves devoted to flight. The three small triangles represent three different starting points on the same reproductive gain curve and show that a particular increase in investment in flight and its associated increase in the survival rate of the migrant may be advantageous at low initial levels of investment, but not at high levels (i.e. the reproductive gain is reduced in the first case, but not in the second). (c) The same basic curves as in (b) here show changes in the optimal investment strategy with changing expectations of migrant survival. The five, thin, exponential decay curves (A,B,C, and two others) represent the average realized relationship between increased investment in flight and the consequent increase in survival rate for five different species. The thick arrow on each curve represents the optimum investment level for that particular species. Note how this level changes with the average expectation of survival.

of investment stabilizes. The reverse arguments apply to populations of individuals that initially over-invest in migration: the investment level will fall to the same equilibrium value.

There will thus be some intermediate value of investment in migration that reduces to a minimum the reproductive gains necessary to make migration advantageous. Ths equilibrium level will in turn depend upon the level of survival that migrants tend to experience, and more especially on how, if at all, an increase in investment can actually bring about an increase in survival (the cost being in terms of a reduced eventual reproductive output).

For the sake of simplicity we assume that at zero investment in migration, survival of migrants is also near to zero. As more investment is made, survival increases, but at a decreasing rate, and then becomes asymptotic to a value reached (at 100 per cent investment) which is dependent upon the type of transport used. For example, a wind-borne migrant such as an aphid would be expected to have a lower maximum survival rate than a self-powered, orienting migrant such as a bird. These two extremes and one intermediate are labelled curves A, B, and C in Fig. 9.2(c).

The optimal investment strategy for each type of migrant is now the one that reduces to a minimum the reproductive gains shown by the indifference curves. These optimal points are indicated by the arrows against each curve in Fig. 9.2(c). Migrants with an intermediate expectation of migration survival should invest more in the migration process than those with either higher or lower expectations. If a migrant has a very low expectation of survival (curve A), it reduces to a minimum its investment in migratory activity, and relies upon contacting a particularly suitable habitat unit (relative to its origin) where it can devote all its reserves to offspring production (high migration risks, high potential gains, but with a low probability of occurrence). Migrants with a higher probability of survival (curve C) make more or less the same investment, but for a different reason (lower migration risks, lower potential gains, but with a higher probability of occurrence).

There can be variations in the optimal strategy in each of curves A, B, and C. For example the individual which tends to invest less than the average for the species in migration has a lower probability of surviving the migration period, but may be fortunate enough to find a high-yielding patch for its offspring. This is a 'risk taker'. Another individual which invests more than average to increase its survival probability may also be fortunate enough to find a good patch, but will be able to lay fewer eggs in it. Such an individual is 'risk averse'. The nature of the total species' range will determine just how varied the individuals of each species are in their investments in migration. The very unpredictability of many migration routes suggests that a broad-based investment 'portfolio' will characterize many insect species (e.g. the various *Oncopeltus* species studied by Dingle 1978).

Fig. 9.2(c) can also suggest a reason for the falling settling thresholds for migrants such as aphids which test a series of plants for suitability before finally settling (Kennedy 1966). If a migrant lands before its optimum investment in travelling (i.e. to the left of the arrows in Fig. 9.2(c), then the reproductive gains necessary to make migration a better strategy than residency have to be high: hence the settling threshold will be high. As time passes, and the insect continues periodically to settle, the necessary reproductive gains fall, and hence the settling threshold may also fall.

Before passing to the next section we must add a number of caveats. Figs. 9.1 and 9.2 apply only to individuals and should not be regarded as representing any reproductive strategy on the part of a parental group that involves producing both migratory and nonmigratory offspring (Hamilton and May 1977).

Secondly, I have assumed that migrants contain a fixed total amount of reserves at their point of origin, which can be used either for migration or for reproduction. A number of migrants feed during migration, and this behaviour may increase their survival without reducing their productivity.

Thirdly, one axis of Fig. 9.2 allows migrants to have survival rates between 0 and 100 per cent of those of residents, but no higher. If migrants are escaping from deteriorating local conditions, it is possible that even allowing for migration losses their average survival exceeds that of residents. In this case many habitats are more suitable for the migrant's offspring, and the balance swings very definitely in favour of migration.

Walking or flying: which alternative?

Movement in insects may be by walking or by flight – both particularly well developed in the class.

Manton (1977) indentifies selection for speed as one of the most important forces in insect evolution. The six-legged nature of insects provides sufficient mechanical stability whilst also allowing for greater walking speed than is shown by many-legged creatures such as myriapods. Such was the advantage of the six-legged plan to small, terrestrial invertebrates that it appears to have evolved at least five times (Manton 1977).

Long-distance movement is, however, more efficiently accomplished by flight. At any particular body size, flying is energetically cheaper than walking but, for all forms of transport, movement becomes proportionately more expensive as body size decreases (Tucker 1969). A rat of about the same body weight as a gull uses ten times as much energy to transport itself over a similar distance. If the line for walking animals such as the rat is extended to body weights of the order of those of insects, the energetic costs of transport become prohibitive: flight is an energetically cheaper alternative.

Flight does nevertheless involve the additional cost of flight muscle maintenance when the insect is not flying. To escape this, insects that need to fly often do so at some particular stage of the adult period, after which flight musculature is histolysed often for reproductive purposes. Insects that never need to fly (i.e. can find suitable habitat units by walking) dispense with wings altogether. Where flight behaviour of any species is variable and expressed as a structural polymorphism, those individuals that do not fly (i.e. are apterous or with atrophied wing muscles) tend, on average and under ideal conditions, to produce more offspring than those that do.

Despite the higher costs of maintaining the flight apparatus, this is occasionally a very necessary expenditure especially when the insect is starved. In such cases, wing muscles are retained until more food is discovered and histolysed thereafter (Dingle 1979).

The evolution of flight in insects

Amongst all the terrestrial invertebrates, flight is the unique achievement of the insects. Of the two major ideas of the evolution of wings in insects (see Wigglesworth 1976) the 'gill theory' emphasizes the patchy nature of the environment of the first insects and the need for them to move between patches. It is thus reminiscent of Southwood's ideas of the importance of the abiotic part of the environment.

The alternative, 'paranotal theory', with its emphasis on assumed predator pressure which encouraged insects to escape enemies by gliding away from danger, highlights the biotic mechanisms that the Taylors emphasize.

Life-history consequences of the evolution of flight Whichever of the two theories is correct, we cannot escape the conclusion that the insects dramatically altered their life-history patterns when they evolved wings. Among the present-day apterygotes (wingless insects such as bristle-tails, spring-tails, etc.), the pre-adult development period is short in relation to the adult (reproductive) lifespan which can be up to seven years (and hence similar to, for example, marine Curstacea).

Mayflies which retain many primitive features, including the soft bodies of their ancestors, have an adult lifespan (a few days at most) which is only a fraction of the pre-adult period of at least a year. The reason for this appears to be that moulting which is a *sine qua non* for survival of wingless invertebrates (to replace damaged cuticle) is impossible for winged insects in which the wing itself is entirely membranous and non-renewable. Amongst adult, winged insects alive today, only the mayflies moult more than once and they are able to do so because both adult cuticles are laid down in the pre-adult stage, before epidermal tissue is withdrawn from the wing rudiment.

Many adult, winged insects die after relatively short periods simply because their wings wear out through fraying and cannot be replaced. Winged individuals which escape the necessity to fly (or dispense with the flight mechanism) can live for as long as their apterygote relatives (e.g., queen ants, bees, and termites). But for all other adult-winged insects the wings represent the 'Henry Ford element' of evolution: with wings, for the first time in insect evolution, came built-in obsolescence.

Evolutionary consequences It follows that there must have been considerable advantages to outweigh the disadvantage just discussed – and the major advantage was the considerably increased mobility that wings allowed.

As will be shown later, this increased mobility is obviously successful on an ecological time scale, but has been equally successful on a geological time scale. Over the 350 million years of the evolution of winged insects only nine out of a total of 39 orders have become extinct – a remarkable survival record (F. M. Carpenter 1977). The reasons become clear when we look in detail at the more recent fossils. Coope (1978) has demonstrated considerable continental movement of insects in the Quaternary era, which coincide with the know movement of habitat types to which the same species are adapted at the present day. Instead of being at the mercy of locally changing conditions that their less mobile competitors face, insects appear to track environmental fluctuations and survive by movement. The consequence was that, once evolved, insects tended to retain their structural form. Permian and even Carboniferous Palaeopteran insects can be placed in their correct orders with a knowledge of only extant types (Jeannel 1960). Rather than change *with* their local environments, insects change their local environments – by moving.

Diversity explained?

It is not necessary here to discuss the diversity of insect types except to the extent that diversity may be relevant to the problem of movement. In this, there seems to be a paradox. Insects are the most diverse of all animal groups (reviewed by May 1978) and yet it could be argued that their large-scale movements, by diluting locally developed adaptations, should have decreased the rate of speciation and prevented such diversification (Dobzhansky *et al.* 1977) for the same reason that migrants today show less intraspecific genetic variability than more sedentary species (e.g., Eanes 1979). It seems, however, that increased movement has allowed the insects to specialize in habitats that would not be available to more sedentary creatures. A single dung pat in a field, per year, would not be a sufficient resource for an animal restricted to movement within that field: the dung pat would be removed by non-specialist saprophages. However, the same event, repeated over several fields and exploited by an aninal capable of moving between fields could result in dung pat specialization through competition with others for the same resource. Movement, rather than prohibiting specialization, may encourage it: the inability to move precludes it.

We therefore arrive at the prediction that specialists move more than generalists not because the habitats to which they are adapted are more spaced out (i.e., the usual argument which assumes that specialization occurred for reasons other than ones associated with the distribution of resources) but because spaced out, temporary habitats select for movement which in turn provides a continuous supply of a particular resource on which specialization can occur. It is both the winged invertebrates (insects) and the winged vertebrates (birds) that provide the greatest diversity within their respective terrestrial groups.

Aquatic environments are more stable and provide an alternative mode of individual transport. Teleosts are more diverse than any other group of marine vertebrates both because they are able to exploit small patches of stable communities (e.g. coral reefs) and because of their generally planktonic egg and larval stages — allowing a movement analagous to that of insects and birds.

The above can be regarded as the general case for specialization: the habitat unit and its temporariness, however, determine the limits to specialization that the exploiting groups can show. Dung-pat insects have to be able to face a far wider range of abiotic conditions during their larval development period than do the larvae of insects exploiting more long-lasting food supplies (e.g. large pieces of rotting wood). In the former case, despite the rapid generation time, there is an equally rapid change in the local environment (the dung pat dries out) to which the larvae must be adapted. This prevents extreme specialization (e.g. on food of a particular water content) that can be found in insects of the more permanent patches, and this limits the diversity of the exploiting taxa.

Without the ability to move there would be no dung-pat specialists: despite the ability to move, the temporariness of the resource determines the level at which further specialization ceases.

Subjective stability of resources – a definition We can define the 'subjective stability' of a habitat as the stability of resource supply as perceived by the individual. The subjective stability will be high in both a resident animal which is unable to move but requires only a single, continuously available resource, and in a highly vagrant species able to seek out small, widely scattered or temporary patches of its required resource.

Resumé – the story so far

To recap this section we conclude that throughout their evolutionary history insects have survived by movement on ecological and geological time scales.

The evolution of wings, the third nodal point of insect evolution (Hinton 1977), is the only one peculiar to insects (the fourth, the evolution of a pupa, is a consequence of the third; and as the pupa is a non-feeding quiescent stage it finds parallels in many other animal groups). It is thus to the possession of wings that we attribute both the insects' remarkable powers of survival (by movement between habitat units) and their diversity (through specialization resulting from competition with others on the same, more or less temporary resources).

The ability to fly also has associated costs. Wings are non-renewable (unlike the cuticle of non-winged arthropods, including adult apterygotes), and therefore limit the active life of adults dependent on them. In addition, flight muscles are used and maintained at the expense of reproductive output. Their *use* can bring a distinct advantage by increasing the survival prospects of the reduced number of eggs that are laid. But their *maintenance* alone, at considerable cost, has no advantage except the rather dubious one of not precluding further use. In such cases, the habitat is subjectively stable only through the insect's ability to move.

If insects live in a stable, permanent habitat, wings are either not developed at all, or are functionless. The habitat is subjectively stable because of the constant conditions.

Polymorphism for wing development, especially common in some groups (e.g. Heteropteran bugs) is an obvious compromise between these two extremes. For all morphs their habitats may be subjectively stable, but for different reasons. It seems likely that in such cases environments are locally stable for only a few generations at a time.

The energetic disadvantages of the possession of wings suggest that the tendency must always be to use the flight mechanism whenever possible: 'if you've got it, flaunt it' is more appropriate for the winged insects than for any other group. Or, put another way, it would be wise for an ecologist faced with a winged animal to assume (rather than be surprised by the fact) that migration is vital for its survival. Since winged insects comprise 70 per cent plus of the animal kingdom, the lessons to be drawn are obvious.

Environmental hostility – alternative survival strategies

From an animal's point of view, environments may be 'spatially hostile' or 'seasonally hostile'.

In spatially hostile environments the insect may move either to a suitable microhabitat within its present habitat unit, or to another habitat unit (most examples of insect migration).

In seasonally hostile environments the survival strategy is often that of diapause. Diapause is a state of arrested development which, unlike for example immobility brought on by chilling, is not reversed by a reversal of the conditions which induced it. Instead the animal has to undergo a period of 'diapause development' which proceeds in temperate species of insects more rapidly at lower temperatures (Lees 1956).

Kennedy (1961a) was the first to point to interesting parallels in the behaviour of insects during migration and diapause, and many others have followed (Dingle 1972, 1978). In both migration and diapause there is an initial lack of responsiveness to stimuli which are attractive to the non-diapausing or non-migrating insects (Kennedy's 'vegetative stimuli'). The migrant will move for a certain minimum time before being prepared to stop on the presentation of appropriate stimuli: the diapausing insect will similarly be unresponsive for a minimum period of time, and then show increasing responsiveness.

In adult insects, the parallels go further. Diapause and migration of adults, when they occur, are both generally pre-reproductive: once they end, maturation of the ovaries commences (indeed this is often the first detectable sign of the end of adult diapause). The association of reproductive inactivity with the migrant phase has been called the 'oogenesis-flight syndrome' (C.G. Johnson 1960). Diapause has never been called the 'oogenesis development syndrome': if it were, the parallels would be appropriately clearer. A less analytic but more catching phrase is that diapause is 'migration through time'.

We conclude that space and time are, from the animal's point of view, almost equivalent dimensions into which it fits its survival strategy (Southwood 1977; Solbreck 1978; L. R. Taylor 1979; R. A. J. Taylor and Taylor 1979), the aim being to increase the subjective stability of the environment. It is also suggested that migration and diapause should both show similar lacks of responsiveness to vegetative stimuli only if there is some minimum space or time over which conditions are adverse (i.e. where movement or diapause occur because there are better conditions at some minimum migration distance away or at some minimum time in the future).

Facultative migration should not show the same lack of responsiveness since it is evoked by a deterioration of local conditions below the average of the environment as a whole – often of the animal's (rather than environment's) own making. Similarly, temporarily arrested activity (e.g., through chilling) should not be followed by unresponsiveness to immediate improvement in the environmental conditions, since the average conditions over time are better than those presently experienced.

Migration and diapause – a synthesis

In this section some simple ideas are examined concerning habitat suitability, migratory survival rate, and the optimum migration distance. By changing the space axes to time axes we extend the ideas to include diapause, and anticipate the next section where the evolutionary choice between migration and diapause is investigated.

Habitat suitability We define habitat suitability in terms of the potential reproductive rate of one organism in any one unit of it. If the suitability is low where the animal currently finds itself, and increases elsewhere, the average suitability may be imagined to change with distance as shown in Fig. 9.3(a). In Fig. 9.3(a) the dotted line shows the suitability along any one transect from the origin; peaks indicate resource concentration. An insect's sensory system must work over the average distance between peaks if it is effectively to exploit the environment (hence pheromone systems and the high searching efficiency of dung-pat feeders, etc.). Beyond a certain distance we reach the edge of the insect's range, and average suitability again falls.

Migrant survival rate With a constant survival rate, the number of migrating insects will decline with distance from the origin. Curve A on Fig. 9.3(b) represents a hypothetical, self-powered migrant in which the farther it moves, the lower its probability of survival. Curve B represents a wind-assisted migrant

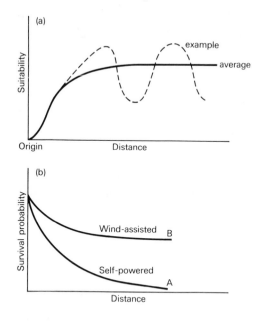

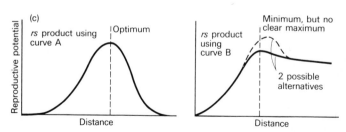

Fig. 9.3(a) Hypothetical relationship between the habitat suitability and distance from an animal's present abode (the origin). (b) Survival probabilities for two types of migrant are related to distance from the origin. (c) Curves A and B of (b) are multiplied in turn by (a) to give the reproductive potential expected at different migration distances. There is a clear optimum using curve A, but only a rather poorly defined minimum using curve B.

that moves as a swarm in which losses are initially large (entering the wind mass), but fall with distance.

The variable survival rate makes curve B different in form from that of curve A; wind assistance also makes it shallower.

Optimum migration distance The two curves in Fig. 9.3(b) can be multiplied in turn by the habitat suitability curve of Fig. 9.3(a) to give the product curves of Fig. 9.3(c). These represent the reproductive potential of individual migrants at increasing distances from their starting points, and will be called '*rs* curves'.

Using curve A of Fig. 9.3(b) there is a definite peak in the *rs* curve at a certain distance. From this we conclude that migrants that are always self-propelled should migrate a fixed distance (related to the proximity of the next suitable habitat unit) and then stop. It is tempting to equate such behaviour with that of obligatory migrants from temporary habitats such as water pools and dung pats in which we would not expect any lack of responsiveness to vegetative stimuli, i.e. they might be expected to use the first pool or first dung pat they encounter on their travels.

The *rs* curve for curve B of Fig. 9.3(b) can be of quite a different shape. For such migrants there seems to be a minimum distance below which migration should not cease, but no obvious maximum distance: for them, the rule might be 'fly for a minimum distance and then keep going for a variable time'. It is for these animals that we can imagine an initial lack of responsiveness to vegetative stimuli (oogenesis—flight syndrome).

The difference between the two types of animal is however more one of degree than of kind. For both types of animal there appears to be an optimum migration distance, which is more clearly defined for the self-powered than for the wind-assisted migrant.

Selection for migration now becomes a process of determining the behaviour that optimizes displacement distance.

A pause, for diapause For each of the three parts of Fig. 9.3 we could replace the horizontal, distance axis with one for time: in that case we would have been discussing diapause rather than migration. In the time-equivalent of Fig. 9.3(c) we arrive at the rule 'diapause for a minimum time and then keep going for a variable time, until conditions improve' (this incidentally is much more akin to the modern view of diapause than to earlier views — see Tauber and Tauber 1976).

To diapause or migrate: which strategy?

The similarities between diapause and migration have been continually emphasized, and in this section I suggest rules by which animals might make the choice between the two (by now it should be obvious that in many insects the choice has to be made).

Consider a three-dimensional graph with the time and distance axes of the previous figure, and erect on each axis the *rs* curve appropriate to the alternative behaviours of diapause and migration (Fig. 9.4). This is a graphical presentation of ideas to be found in matrix form in Solbreck (1978) and Southwood (1977).

Initially a combination of the two axes, time and distance, seems impossible until we realize that they are inter-convertible by the animal's own activity of movement. To move a certain distance d, the animal requires a certain time t. This time is obviously dependent on the animal's size and its ability to move (self-powered, wind-assisted, etc.).

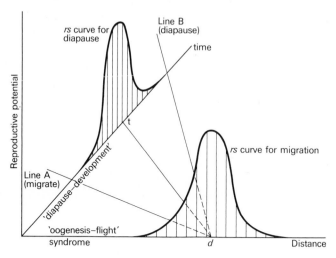

Fig. 9.4. The reproductive potentials are represented on the time and distance axes for the alternative strategies of diapause and migration. The animal is able to translate time into distance by migration, but whether or not it does so depends on its powers of movement. A rapidly moving species represented by line A will choose the migration option: a slower species (represented by line B) will diapause.

If t and d are related in the way shown in Fig. 9.4 (i.e. each coincides with its respective *rs* optimum), the choice between diapausing in one place ('time travel') or migrating to another ('space travel') is evenly rewarded. If however the situation is as shown by line A, i.e. it takes a shorter time to move distance d, then selection will favour migration rather than diapause. A slower-moving animal is represented by line B, for which diapause is the better alternative.

It becomes clear that the decision to travel in time or space depends not only on the period of time for which the local conditions may be adverse, and on the minimum distance to the next suitable habitat unit, but also on the animal's own speed of movement (effectively changing time into space). Creatures that are small in relation to habitat unit dispersion will generally adopt the 'diapause strategy' (best developed in the smallest organisms such as viruses and bacteria with their long quiescent stages), whilst creatures which are larger in relation to their equivalent units will migrate (best shown by the largest mammals, the whales and elephants).

Insects as a class, with sizes intermediate between these two extremes, have adopted a mixed strategy: some species diapause, others migrate. This mixed strategy may itself have contributed to the diversity of the insects, since viruses on the one hand do not have the movement option, whilst most very large mammals (with an exception pointed out below) do not use the diapause option. This suggests that the diversity of the insects relative to smaller organisms may be real, rather than an artefact of our inadequate taxonomic understanding of the latter groups (May 1978).

The life-history polymorphism of each insect species is characterized by stages of differing degrees of mobility: for the reasons outlined above it would seem that if a larval stage requires to time or space travel, it will generally adopt the former (diapause) because of its reduced powers of movement, whereas the adult stage will more frequently space travel (i.e. migrate).

Whatever option is, in fact, adopted will depend on the feasibility of the alternatives. We can imagine a whole succession of *rs* curves linking the time and distance axes of Fig. 9.4, to create a three-dimensional relief of what might be called the 'migration terrain' that the insect *must* traverse in some way (Fig. 9.5).

If the *rs* curves are all more or less of the same height (creating a smooth migration ridge, Fig. 9.5(a), then the insect may, if it migrates, stop periodically en route, to lay eggs or feed in response to vegetative stimuli (e.g. the Monarch butterfly during its southerly flight in North America: Eanes 1979). If, however, there is a deep saddle on the migration ridge (Fig. 9.5(b)) the options are more polarized, and the migration terrain more hostile, a fact that will select for a much more definite suspension of responses to vegetative stimuli during migration until a considerable distance has been travelled (e.g. as in locusts migrating from their outbreak areas). Finally, in some parts of the species' range one or other option may not be available, and the migration terrain takes the form of a single hill on one or other axis (Fig. 9.5(c)). At the northern edge of the range of the milkweed bug, *Oncopeltus*, those individuals unable to complete development to the migrating adult stage are killed by the first severe frosts of autumn (Dingle 1978).

Before concluding this section we draw two other lessons that extend the arguments to other groups.

If seasonal shifts in habitat units are large in relation to an animal's ability to move, even physically large animals may 'diapause' (hibernate) during adverse conditions (e.g., polar bears in extreme environments).

In a group of animals such as the birds which have obviously evolved an ability to move rapidly between habitat units, we would generally expect the migration option to be taken. But the smallest of birds (humming-birds) even in tropical habitats 'diapause' on a daily basis. This may be because the movement option (to other places where food flowers might still be available) would involve too much travel time for such small creatures between what are for them widely scattered habitat units. It is better to wait for another day in the same place.

Abiotic and biotic life-history components and the migration syndrome

In this last section we revisit the two major schools of migration theory and try

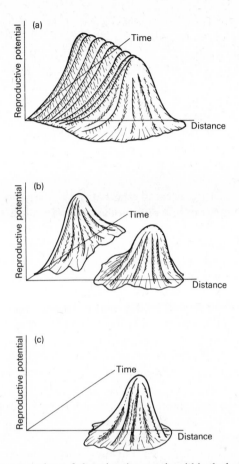

Fig. 9.5. Three representations of the migration terrain within the body of Fig. 9.4. In each case the insect has to traverse the terrain somehow, starting at the origin, and must pass with time into the body of each graph. (a) represents a benign terrain. If the insect migrates, it can feed and reproduce en route (e.g. Monarch butterfly). (b) represents rather more polarized options. If the insect migrates, we might expect a fully developed oogenesis-flight syndrome (e.g. locusts migrating from their outbreak areas). In (c) the insect *has* to migrate, since it will perish if it remains where it is (e.g. *Oncopeltus* species – see text)

to determine simple connections between them. There are many things that migrants achieve by moving, but perhaps the most thought-provoking is that they exchange one sort of mortality for another. Mortality during migration is generally thought to be density independent, and may also be lower than usually believed (e.g., Farrow 1974).

Southwood's ideas mean that migrants exchange a high local abiotic mortality (the temporary habitat coming to an end of its useful life) for a variable but on average lower abiotic mortality. L. R. Taylor and Taylor (1977) emphasized that migrants are escaping a potentially high local biotic (density-dependent) mortality due to competition for limited resources, for a density-independent mortality. This causes these authors to question the field importance of competition (Taylor and Taylor, Chapter 10) which has so far been seminal in all population regulation theories, many based on laboratory experiments on confined populations: in nature, the migrant avoids competition by moving and, given the choice, will always do so. This option that potential competitors have serves to explain to these authors (Taylor and Taylor Chapter 10) why competitive mechanisms, especially in vertebrates, are ritualized and do not lead to catastrophic escalation of both competitive conflicts themselves, and the weapons with which these conflicts are fought. Epideictic displays (Wynne-Edwards 1978) are seen as a signal to others of the local population density, and determine not the number of offspring produced per individual, but whether any offspring are produced at all 'here and now', rather than 'later and/or elsewhere'.

The balance between the attraction of other members of the same species (e.g. for mating purposes or as indicators of suitable habitats) and repulsion due to competition sublimated in the way just outlined is encapsulated in the Δ-response (L. R. Taylor and Taylor 1977), two power functions of density with the impressive descriptive abilities mentioned earlier. What is of particular interest is how the Δ-response might have been tuned during evolution to take account of the type of habitat the animal exploits. A temporary-habitat species would be expected to have a different Δ-response from a permanent-habitat creature, and there should be consistencies between animals belonging to these two groups.

In this section we make several observations on the level of density dependence in natural populations that help to direct where to look for Δ-responses.

Life-table studies are usually carried out on creatures not especially noted for their migratory tendencies. In many but not all of the examples reviewed by Podoler and Rogers (1975), the original authors tested for the density relationships of each life-history sub-mortality, or provide enough information for this to be done. If the mammals, birds, and insects are treated separately, and their life-history mortalities assigned to density-dependent or independent causes, the pattern shown in Fig. 9.6 emerges. Total mammalian life-history mortality is lower and less variable, and a larger proportion of it is density-dependent in nature in comparison with the life-history mortalities of insects. The birds are intermediate between these other two groups. Fig. 9.6 also gives the mean fecundity per generation of each animal group, estimated as the antilogarithm of the mean total life-history mortality (K of Varley and Gradwell 1968).

Because density-dependent mortalities (or, more correctly, losses) are proportionately more important for the vertebrates, it is here that a search could be made for density-dependent movement as a major population-regulation mechanism.

A second observation concerns the type of density dependence shown by

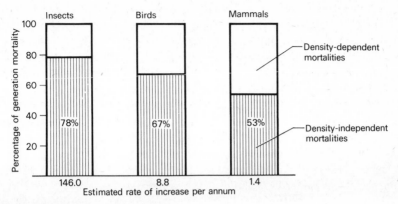

Fig. 9.6. Allocation of total life-history mortalities to density-dependent or density-in-dependent causes reveals a difference between the mammals, birds, and insects, which can also be related to their differing fecundities (lower figure). (Original data from Podoler and Rogers (1975)).

creatures of temporary as opposed to permanent habitats. In a review of more or less the same life-table information, Stubbs (1977) showed that creatures of temporary habitats not only had higher rates of increase (a long acknowledged feature of such animals) but also had density-dependent mortalities of higher terminal slopes than permanent-habitat creatures. This is shown diagrammatically in Fig. 9.7 with, superimposed on each graph, a density-independent mortality that is meant to represent the mortality costs of evading such competition mortality by migration.

It is clear from the two graphs in Fig. 9.7. that the penalties of not migrating are greater for the temporary than for the permanent-habitat creature. Thus temporary-habitat creatures might be expected to show a greater sensitivity in their Δ-responses.

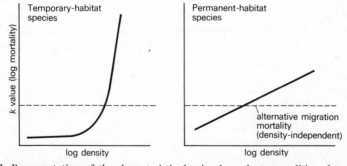

Fig. 9.7. Representation of the characteristic density-dependent mortalities of temporary- and permanent-habitat species, related to a hypothetical level of migration mortality which is assumed to be density-independent. The penalties for not migrating appear to be more severe for the temporary-habitat species.

These two examples, of the vertebrates on the one hand, and the highly *r*-selected species on the other, seem strange companions in a search for Δ-responses since they are almost at opposite ends of the *r*–*K* spectrum. We conclude that in vertebrates Δ-responses are likely to be shown more frequently, whereas they are likely to be more dramatic in temporary-habitat creatures throughout the animal kingdom. For insects the simplifying assumption that the life stage that potentially suffers the extreme density-dependent mortality (i.e., the larval stage) is also the one capable of migrating to escape it, is clearly not justified. In them the adult has to recognise the symptoms of future over-crowding (e.g., eggs and larvae already present) and act accordingly (e.g., there is an exponential drop in the arrival of female dung-flies at freshly deposited dung pats, G. A. Parker 1978 and see also Chapter 6).

This discussion brings us to the single most important difficulty of distinguishing the proximate role of the abiotic versus the biotic environment in determining migration tendency. All authors acknowledge the evolutionary importance of migration, but as Brady (1979) has recently pointed out, evolution is a process of the optimization of characteristics of which we are presently only dimly aware. The fact that predictions generated by our hypothesis are contradicted by experiments in which we attempt to hold all but the crucial variables constant, does not necessarily cause us immediately to abandon our hypothesis, since we can easily invoke *certeris paribus* (all other things being equal) clauses that produce an immediate philosophical stalemate. Any evolutionary phenomenon must be universal: is it therefore disprovable? It is easy to suggest non-Popperian approaches to biology (e.g., Southwood 1977) but this still does not remove the difficulty of studying an optimization process in which the constraints are not only unknown for their quantity, but also their type.

There is no doubt that the Δ-model of L. R. Taylor and Taylor (1977) can generate many biologically realistic results: but until we can correlate the parameters of the Δ-response to meaningful (sic) aspects of an animal's life-history strategy, and furthermore show a degree of consistency among results from animals with similar strategies, this exciting framework of ideas will be no more than a stimulus to sensible research work — perhaps, in Southwood's (1977) challenging metaphor, the ecologist's periodic table, waiting to be filled?

10

Insect migration as a paradigm for survival by movement

L.R. TAYLOR and R.A.J. TAYLOR

And many lines of organisms must have perished then, and been unable to propagate their kind. For whatever you see feeding on the vital air, either craft, strength, *or finally mobility* has been protecting and preserving that race from earliest times. [Haldane's translation from *De Rerum Natura, Book V*. Titus Lucretius Carus, 95—55 BC]

Movement links the behaviour of the individual with the biology of the population. Without movement the individual has no behaviour and the population has no cohesion so that distribution in space is isolated from distribution in time and there is no survival. Animal movement is therefore common ground in the study of spatial behaviour and temporal dynamics. But the disciplines of animal behaviour and population dynamics have different approaches. The behavioural approach tends to be analytical and introverted, isolating the organism in its search for motives and chemical mechanisms. This approach may overlook the fact that organisms cannot survive alone and out of context, that its samples are small and the world more varied than its experiments. In contrast, the approach of dynamics is synthetic and extrovert, assuming randomness where there is only specific and individual behaviour so that, as J. H. Lawton and McNeill (1979) expressed it, 'bizarre ecologies with low probabilities of occurrence – with which the naturalist is so familiar – confound the unwary theoretician'.

It is proper therefore, to integrate behaviour and dynamics, but difficult. Although we are not the first to try, we approach the task with trepidation knowing that both disciplines need to modify familiar concepts. Behaviour must surrender some descriptive niceties and dynamics some mathematical facility. Also the experimental approach alone is too narrow and too time-consuming to cope with the whole gamut of 'bizarre ecologies'. Only a comparative spatial dynamics is likely to coordinate the diversity of data needed for a realistic dynamic theory.

Insect migration has a history long enough to illustrate some of these pitfalls and perhaps to provide a model for a more general concept of the role of movement in survival.

Introduction

Kennedy (1969) has given us the most useful working definition for animal behaviour: *the integrated functioning of the whole animal in its environment with special reference to its movements.* We interpret this to mean the activities of an intact, living individual without undue reference to the internal mechanisms that come within the realms of physiology and anatomy. The qualifications 'in its environment' and 'with special reference to its movements' ensure that behaviour in uniform arenas or constrained by enclosures are treated with reserve. Unconstrained movement in an environment that changes constantly in space and time is thus the essence of the behaviour of an individual and, because individuals do not exist alone, one of the major components of these environmental fluctuations will be the changing numbers of other individuals of its own species. Behaviour can therefore only be defined in relation to population density. It is not the same in isolation as in a crowd and it determines spatial distribution because the degree of crowding is not just a product of multiplication in animals, but also of movement (L. R. Taylor 1961, 1971).

It follows that the behavioural definition of a species then becomes 'the potential range of behaviours of all the individuals at all densities'. This is the complement of Southwood's (1977) 'environmental templet' which in turn becomes 'the integrated environmental continuum that has moulded the species behaviour in space and time'. Again, population density is a factor in the species' own evolution whilst the templet's rate of change will have conditioned the species' spatial mobility (Southwood 1962). Whether we should continue to treat these two interacting entities – the species and its environmental templet – separately, or fuse them into some abstraction like Uexküll's (1957) *Umwelt* remains obscure until behavioural responses can be quantified in some common measurable property. It seems to us that a behavioural definition of a species would be useful in ecology because other definitions have no environmental dimension. Genetic and taxonomic definitions identify the species in isolation from the environmental flux that evolved it. For this reason we have previously suggested (L. R. Taylor and R. A. J. Taylor 1977) that the property required to relate behaviour to population dynamics must be a vector quantity common to all species. It must also complete the dynamic cycle of growth, division, and redistribution (R. A. J. Taylor and L. R. Taylor 1979). The movement visible in behaviour provides this essential quantitative component of dynamics and so it is behaviour that provides the spatial redistribution that is often more characteristic of the species than the familiar temporal dynamic properties of rates of growth, reproduction, and mortality (L. R. Taylor and R. A. J. Taylor 1978).

For instance, whilst they are alive, the males of many aphid species such as the black aphid, *Aphis fabae* Scop., are indistinguishable from males of related *Aphis* species except by their behaviour in response to the environment. This behaviour is made visible by their movement in search of a host plant or mate. Again, the ecologies of gynoparous clonal summer siblings of the same aphid species can range widely. These differences between siblings reside partly in unexplained characteristics of reproduction and longevity (L. R. Taylor 1975; Kempton, Lowe, and Bintcliffe 1980; Wellings, Leather, and Dixon 1980). But it is their spatial behaviour that provides the most striking evidence of the potential range of ecologies within these clones, for genetically identical sisters can range from almost sessile neotenous individuals to some of the most mobile of all insect travellers. Quantitatively the behaviours of clonal morphs differ more than those which separate the males of different species and, having no immediate genetic or sexual origin, these behaviours are determined largely by environmental constraints, mainly population density.

As Kennedy (1972) emphasized, behaviour does not exist so long as it remains only a part of the individual's unknowable intention. It comes into existence in the act of performance. Movement is, therefore, the only direct visible and measurable evidence of behaviour. It is also the only way in which this ecologically vital property can be incorporated into a quantitative discipline such as population dynamics that has fixed geographical coordinates. For example, in order to interpret the population dynamics of the white butterfly, *Pieris rapae* Linn., the spatial coordinates of individual host plants are an essential prerequisite. The sequential oviposition movements of females with respect to individual plants then account for the frequency distribution of eggs on plants. These movements also elicit the curious migration pattern – linear flight within each day but with no correlation in direction between days – that determines the spatial distribution of the eggs and hence the survival prospects of the next generation (Jones, Gilbert, Guppy, and Nealis 1980). The movements also make intelligible the differences between the oviposition behaviour that evolved in populations of this species in Canada and Australia (Jones 1977) whilst the butterfly was adapting to new environments after its introduction by man. Although they share a common recent ancestry, these two spatially separated populations have become effectively different ecological protospecies. Such spatial, or behavioural, adaptation to new environments may explain why most immigrant species eventually adjust their population levels and become absorbed into the existing complex. It also explains why population dynamics is misleading without the spatial dimension. Solely temporal mechanisms of stability are not relevant to evolution in the real world where nothing is spatially stable (Steele 1979). The total environment, comprising food, mates, other organisms, and the internal environment of the organism itself, as well as the physical environment, is in a constant state of flux so that adaptation, and consequently survival, is mostly a matter of movement either to escape misfortune or to find resources.

What especially characterizes this behavioural motion in animals is its intrinsic initiation. Although the presumptive intention to move is not measurable in nature, and is ecologically irrelevant until translated into effective action, the motion itself originates in an internal drive that separates the animals from the plants and the physical world. Contrary to Newton's first law – *Every body perseveres in its state of rest, or of uniform motion in a right line, unless it is compelled to change that state by forces impressed thereon* – the motion of *animal* bodies is never linear, never uniform, and rarely impressed from outside. In animals the initiation, orientation, and the motive force itself are intrinsic, even though environmental forces have shaped them in the past and may influence them in the present. Unlike in the physical sciences, motion is not predictable *a priori* in either the individual or the species. Owing to the physical origins of mathematical thought, this is easily overlooked in both analysis and modelling. Caught in a storm, an animal may be obliged to submit momentarily to overwhelming extrinsic forces, but control over its immediate bodily motion is only rescinded permanently by death. It is therefore necessary to distinguish the active use an animal may make of free environmental energy for motion, flotation down a stream, say, or a ride on a camel, and the resulting illusion of random motion that is implicit in much population dynamics. If animal movement is defined as the spatial component of intrinsic behaviour, it is *by definition* nonrandom. Loss of even partial control is only temporary, although animals and plants may actively use the environmental medium or another organism for transport which magnifies their ambit of movement (L. R. Taylor 1958; R. E. Berry and Taylor 1968; L. R. Taylor, Woiwod, and Taylor 1979). In summarizing insect dispersal Kennedy (1975) recognized specialized behaviour as its most controversial aspect. He wrote, 'Plant seeds undergo involuntary dispersal but we now know that the displacements even of the feeblest insects, thrips, aphids, leafhoppers, and small members of many other orders, do depend on a specialized behaviour pattern', although L. R. Taylor (1958) had already emphasized the dynamic parallel that even fungal spores may be actively projected. Apart from this uncertain behavioural border-line between plants and animals, the behavioural criterion used by Kennedy is not peculiar to dispersal. To define animal behaviour as 'voluntary' is tautological. The deficiency he recognized lies, not in past failure to credit fully the active nature of animal migration *per se*. It lies in the failure to distinguish between controlled movement as being peculiar to animals, and migration as being the component of movement peculiar to population dynamics whether controlled or not. In evolutionary ecology the historical 'accident' is as important as the most complex behaviour. The probability (of success) may be less, but it is the outcome that counts.

Again, the remarkable ability to orientate possessed by many animals may be used during prolonged movements but, unless we know the outcome for both the individual and the species, the ecological function of orientation is not so easily identified. Attempting to divine the ecological function from the physiological mechanism is hazardous whether short-term ecological, or long-term evolutionary, success is the criterion. Survival by movement may be achieved by a multiplicity of means including those where, as in the aphids just mentioned, the appropriate genetic model is for minimal interference with current behaviour as in Mayr's (1976) 'open memory'. In the genetically identical clones of aphids, environmentally controlled polymorphism may respond more rapidly to changing conditions and be more effective than slowly operating genetic polymorphism (R. J. Berry 1979). Details of this mechanism are now well documented and the relevant aspects have been summarized elsewhere (L. R. Taylor and Taylor 1977). All the subtle intermediate gradations of direct and indirect genetic involvement, and hence of response time, must be anticipated in different species to make allowance for the many alternative solutions to the single problem of surviving. It is necessary to allow for the potential range of spatial as well as temporal solutions adopted by different individuals of the same species to the same survival problem (Wellington 1977; Łomnicki 1978). For this reason it is becoming evident that current dynamic theory is too narrow, even when mathematically self-indulgent (Lawlor 1978; Wangersky 1978), to cope with the spatial or behavioural solutions to dynamic problems. The discontinuities and non-linearities introduced by behaviour into real data make their analysis tedious and time-consuming, but ecologically essential. Behavioural 'accidents', like the colonization of oceanic islands such as Surtsey or the Galapagos, need to be incorporated into theory without allowing random concepts to dominate the model and this makes expectations based on commonly-observed behaviour inadequate.

For example, the normal behaviour of a chrysomelid beetle, *Gastrophysa viridula* Degeer, yields a picture of continuous but very small-scale movement resulting in redistribution only within the observed 'population'; it would not usually be regarded as a migrant (R. W. Smith and Whittaker 1980). Nevertheless, an island frequently subjected to flooding that usually eliminates the beetles, is always subsequently repopulated (Whittaker, Ellistone, and Patrick 1979). This must require immigration. Evidently the emigration that occurs is not easily distinguished from internal movements. The borderline between eventualities that may be regarded as 'accidental' or due to deceptively commonplace behaviour is difficult to define.

More extreme examples occasionally arise. Flight and its limitations have been very well investigated in the desert locust, *Schistocerca gregaria* Forskål, and the probabilities of migrants arriving in the British Isles from North Africa are extremely remote. Nevertheless, this does happen occasionally owing to an accidental chain of coincidences in the movements of a few individuals which results in their flights falling within the long tail of a highly-skewed frequency distribution for individual flight times. The event can only occur in practice with extremely large starting populations and is quite undetectable under experimental conditions (R. A. J. Taylor 1979*b*). Nevertheless, such eventualities may be crucial when environments change and apparently permanent habitats become no longer tenable. Such alternative behavioural, or spatial, solutions to the environmental changes that occur during the lifetime of a species must be available to all species that survive.

Occasionally, there is clear observational evidence of a change in behaviour in relation to population density that clarifies an issue. Conventional dynamics, mainly larval and pupal mortality and the effects of resource limitation on subsequent adult fecundity, appear to explain temporal changes in numbers of the cinnabar moth, *Tyria jacobaeae* L. feeding on ragwort, *Senecio jacobaea* L. most of the time in Britain (Dempster and Lakhani 1979). However, there are times when, to become realistic, the model demands a density-dependent loss of adults. This is occasioned by emigrations seen in the field only at times of very high density; but density-dependent flight activity is known from laboratory experiments. Emigration may not be noticeable at the usual level of density in the field, but sets an upper limit to population increase in the model (Lakhani and Dempster 1981).

A more striking development in analytical models for real data that are intensive and extensive enough to identify spatial components, has recently taken place in the dynamics of the larch bud moth, *Zairaphera diniana* Gn., recorded for twenty generations in 20 regions, covering 20 000 km^2 throughout the Upper Engadine in Switzerland. Fluctuations in abundance in time and space were formerly attributed to classical density-dependent resource depletion acting selectively on two genetic polymorphs in optimal and suboptimal areas, differentially affected by climate (Baltensweiler 1970). Internal migration had been reported frequently but treated as anecdotal and hence not incorporated into the model. Such migration has now been systematically observed to occur both vertically up the mountains, and horizontally over distances of 100 km. These movements are density-dependent and together appear to account for much of the cyclical synchronization and singularity of the species dynamics (Baltensweiler and Fischlin 1979). Survival by movement becomes the positive dynamic alternative to the negative view of mortality due to adversity.

A Canadian problem with a forest lepidopteron, *Choristoneura fumiferana* (Clem.), has an almost diametrically opposite history. For years conventional analysis, by life tables, of the carefully sampled successive generations of spruce budworm in north-western Canada showed large discrepancies of measured oviposition from that expected from observed pupal emergence (Greenbank 1963). The gains or losses could only be attributed to supposed immigration or emigration of egg-laden moths (Henson 1951; Greenbank 1957). But only recently has the required massive aerial migration been verified by high-tower visual observation, radar observation, and aeroplane sampling (Greenbank, Schaeffer, and Rainey 1979). These flights are regular and strongly active, but the moths are not orientated to or from recognized coordinates; they are wind-borne into zones of convergence, moving with wind-shift fronts, and affected by storm cells.

The importance of these examples lies in the fact that, depending on the spatial scales of the movement and the so-called 'population' observed, purely temporal dynamics may appear to provide an adequate explanation of short-term population function. Given a large enough area, all external migration is seen to be fatal, whilst all successful migration is internal and, though essential, may be overlooked. Attitudes to migration have often been conditioned by this and by the interpretation of 'migration'.

The philosophical issue of cause and effect, accident or design, and the scales of observation that lie behind these changing attitudes to movement as a dynamic function may seem of little direct ecological concern; yet they have been the most contentious aspect of the study of insect migration (C. G. Johnson 1965; Kennedy 1975). This protracted controversy has been both a distraction and an incentive to thoughtful analysis of the problem of what the function of migration might be and how behaviour and physiology are involved in it. It provides an excellent model for assessing the ecological property of behavioural motion as a paradigm for the survival value of animal movement.

Migration in the Insecta

The ecology and taxonomy of the insects are more diverse than those of any other group of organisms. The species are numerous, have ranged over four orders of magnitude in size, are aquatic, terrestrial, and aerial, occur in every region from the equator to the poles, and live in extreme and curious environments. They feed everywhere, from inside sponges to high in the air; they are herbivores, carnivores, parasites, hyper-parasites, and scavengers. Their generation times range from hours to decades; some species can evolve with extreme rapidity, new ecotypes becoming established within a year; yet others have remained unchanged for millions of years. They utilize almost every known reproductive system, including many peculiar to themselves, and have evolved the most highly organized social systems. They range from the commonest to the rarest of species, and in addition to the diversity of their population structures, genetically identical siblings can range from sessile individuals to the most adventurous travellers over thousands of kilometres, as already mentioned.

If a general theory of the ecology of movement is possible, the Insecta will provide the most varied examples and hence the most critical tests. Those same attributes that generate the diversity of the class also provide the large and varied samples needed to measure individual variability and specific population parameters, especially since the offspring may range from a single near-adult to thousands of minute eggs that must pass through a variety of ecologically different development stages. Also, the protracted philosophical controversy on insect migration illustrates many of the problems of identifying the role of movement as an ecological property and offers a foundation for further development.

The first coordinator of the study of insect migration was J. W. Tutt who edited the *Entomologist's Record and Journal of Variation* from 1890 to 1910. In it he published a series of papers (1898–1902) critically reviewing the literature on insect movement during the previous two centuries. These papers were collected and enlarged in 1902 under the title *The migration and dispersal of insects*. In choosing his title he began a bifurcation of the discipline that continued until the final dominating statement by C. G. Johnson in 1969, *The migration and dispersal of insects by flight*. Johnson's restriction to movement by flight was imposed by the rapid growth of the subject and lost little, save one important feature of the time scale of movement at the borders of paleontology and neontology (see later), because Tutt's coverage was also largely confined to flight. Tutt opened almost every door to the chambers explored later and his opening paragraphs still provide an excellent introduction

It is well-known that many insects, at certain irregular periods, leave the district in which they come to maturity and fly to other localities. Sometimes these flights extend only a comparatively short distance, at other times, hundreds of miles are covered. The term migration is usually applied to these movements of insects, in common with the more regular periodical movements which are carried out by birds and fishes. The application of the term to the movements of insects must therefore, be considered as referring to irregular dispersal movements and by no means to regular movements to and from a given locality. As a matter of fact, it has never been thoroughly shown that the progeny of any immigrants, which have settled in new quarters, have returned to the home of their ancestors, although it has been considered highly probable in the case of certain locusts, and suggested also in the case of one butterfly, *Anosia archippus*. [Now *Danaus plexippus* Linn., the Monarch butterfly]

He went on to point out that the lightness of insects and their ability to go without food (i.e. to store large reserves) enables them to fly immense distances over oceans, aided by the wind. Also, that more or less accidental dispersal by man and on floating trees, etc. has introduced them into countries thousands of miles distant from their native habitat, quoting A. R. Wallace (1876) for evidence. However, their sedentary range is often small despite their great natural powers of dispersal, mainly because their food and climatic tolerance restrict permanent establishment geographically, whilst there are few geographical barriers to 'migration and dispersal'. He then considered their great antiquity, present genera existing when 'Ichthyosaurus ruled the sea, and Pterodactyl the air', and in his closing chapter speculated that the 'migration instinct' is a survival of a very ancient habit, subsequently preserved for any advantageous purpose. This speculation was prompted by his list of potential advantages of migration. They are: to increase the number of communities; to promote cross-fertilization; to 'dispose of superfluous individuals'; to seek new feeding grounds or breeding grounds; to extend the range of the species by adapting to new environments through repeated trial invasions until successful modifications occur; to avoid 'absolute extinction' by continually seeking new grounds'; possibly leading ultimately to new species by adapting to new conditions; to avoid competition in an old home by finding a new one as in *Pieris rapae* which, in the course of the previous quarter of a century, had already covered Canada and more than half of the United States where butterfly species of similar ecology were few.

Of the possible causes, he listed local overfeeding which forces individuals 'to leave their native homes and seek pastures new elsewhere' and some unexplained sexual condition. He also distinguished between local and migratory flights, within and between breeding-grounds, but emphasized that 'comparatively short local flights in one species may be just as truly migratory and undertaken for exactly the same benefit to the species as are the longer flights of other species'. He noted a general north—south trend in migration due to a spread pole-wards from the tropics and that swarming is sometimes associated with migration but often not; and he was not convinced of return migrations. He wanted 'more exact data before a return can be accepted' for the Monarch. Some species appeared to him to have a 'constant migratory instinct' and 'certain individuals' apparently set off to fly 'as far as they can in a more or less definite direction' possibly because 'in some way they recognize that the ground is already occupied', and the excess individuals begin a 'migratory quest' for unoccupied ground. A species in a very favourable habitat only avoids extermination by such migrations, he claimed, so avoiding 'eating up every available scrap of food before the larvae are half grown', and offspring would then have to work back to the favourable areas when resources were restored there. All species do their best to extend their range but causes of dispersion were still very obscure to him, especially in moths and butterflies whose only food is nectar but who leave a 'flowery Eden' and take vast journeys to places with leaves but no flowers 'for the sake of their yet unknown progeny'. He found nothing in parallel between migration in birds, which is 'regular, systematic and purposeful' and in insects where it is 'spasmodic, irregular, uncertain and undertaken solely on account of the absolute necessities of the time'; more observation was needed.

The problem

Certain features in this, admittedly inadequate, resumé of Tutt's detailed review require notice, setting aside the unconcious altruism and vitalism. Amongst more than forty identifiable components in his approach, centres of population are implicit. He suggested quite strongly that depletion of the population by emigration required restoration by immigration, and this, in turn, implied a return for which he had as yet no convincing evidence. This point was subsequently taken up by C. B. Williams who regarded it as a serious evolutionary difficulty (C. B. Williams, Cockbill, Gibbs, and Downes 1942; C. B. Williams

1951) as did Lack and Lack (1951). Kettlewell (1952) sought a genetic mechanism for it. More recently, MacArthur (1972) argued that the loss of genes for migration created an impassable block to continued emigration, a view not accepted by Wiltshire (1946). It may be noticed that such an argument is based solely on the concept of migration as a product of classical population dynamics and genetics, that is, repeated resource depletion due to over-population and the correspondingly repeated loss of potential migrants *from a fixed residential centre.*

Tutt also admitted a long evolutionary history for compulsive migration, with obvious behavioural and orientational ancillaries modified for the requirements of the particular species. Although this showed great insight, there is some apparent anomaly which he compounded with his assertion that migration is a response to current adversity. He also, unhappily, suggested the locust as the most likely example of a returning migrant instead of the Monarch. Much of the subsequent growth of population dynamics hangs on this implicit concept, that species have a centre of population, where the dynamics are concerned only with numerical changes in time, whilst the migrant species is the exception that requires special explanation. For this logic to be viable, such centres need to be clearly demonstrated. We return to this issue with C. G. Johnson and R. C. Rainey.

Tutt took for granted that migration is a behavioural phenomenon and only a very few authors have denied that, as he put it, insects may be aided by the wind in their migrations; they are not forced to migrate by the wind but make use of it. Nevertheless, he did allow some 'accidental' transport by other organisms, such as man and floating trees, so that migrations may be, like many other activities including research, fortunate accidents that only occur to the prepared 'mind'. Alternatively, all organisms migrate in one way or another, so all 'minds' are in some degree prepared and the most obvious kinds of migratory behaviour are not the only kinds. This issue was raised by C. B. Williams and J. S. Kennedy.

When Tutt wrote of migration as both an individual and a species phenomenon, he did not distinguish clearly between migration and dispersal, except to say that insects do not migrate like birds. Modern population dynamics had not then developed very far and Tutt was not clear what the ecological difference is between an individual and a species, nor how the two are related. This fault remained with the subject for a long time.

What he did *not* do was to try to define migration; he described its attributes and wondered about its function. Subsequently, much confusion arose because definitions were produced without recognition of their logical limitations. Tutt reported what had been observed in a wide range of species of differing ecologies in different latitudes. Most of the evidence he presented has since been confirmed and extended. Mass mortality of larval stages due to predation, parasitism, disease, and starvation has been recorded in very high-density concentrations. Mass emigration of adults has been observed from similar concentrations. These situations, if not peculiar to the Insecta, are aggravated by their synchronous, short-cycle life-histories with flight restricted to the adults. Compass-orientated flight is definitely established especially in butterflies, several species flying through the same place on different bearings at the same time. Surprisingly, no adequate frequency distributions are available for individual track bearings at such a point. Less surprisingly, no full-flight journeys for individuals are known, nor any complete courses for the streams of individuals in one of these migrations extending over several weeks. Mark-recapture data show two-point tracks for the Monarch converging southward in autumn and diverging on the return flight northward in spring in North America, reflecting the shape of the continent (Urquhart and Urquhart 1978, 1979). Marked Oriental Army-worm, *Pseudalatia separata* (Wlk.) recoveries, few in number, show roughly parallel north-east/south-west tracks in China (Li, Wong, and Woo 1964).

As Tutt surmised, two-way flights do occur in some species but there is little evidence of the same systematic, narrow channelling of flight in both directions that characterizes some bird migration and is partly learned.

The high-density emigration observations tend to be in arid, unpopulated regions and observations are few and not sytematic but, if such 'centres' of emigration are presumed to remain fixed, then all this evidence is coherent and explains much of Tutt's speculation. The resultant picture is one of pulsating population centres, such as conventional population dynamics has dealt with, having either regular or irregular population explosions that are controlled by infant mortality due to enemies or disease. Or, when enemies are out of phase, death is by starvation. These are the familiar Malthusian controls. 'To escape the final catastrophe' in such centres, high densities produce emigrant adults which fly northward in the northern hemisphere, southward in the southern hemisphere, to arrive at later seasons and lower temperatures at higher latitudes where a less synchronized and less explosive population growth produces a less concentrated return migration.

By this means, individuals benefit from reduced competition along with the exploration of new sites and populations achieve cross-fertilization with the selection of variations for changing climates and the avoidance of ultimate overpopulation and final extinction. New feeding and breeding grounds become colonized, 'satisfying the species requirement' for increased range. Because the emigrants need to be able to establish new colonies, mating before migration and delayed oviposition are advantageous. Also, because migration is compelled by the immediate threat, it is likely to occur immediately wings become available. Aggregation will result from immigrants returning to the population centre and 'excess individuals' are either left behind to starve or fly away to be lost.

Despite the tidiness of this theory, Tutt remained uncommitted. Much later, and with a better understanding of genetic mechanisms and the problems of altruism, Comins, Hamilton, and May (1980) still showed such a subjective view of migration when they wrote 'it is by no means obvious what advantage an individual organism gains by undertaking a perilous dispersal movement instead of staying back to compete more safely in the locality where it was reared'. In reality, there is no guarantee of safer competition, no stable locality, and, because neither size of locality nor distance of migration are specified, no specific peril can be attached to migration. Evolution does not condemn the coward if he survives and it is more rational to expect migration than competition *unless the concept of migration presumes great distance and great risk.* Neither of these is synonymous with most such movements, unless the definition excludes the short, safe movements and so creates a circular argument.

In fact the population-centre model leaves several crucial pieces of information, already known to Tutt, unexplained. If population centres are the rule, why are they not more frequently found? If centres exist, not only is return migration a necessity, it requires accurate navigation as is found in the Monarch but is impossible for most insects as small as aphids, which constitute the majority af all known species. Tutt accepted directional (orientated) flight as a possibility, but he also recognized 'wind aided' migration. How could a small insect such as an aphid, dependent upon the aid of the wind to migrate, be sure of returning to a population centre? Again, it was usual to classify insects as migratory or nonmigratory. As late as 1971, van Valen could state that selection within populations is against dispersal in 'equilibrium species'. If migration is essential in one species, why not in another? If migration exists, in numerical terms it is 'equilibrium species' that constitute the exception, not the rule.

Also, if migration had all the appearance of an ancient habit, deeply ingrained genetically as one must suppose, so that some individuals take off and fly 'as far and as straight as possible', how can this be a response to current adversity at the centre where the habit developed? The population-centre concept is at the heart of the migration dilemma.

Tutt had no theory of migration by which to classify and judge his data. Because most of them were quoted from other authors, his comments often seem self-contradictory and some of them remained so for over half a century. Selection from his evidence could be used to support most reasonable hypotheses and subsequent history was, in effect, a search for a comprehensive theory of migration. First, there was a need to respond to his appeal for more data on the basic facts about whether insect migration is orientated, whether there is a return migration, whether swarming and sex are properties integral with migration, and what is the status of this ancient habit? The questions he posed are genetic, physiological, behavioural, and the final question about the function of migration is ecological. It is not possible to follow the subsequent discussion of these problems chronologically in reasonable space, and it is equally difficult to define each author's position and contribution because views evolved with time and, in retrospect, are not always consistent. We have tried to take the issues as seen through the written contributions and present these in a logical sequence. If we have misrepresented people, it is not intentional.

Tutt pleaded for more evidence of both migration mechanics and proof of return flights, before the function of migration could be properly evaluated. The first to take up the challenge was C. B. Williams.

The controversy

Williams began to publish evidence of linear flight — obviously migratory by any criterion — in 1917. By 1930 he had enough observations to fill one book and had already established the existence of migration, however interpreted, and also of an orientation mechanism, at least in day-flying Lepidoptera. Williams continued to collect evidence until 1976, publishing in all about sixty technical papers and two books. His amassed data leave no room for doubt: linear flight is a fact in many species and directions are highly specific, several species often migrating in different directions at the same place and time. Return migrations, when they can be recognized, are diffuse and rarely along the same narrow route. He tried to define migration physiologically, behaviourally, and ecologically. This was a logically impossible task, for a *definition* cannot be valid at three different levels of organization simultaneously, and much of the controversy centres around this problem.

In 1928, two years before Williams' first book, Felt published 'Dispersal of insects by air currents' (*N.Y. State Museum Bull.*), an equally convincing mass of data on the downwind migration of many insect species. Later, these anecdotal records were to be confirmed quantitatively by Hardy and Milne (1938), Glick (1939), Freeman (1945), C. G. Johnson (1957), and L. R. Taylor (1974) who found vast numbers of individuals and species of insects so high in the air and so small — some less than 1 mm in length at thousands of metres a.g.l. — that total control over orientation was inconceivable during the flights sampled. Nevertheless, L. R. Taylor (1960a) showed the insects were alive and healthy and therefore in control of their immediate life functions, including flight. Meantime, Felt (1925, 1926, 1928) had insisted that all movement was downwind and C. B. Williams (1926, 1930) had rejected this involuntary transport as migration. Orientation had become a criterion. By implication, migration was being defined physiologically, not ecologically, and the controversy had been started.

Williams' interest lay mainly in the mechanism of orientation which he investigated in a mass of observational data of prolonged unidirectional flights of day-flying Lepidoptera along fixed compass bearings in the tropics. He finally concluded that orientation was mediated by light or the Earth's magnetic field and this now seems likely to be right (Gould 1980). The actual compass bearings could be given no specific relevance because no migrations were followed from start to finish, and the importance of this lay in the necessity for a return migration along the same route if the population centre remained at geographically fixed coordinates. Of this, C. B. Williams (1958) wrote that the absence of evidence was an expression of our ignorance rather than our knowledge. However, migration could legitimately be described as 'purposive' in the sense that an essential goal existed and the mechanism necessary to find it must also exist.

Systematic observations of a butterfly, *Ascia monuste* (L.), in Florida showed the ontogeny of migration to be a progressive build-up of flight to fully developed exodus, but it was not a regular out-and-home movement (Neilsen 1961). Mass exodus from very-high-density centres of populations of another butterfly, *Pyrameis cardui* (L.), probably associated with overreaching resource supply, was observed on rare occasions but only in semidesert areas where continuous observation was impossible (Skertchly 1879; Egli 1950), but such occasional observations provided the impetus needed for further search. If population centres were geographically fixed, it followed that migration routes were proscribed and travelled over year after year; and there was considerable evidence for channelling of routes at least in one direction (Beebe 1949; Lack and Lack 1951; C. B. Williams, Common, French, Muspratt, and Williams 1956).

However, this could not explain the movements of aphids and locusts which appeared to be equally concerned with survival after mass exodus from a ravaged environment, but were not compass orientated. It was also significant that the only moths to produce convincing evidence of oriented flight were the day-flying *Urania* species (C. B. Williams 1930). Subsequent observation of night flight by moths increasingly gave evidence of downwind migration from trapping (L. R. Taylor and Brown 1972), by direct visual observation (E. S. Brown 1970), and much more extensively by radar (Schaeffer 1976; Roffey 1972; Riley and Reynolds 1979; Greenbank, Schaeffer, and Rainey 1979; Riley, Reynolds, and Farmery 1981).

After the exchange between Williams and Felt the fact of orientation during some migrations was no longer in doubt. Williams' evidence was conclusive. Equally beyond doubt were Felt's observations of the downwind movement of large numbers of insects many of which had left their birthplace irrevocably. The first thousand metres of the atmosphere are daily filled with such insects, too small to control their flight direction but still possessing the behaviour that initiates such flights. Their presence, therefore, demands a common ecological explanation for the evolution of the behaviour and this makes segregation of migration into 'active' and 'passive' highly artificial and misleading (L. R. Taylor 1957a). Except for a few specialized groups, accurate flight control is needed ultimately even by the smallest insect species for mating and oviposition. The relevance of the segregation between compass-oriented and downwind flight is therefore not its ecological, migratory function, but a physiological one based on the evolutionary history of body size and the resulting functional relationship between flight speed and environmental factors, mainly wind speed.

L. R. Taylor (1958, 1960b) proposed and finally demonstrated (1974) a 'boundary layer' of air near the ground, varying in depth with wind speed but deeper for large than small insects, within which flight speed exceeds wind speed and control of directional flight is absolute. For large insects, whose flight is comparatively powerful, compass-oriented migration may be wholly within this boundary layer which is deep. Outside the layer, direction of any flight including migratory flight, must have a downwind component although this does not mean total loss of control; upwind flight is, however, impossible. He also showed that the flight above the boundary layer is not 'random', because the mean profiles of density with respect to height are highly specific and differ from inert particles; in other words even small insects are not blown accidentally and uncontrollably by the wind, although it dominates their direction if and when they choose to fly in this 'free air', which measurement showed that many do. Some species reduce the risks attendant upon partial loss of directional control by flying at dusk when the boundary layer is deeper, for example, or only when wind speed is low. The problem facing the insect is to manipulate its flight so as to make use of the boundary-layer interface in satisfying the ecological demands made upon it. Crossing the interface is usually a deliberate migratory act, making possible the use of the free transport energy available above it, but with increased survival risk due to partial loss of control of co-ordinates in space and the difficulties of subsequently regaining position, when necessary.

L. R. Taylor (1965*a*) recognized four potential categories of migratory flight in aphids: oriented flight within the boundary layer, limited in extent by flight reserves at low flight-speeds; stratiform drift in massed flights that remain passively cohesive in low-velocity, laminar air-flow for up to one or two kilometres, also within the boundary layer and often in the early morning and evening; cumuliform migration, usually in the middle of the day when aphids, scattered by turbulent convection outside the boundary layer, often reach thousands of metres and are distributed over wide areas up to tens or hundreds of kilometres; jet-stream transport at high altitudes when aphids are carried at high velocities in linear motion that can continue over night and reach great distances (thousands of kilometres). He did not regard the migrant as a biologically inert particle, although this conclusion has sometimes been drawn from the evidence that some aspects of its aerial motion can be so treated.

With compass-oriented flight and wind-directed migration established, the main effort focused on finding a behavioural definition for migration (Kennedy 1961*a*; C. G. Johnson 1962) because behaviour during migration is often visibly specialized. The behavioural components most emphasized were active initiation of flight, (C. G. Johnson 1960) persistent linear flight (Kennedy 1961*a*; Neilsen 1961) and non-appetential flight (Provost 1953) but, as ideas developed, positions constantly changed.

C. G. Johnson was interested in aphid migration and, having discovered that daily changes in numbers in the air were not well explained by current environmental conditions, sought an explanation in terms of the exodus of new migrants produced each day. The resulting model for daily cycles of migration in *Aphis fabae* Scop. (C. G. Johnson and Taylor 1957) showed that migration was a brief event at the beginning of adult life. It was then found that the initiation of take-off could be defined by the end of the teneral period, a developmental stage during which wings are hardening (L. R. Taylor 1957*b*), making due allowance for thresholds causing take-off delays in inclement weather. This 'post-teneral' migratory activity was compulsive in certain individuals so that, for them, reproduction was inhibited until migration had taken place (B. Johnson 1958).

C. G. Johnson (1960) used this evidence to argue that compulsive migration could anticipate adversity rather than await the event. This opened the door to a spate of work on factors affecting alate morph production and activation in aphids that has created a different approach to migration and was a turning point in the study of insect migration (Kennedy 1961*a*).

The general system for migration and dispersal by flight proposed by C. G. Johnson (1960) was based on the premise that migration was not a current reaction to adversity but, as Tutt had surmised, a fundamental function with a long evolutionary history which expressed itself differently in different species. The only common feature is an active exodus by new adults, 'post-teneral' and pre-ovipositional, which he later developed into the essentially physiological 'oogenesis—flight' syndrome (C. G. Johnson 1963, 1966). He also commented (1961) that 'it is difficult to characterize migration by one special kind of flight; but the ecological criterion of displacement from the breeding site and the acquisition of another is universal'. However, he distinguished migratory from non-migratory insects behaviourally, by their appetential flight (C. G. Johnson 1962), and his aim in 1966 was to 'characterize migration and adaptive dispersal physiologically and ecologically, as well as behaviourally'. Heape (1931) had earlier classified migratory movement as nontrivial, a behavioural criterion. This requires a specification for trivial, and Provost (1960) interpreted it in classical ethological terms as 'appetential flight' in mosquitoes, meaning that trivial flights had an immediate objective. Migration then becomes non-appetential or an exodus flight that has no immediate objective, being essentially self-satisfying travel for travel's sake – a negative definition from the ecological point of view.

Kennedy (1951) attributed long-range movements to 'both insects and wind acting together'. He emphasized that 'active' and 'passive' migration are not logical alternatives, 'passive migration' being a contradiction in terms because the insects' contribution is always actively persistent. The means of traversing the ground he considered secondary but the geographical trend of locust swarm movement was not created by either the locust or the wind, but by their conflict. This is again an essentially behavioural view of migration and was later (1961*a*) focused on the alternation of motivation and inhibition. He recognized ecological laws 'on a higher plane than behaviour' (1951) but regarded them as the 'interaction of a whole population of insects and a whole system of climatic and geographical circumstances'.

Kennedy's main concern with motivation led to a search for behavioural characteristics that would embrace both compass-orientated and wind-orientated flight. He found the answer in persistent, undistracted, straightened-out, loco-motor activity, whilst vegetative functions of feeding and reproduction are inhibited. This view of the control of migration by the central nervous system through thresholds of mutually antagonistic fundamental drives, has much in common with Sherrington's successive induction model of spinal reflexes and became known as the Sherrington—Kennedy behavioural model (Southwood 1962). Kennedy's view does not distinguish between linear flight intrinsically directed or extrinsically imposed; it is the continuation of drive that counts. But there are many examples such as the Monarch (see also C. B. Williams 1949) in which flight that eventually results in linear transition over great distances, does not inhibit feeding, oviposition, or nocturnal roosting in circum-stances that involve quite marked deviations from linear flight. Behaviour that is obviously social intervenes prior to settling down at sunset and behaviour that must often be considered as 'trivial' by other standards is only recognizably migrant during spasmodic intervals.

Heape's (1931) separation of flight into mutually exclusive categories of migratory and trivial can also be interpreted functionally as well as behav-iourally. Cavalli-Sforza (1962) used these categories for movements by man. Trivial behaviour then consists of feeding, sexual, reproductive, or other 'loops' which result in no final change of place and are thus, *by definition*, random with respect to the underlying trend of movement. The trend is then migratory because it results in a permanent shift in position. This is no longer a behavioural definition. No implication of continuity of behaviour, compulsion, orientation, origin, or mode of transfer is made. Migration is a purely spatial concept; it is the persisting change in place that is left when all other minor excursions are removed, whatever their cause or function. It includes the spatial outcome of Williams' 'migration', Provost's 'non-appetential behaviour', Johnson's 'oogen-esis—flight syndrome', and Kennedy's 'drives'. However, it differs fundamentally from all these in that it does not exclude 'accidental' migrants such as a foraging bee that is blown off course and unable to recover. It is an ecological criterion judged solely by results. If the bee is solitary, female, and fertilized, and estab-lishes viable offspring in the new place, in dynamic terms it has migrated what-ever the cause (L. R. Taylor and Taylor 1978). Migratory physiology and behav-iour are real, but they do not *define* migration.

A special case

The Desert locust *Schistocera gregaria* Forskal has been the subject of intensive genetic, physiological, morphological, and behavioural study, and its ecology has been monitored more extensively and for longer than any animal except Man. In particular its movements have been followed systematically for many years over an area of 30 million km^2, from 15°W to 90°E and from 40°N to 10°S. The morph complex is density-dependent and largely genetically independent, but carried over accumulatively for several generations. At their extremes, morphs differ more – morphologically, physiologically, and behaviourally – than some taxonomically recognized grasshoppper species. The extreme morphs are sedentary, unaggregated *solitaria* and swarming, day-flying *gregaria*. The differences in behaviour are ecologically relevant and it has been suggested that the habit of aggregation developed in flightless nymphs as a protection against predators (Gillett, Hogarth, and Noble 1979). But it is the movements of flying swarms that most concern us here. Swarm cohesion is maintained by aggregative behaviour against the strong forces of atmospheric turbulent diffusion and against the fast (up to 23 km h^{-1}) flight of groups of individuals that is visibly linear when seen from below, but is directed apparently at random within the swarm as a whole. Escape from the swarm is prevented by orientation towards neighbours at its surface, a tenuous skin effect. Swarms vary in height from low-flying stratiform structures to towering cumuliform masses extending up to 2000 m.a.g.l. that may become detached from the ground in very active convection (Waloff 1972). The skin effect at the surface converts the powerful linear flight of the randomly orientated groups into a rolling motion of the whole aggregation (Sayer 1956; Waloff 1958) that accurately tracks wind direction at a speed equal to or less than the wind (Rainey 1963). Deviations of swarm direction from wind direction are minimal (Draper 1980) and reduction below wind speed is accountable largely to errors of measurement due to local variation in swarm structure and to surface drag caused by locusts settling out in the van and delaying take-off in the wake. High fliers tend to be more downwind orientated than those in lower strata where wind speed is less, and this is attributed by Waloff (1972) to Kennedy's (1951) optomotor reaction. There is some loss of swarm cohesion in high winds but this is restored when wind decreases (Sayer 1962).

Rainey's interests lay mainly in the association between locust migration and meteorology and his 'new hypothesis' (1951) was designed primarily to explain the mechanism of swarm motion. As a result of swarm cohesion, a powerful linear flight that could make headway against most windspeeds, if associated with any orientation mechanism that may exist (Ellis and Ashall 1957; Riley and Reynolds 1979), is converted by behavioural mechanics into superficially aimless drift down seasonal winds in eastern Africa (Waloff 1946), in the Indian sub-continent (Rao 1960) and elsewhere throughout the species range (Waloff 1966). This drift performs an ecologically sound function. Over-all movement is towards zones of wind convergence, the I.T.C.Z. that move seasonally within the tropics where rainfall is most likely to provide vegetation for the next generation. Migration, however complex and counterintuitive its mechanism, is therefore basically a prolonged oviposition flight, terminated by environmental cues that anticipate improved prospects for survival of the offspring; where the winds meet, the prognostication is rain and therefore food.

A revelation

The life-style of *Schistocerca gregaria* has proved to be an ecological revelation. It has been studied intensively and meticulously. The insects' migrations have been monitored with patient care for so long that major systematic revison of their interpretation is not now likely. C. B. Williams' (1930) and MacArthur's (1972) fears were groundless. Although it may take hundreds, or thousands, of generations before the progeny of a given individual will return to its precise ancestral home, there is no potential loss of migratory genes. Indeed, the reverse is true. The greater the motion, the more likely it becomes, for it is by migration that survivorship is assured.

The main components in this dynamic locust system are most easily recognized by considering all the individuals of the species as a single, geographically unrestricted but internally cohesive, yet highly fluid, population system. All the recognized behavioural and physiological attributes of migrants are present in a large proportion of individuals: post-teneral, compulsive, non-trivial, appetential, linear, long-distance, oriented, massed, persistent, pre-reproductive, flight. However, because the migratory drive is balanced against an equally vital aggregative compulsion that restrains indefinite separation of the individuals, the resultant geographical displacement is totally different from expectation. All the expenditure of energy achieves nothing except to keep the locust in the air; the wind does all the real work as it does for aphids. Migration remains essentially internal, both to the local population element, the swarm, and to the whole species population. The concept of population itself has changed. There is no centre. There are mobile semi-coherent elements that remain a part of the whole population of the species in a space—time continuum, by a common interest in each other that balances the tendency to separate, and by a common interest in their environment. Together these two motivations prevent population disintegration, the greatest risk in evolutionary history.

Stated in this way, the concept of population and the role of migration and aggregation are interdependent and can be generalized. The population is not an isolated group of individuals chosen for observational convenience, like a single locust swarm. It is all the individuals of the ecological species and forms a continuum in space and time. Survival is by movement within the population and, if it spatially separates the birthplace of generations, we call it migration. Unsuccessful migration leads to loss from the population by death, but migration is an essential component of population structure undertaken by nearly all individuals; it only becomes strikingly obvious when the spatial scale is large and the dynamic structure of the population is very fluid because its environment is especially mobile. It is always balanced by the restraining influence of individual attraction, often sexual, which we call aggregation (L. R. Taylor and Taylor 1977).

Because so much is known of their ecology, the aphids have been used previously (L. R. Taylor and Taylor 1978) to characterize the vagrant way of life which magnifies the spatial component of dynamics and so demonstrates that population dynamics are not universal or fundamental if they assume a fixed population centre. But the asexual phase of aphid population cycles may seem to justify their treatment as a special case. Indeed aphids were chosen partly to demonstrate the existence of aggregation as a non-sexual, intraspecific mechanism that yields environmental cues (Kennedy and Crawley 1967). No such barrier applies to the locusts: their phase change is neither asexual as in aphids, nor immediately genetic. It merely makes visible the changes in spatial behaviour associated with population density that are well known in all organisms, although not always taken into account in population dynamics, and emphasizes the integrative role of aggregation as an essential counterbalance to migration. The vagrant way of life avoids adversity by movement without altruism but more geographically concentrated populations, like the chrysomelid beetle mentioned earlier, give an illusion of 'migration' being disgenic because much internal movement is discounted so that the occasional long-distance migrant is seen out of context. In reality all populations are spatially mobile in some degree and on some time-scale.

Integrative levels

The search for a definition for migration in insects illustrates the necessity to understand clearly at what level a recognizable function operates. Early attempts to define migration failed because the demands made of the definition were too high. An ecological property can only be defined ecologically. Once defined, it can be *characterized* behaviourally, physiologically, or genetically for individuals or species, but not *defined* at lower levels of organization. There is no logical reason to expect the same behavioural mechanism to be used by different species to achieve a given ecological end; it is evolutionarily unlikely. The behaviour of a given species can be separated into the three basic life categories of growth, reproduction, and redistribution, as has been done by Caldwell and Rankin (1974), but these specific behavioural equivalents of the dynamic functions are not transferable to another species. The redistribution behaviour for a species is migratory behaviour, but the definition of migration in ecology is change of place between generations which has no equivalent in behaviour or physiology. In the same way, feeding has no common behavioural definition for all species, having nothing in common between, say, amoeba and man except its dynamic function, providing material for growth. The attention paid to behaviour *per se* in the discussion about migration directed interest away from its redistribution function to a multiplicity of specific mechanisms. Not all movement is migration, however, and often the same behaviour confuses two essentially different functions. This is especially obvious in what have become known as foraging strategies. The oviposition flight of a parasite is a migratory movement that determines the site and survival prospects of the next generation. It has come to be treated simultaneously with predation, which is a feeding movement concerned with survival of the current generation, under the common heading of foraging, although the dynamic function is vitally different. In some insect species, the two functions do operate simultaneously, the same individual providing both prey and host; the parasitoid *Aphelinus thompsoni*, for example, also preys on its host, the aphid *Drepanosiphum platanoidis* (Collins, Ward, and Dixon 1981). But in other parasites the host is not the prey. Confounding the two *ecological* processes because the *behaviour* looks the same is then confusing issues.

An ecological approach

The continuing dialogue about the meaning and function of migration eventually directed attention to two vital ecological issues. Fletcher (1925) had commented that migration is 'merely an accentuated form of dispersal by flight', and more or less normal. He claimed that the picture of insects confining themselves to a locality 'since there is no incentive to leave' is only partly true. There is a constant diffusion which is only noticed when suitable localities are separated, or when insects move *en masse*.

In 1951, E. S. Brown showed that the migration rates of different species of aquatic Hemiptera in Hertfordshire, England correlated well with the permanence of their habitats. In particular the mainly brachypterous species of *Cymatia* and *Micronecta* were confined to large lakes and rivers where movement by swimming makes flight less compulsive. Loss of wings, flight muscles, and flight behaviour are often reversible, this condition being secondary to the primary winged condition adapted to a heterogeneous environment (Wigglesworth 1963).

Southwood (1962) greatly clarified and generalized this ecological view, re-directing attention to the survival function of migration as a means of keeping pace with the spatio-temporal variations in habitat. Unfortunately, he, like C. G. Johnson, added untenable qualifications; i.e. leaving the 'population territory or habitat' and also Kennedy's 'persistent locomotor activity'. The ecological relationship between migratoriness and permanence of habitat is a species property — proportionate adaptation. Leaving the habitat and persistence are properties of individual behaviour — peculiar to the case in point. The two sibling moths that emerge 10 m inside a forest, one of which flutters erratically 20 m out of the forest whilst the other flies persistently for 20 km further in, illustrate the anomaly. Both are migrants if successful, but neither conforms to both these criteria. Subsequently this developed into the τ/H ratio (Southwood 1971) which was derived from consideration of the length of time a habitat is suitable for occupation in relation to the species' reproductive cycle, a different but parallel approach to MacArthur and Wilson's (1967) $r-K$ continuum and an essential component in migratory function.

Ibbotson and Kennedy (1951, 1959) found that *Aphis fabae* individuals *'certainly use each other as signals of food'*, just as migrant cypris larvae of *Balanus balanoides* are attracted to settled ones (Connell 1971), and Kennedy (1966a) concluded that 'it would probably be better to work on the assumption that the members of a population are not usually quite incommunicado, intraspecific communication being the rule rather than the exception....Indirect *evidence of widespread communication comes from many examples of responses to increased population density'*, including long-distance communication. He went on to say that all insects should be considered as 'both attracting and repelling conspecifics', referring these to mutual aggregation or avoidance *in spatial terms* (Kennedy and Crawley 1967). In particular he considered that sexual and protective behaviours are aggregating mechanisms. Malthusian effects of overpopulation, that result in resource shortage for passive victims, are anticipated by communicating current density, so avoiding the deleterious effects by moving apart (Kennedy 1966a); he was reflecting Tutt's views of sixty years earlier.

This concept, of spatial dynamics resulting from the balance between aggregation and migration, had also been reached earlier by L. R. Taylor (1961) from quite different evidence. The spatial distribution that is the outcome of such behaviour was found to have specific population parameters interrelating the first and second moments, density mean, and variance, and these were attributed to spatial characteristics of aggregation and migration that differed disproportionately at different densities (L. R. Taylor 1965b, 1971). The evidence for such specific spatial and temporal dynamic patterns is now overwhelming in insects and birds and for most animal taxa (L. R. Taylor, Woiwod, and Perry 1978, 1980) and has an equivalence with temporal variations in density (L. R. Taylor and Woiwod 1980) but with wider-ranging species parameters. A conceptual mechanism balancing the association between density-dependent aggregation and migration has also been suggested, the Δ-model (L. R. Taylor and Taylor 1977, 1978; R. A. J. Taylor and Taylor 1979), that can be simulated with some success (R. A. J. Taylor 1981a,b). Dependence of behaviour on density in the model reaffirms Kennedy's (1966a, 1972) assertions. Increasing evidence of density-dependent migration is now appearing in all taxa, often with associated population effects like those examples quoted earlier.

Models for dynamic motion

The first model for animal migration was based on the classic mosquito work of

Ross (1905, 1911) who recognized two kinds of movement, major and minor vicissitudes. The approach was taken up by K. Pearson and Blakeman (1906) who correctly adopted Ross's ecological classification, which they now called 'flights' between breeding sites, and 'flitters' for feeding and mating. Their behavioural approach was less fortunate. They chose the physical diffusion model which treats of identical particles having no intrinsic motion. The definition, as given by Kendall and Buckland (1971), is the basis of much that followed. The process is such that the displacement of the variate (its increment) in time dt follows a Normal distribution with variance proportional to dt. It assumes random orientation as well as distance travelled at constant velocity (D) and results in the so-called 'half-normal' regression equation

$$N = \exp (a+bx^2) \tag{1}$$

of density N on distance x, in which b is inversely proportional to the diffusion coefficient D.

The analogy with dispersing animals is uncertain. Whether the individual differences in behaviour were presumed to follow a normal distribution or the individuals all alike and the displacements imposed on them by the environment were actually like diffusion is no longer clear. It is evident with hindsight that this physical approach introduced into individual migratory motion, and also into the confusing population function of dispersal, the fundamentally unbiological concept of randomness; and the model was unfortunately followed by R. A. Fisher (1922), Dobzhansky and Wright (1943), and Skellam (1951). Their collective authority gave credence to a concept so foreign to the intensely individual and specific activities of animals that it may account for much of the subsequent dichotomy between animal behaviour and dynamics.

As shown by Horn in Chapter 4, theoreticians now incorporate emigration and immigration into population models, but rarely has the individual's motion been realistic or the mobility of the population itself taken into account. The mathematical difficulties of using behaviourally controlled motion in coordinate space and real time have delayed the process. But it has also been held back because interest was directed towards the concept of abstract 'competition' between apparently immobile, plant-like, individuals rather than to their active search for survival in a varied terrain predominantly indifferent to them. During evolution this intrinsic motion offered a prospect of environmental optimization by a mechanism quite different from that provided by maximized reproductive rate. We suppose that the motion led to selection for increasingly sophisticated mechanisms of animal movement, hence to the sensory mechanisms that direct it, and ultimately to the neural system that coordinates it. None of this selection occurred in plants because they are basically static and their dispersal nearly passive and random and correspondingly wasteful. The mathematical theory developed by K. Pearson is too complex for verification but it appears to have influenced R. A. Fisher (1922) to predict that the spread of genes in space would be found to obey the Gas Laws, a highly improbable ecological concept for animals if not for plants.

Much later, attempts were made to measure the amount of directional bias needed to produce the measured recovery rates of marked and recovered long-distance migrants; for example in Saila and Shappy's (1963) cardioid model, salmon were used to verify results. The required orientation bias was quite small, partly because the final stages of salmon migration use widely-diffused chemical environmental clues that are remembered from youth. The memory mechanism would be different in insects that have usually not visited the goal before, although pheromone production suggests that the inherited memory based on sensory responses to other individuals, as in Kennedy's environmental cues, could be equally viable. The diffusion distances in air would be shorter than in water. The other relevant factor is that large losses may sometimes be as acceptable during redistribution as they are during development, such as pre-dation of immature fish and insects, for example. The extreme condition found in the dispersal of spores and fine seeds such as orchids relies entirely on blanket coverage, so a physical model must sometimes have relevance. Nonrandom components are introduced by controlling mechanisms confining release to appropriate seasons and environmental conditions, whilst the complex shapes of seeds ensure that the distance function is not random. The release of Coccid crawlers, with no control whatever over flight, is constrained by light, tempera-ture, and humidity and differs between adjacent bushes (M. G. Hill 1980). Spiders can control only the time of release, the lift exerted by the length of their floating line, and, possibly, their time of alighting (L. R. Taylor 1974).

Although, with certain reservations, the migration of flightless spores, seeds, and wingless arthropods may be approximated by physical diffusion, the vast majority of insects have wings and, however small and powerless they may appear, we can see no justification for equating their movements with random physical mechanics (L. R. Taylor 1965a; and R. A. J. Taylor 1978). Of all the small insects, the migrations of aphids have been most studied and every aspect of the flight, except the major directional component which depends on the wind, is controlled by complex behavioural mechanisms. Not only do thresholds for such simple processes as, for example, take-off in relation to light and temperature, differ betweeen species and biotypes, the frequency distributions of individual responses are not Normal distributions and depend upon substrate and previous experience (Dry and Taylor 1970; Halgren and Taylor 1968).

In all, there are millions of specific solutions required for a migration model and the basic model must be sufficiently flexible to permit migratory processes as different as in so-called 'non-migrant' *Drosophila*, directional migrants like the Monarch, territorial birds, diffuse migrant aphids, and man (L. R. Taylor and Taylor 1977). To date, the only family of regression equations of density on distance that approaches this degree of universality is a generalization of the

Γ-distribution (when $\gamma = 0$), which includes the Weibull distribution ($\alpha = 0$) and the half-normal ($a = \frac{1}{2}$, $c = 2$, $\gamma = 0$) and which has the probability density function

$$p(x) = \frac{c(x-y)^{c\alpha-1}}{\beta^{c\alpha}\Gamma(d)} \; \exp\left[\frac{(x-\gamma)^c}{\beta}\right] \tag{2}$$

which yields the four-parameter density N by distance x regression equation

$$N = \exp\,(a+bx^c+d\ln x) \tag{3}$$

and which describes all the 15 sets of data available at present. Of these, thirteen sets can be simplified to two-parameter special cases and none is fitted well by the half-normal (R. A. J. Taylor 1980).

The inadequacy of the physiological experiments to explain the long-distance migrations achieved by some species, notably locusts, but also aphids (Elton 1925; R. E. Berry and Taylor 1968) suggests that too little attention has been paid to individual variation. From the evidence of a locust migration in which success rate was known, albeit very approximately, the behavioural contribution appeared superficially to be minimal (R. A. J. Taylor 1979b). The orientation appeared to be attributable largely to meteorological conditions, especially the winds associated with convergence zones. Nevertheless, as in aphids, behavioural control of migration was found to be highly sophisticated, determining time spent airborne and the vertical profile of density in relation to height, so regulating the distance travelled and its downwind direction that the density–distance distribution is complex and highly specific (R. A. J. Taylor 1980; L. R. Taylor *et al.*, 1979b).

The directional component of insect migration remains unresolved. Orientation and goal-finding are both convincingly demonstrated; orientation in numerous, mainly tropical, day-flying Lepidoptera and goal-finding by the Monarch in Mexico (Urquhart and Urquhart 1976).

However, no information is available as yet on losses during these flights. Also, the concentration of the Monarch in the south is not directly concerned with host-plant availability. The aggregations occur on non-host trees during the winter and have more in common with aggregations of ladybirds on mountain tops and the Bogung moths in caves (Common 1954) than with host-finding and oviposition. The temporal equivalent of migration, tracking the benign environment in space, is diapause, awaiting the return of the benign environment in time. This has been extensively investigated by Dingle (1978) and his colleagues (see also Rogers, Chapter 9). Aggregations such as those mentioned above have the appearance of relict behaviour fixed to geographical coordinates from some past event in history and seem to be compounded of elements of both migration and diapause.

The compass orientation of insect migrants also sometimes has an appearance of relict behaviour. Long-range, downwind movements of the nocturnal moth *Laphygma (Spodoptera) exigua* Hübner can often, but not always, be explained by continuous flight over land and sea by day and night (R. A. French 1969). Sometimes nocturnal flight over land, combined with resting during the day, provides a better model (Mikkola 1970). Observational records show that normal diel (daily flight) periodicity can break down during migration in many insects including aphids (R. E. Berry and Taylor 1968) and Lepidoptera (C. B. Williams 1930). In laboratory experiments (Macaulay 1972) the migrant Silver-Y moth, *Plusia gamma* L., flies continuously during the first three days of adult life, but only at night thereafter. Field observations indicate that there may be a separation into downwind flight by night and oriented flight by day (L. R. Taylor, French, and Macaulay 1973) although the categories are correlated with sea and land and the data not sufficient to be totally convincing. This could provide a connecting link between oriented day-flying and downwind night-flying Lepidoptera. Nocturnal downwind migration by *Spodoptera exempta* (Wlk.), for example, is highly successful in redistribution (E. S. Brown, Betts, and Rainey 1969) and the selection by migrants of winds in the appropriate general direction could be associated with their moisture content. At night, larger moths fly higher (L. R. Taylor, Brown, and Littlewood 1979a), which again implies that the downwind orientation is being used actively. All this evidence must raise doubts about flight contained within narrowly-defined compass bearings as a general phenomenon of insect migration. Rather the combined evidence suggests, not a positive mechanism towards a specific goal, but a means of increasing ambit in most insects, with a general directional trend which tends to be north/south in high latitudes but less so in the tropics, just as Tutt forecast. The white cabbage butterfly, mentioned earlier, that flies straight each day but in different directions on successive days, is maximizing ground cover or search efficiency. The different orientation by different species at the same time and place still requires explanation and more species like the Monarch may yet be found.

The long-term effects of accumulated small changes in the area of coverage by a species, caused by individuals that persistently seek a more benign environment, are remarkably successful in tracking the area of maximum survival as it moves through space during longer periods of time – over thousands of generations. This aspect was missing in population studies until recently; it is implied by the island biogeography theory of MacArthur and Wilson (1967) but its implications for the general dynamics of the parent population are rarely pursued. The constant succession of immigrants and emigrants implies, not only a numerical balance in time at the point in space occupied by the island, but a constant shifting of the gravitational centre of the parent population itself. The process is less easily visible in long-lived vertebrate populations, although there is increasing evidence of systematic migration of the young in many species as observation becomes more intensive. For so-called autochthonous organisms and where there is strong territorial behaviour in a confined habitat that itself moves slowly, the movement may be largely internal and not easily recognized.

Nevertheless, in insect species that are flightless or seen only in local flights like the chrysomelid mentioned earlier, the end-product of thousands of generations of constant search for available oviposition sites in which to implant the next generation, can produce striking evidence of the effect of accumulated small individual migrations of a few metres into population drift over thousands of kilometres (Fig. 10.1). The evidence now built up by Coope (1979) and his colleagues is incontrovertible. From sites in Great Britain laid down during the last few hundred thousand years, 'nonmigratory' beetles have been recovered that are now only known in geographical locations from North Africa to the Tibetan Plateau and Arctic Siberia. The arrows in Fig. 10.1 show the most direct routes, not the complex sinuous paths that successive climatic cycles have imposed on the motions of the habitat as it has been repeatedly sought and refound by the searching of successive generations. Without those specialized individual migratory movements between birth and reproduction the amoeboid movements of whole populations, tracking the moving resources, would have had to depend upon random dispersal and chance, with levels of mortality and reproduction like those found in plants and some parasites.

The opposing behavioural forces of migration and aggregation sought by L. R. Taylor (1961, 1965b, 1971) to explain the current spatial distributions of organisms, and the corresponding forces seen by Kennedy in aphid and locust behaviour, seem more likely to provide the necessary mechanism. L. R. Taylor and Taylor's (1977) Δ-model proposed a balance of such motions, corresponding to Kennedy's (1966a) repulsion and attraction behaviour in relation to density, that increases and reduces the distance between individuals in proportion to the difference between two power functions of population density. This model attempts to account for the spatial property common to so many organisms. So far it seems to be the only model that treats Kennedy's behavioural properties as continuous functions in space and time. It also treats the resultant movement as a continuous function of distance without imposing subjectively selected 'patches', 'habitats', or 'populations', for which no criteria can usually be given.

Migration and competition

The synthesis of migratory movement and population dynamics changes the emphasis in Darwinian evolution. Darwin's basic premise of environmental selection through differential survival was directed primarily to maximized reproduction through competitive superiority; essentially a concept of the here and now. The tacit acceptance of such a static population with mechanistic

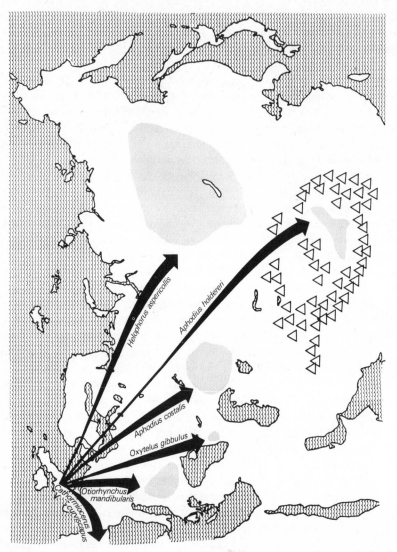

Fig. 10.1. The work of G. R. Coope and his colleagues shows constant change in the beetle fauna of Great Britain over the last 100 000 years. Some of the most sedentary species are now found isolated in sites over Afro–Euro–Asia. The whole species population has been shifted constantly by the niche-searching migrations of its individuals. The arrows cannot indicate the sinuous path of the populations which are not known. (Adapted, with acknowledgements to G. Russell Coope.)

competitive constraints on maximized reproduction, shows the prevailing view of migration as a high-risk enterprise, often treated as equivalent to mortality. Recognition of migratory movement as a density-dependent behavioural response, common to all individuals in some degree whilst avoiding competitive reactions without implicating altruism, provides a different perspective for the evolutionary mechanism. It is one that emphasizes individual choice rather than competitive reaction. To stay or not to stay are alternatives that relieve the one-sidedness of selection that has troubled other naturalists before Wynne-Edwards (1962). If the individual must stay to compete, it must also produce more competitive offspring than its competitors: an implacable cycle leading to increasingly competitive individuals. This is not the observed outcome of evolution which is, by and large, towards more sophisticated behavioural mechanisms that avoid direct competition. Competition between individuals, as distinct from predation, is rarely conclusive: both individuals usually survive and conditions may change so that the initial loser gains ascendancy and survives to reproduce. The highly developed 'social' avoidance of damage to the loser by submission and avoidance behaviour is evidence that cowardice is not penalized in ultimate reproductive success; submission may appear to be a social phenomenon but its antecedents may well have been in purely selfish individual spatial behaviour. Classically, populations increase until constrained by compretition, ostensibly judged to have occurred by the outcome in current reproductive success. There is, however, a long history of dissatisfaction with this indeterminate assessment of competition (Thompson 1929; Dobzhansky 1950; Birch 1957; Williamson 1957; Cole 1960; A. Milne 1961; Grime 1979) because it offers no clearly defined programme of events. Rarely is the evidence of genetic superiority available. What precisely happens during competitive interactions in the unrestrained and heterogeneous environment required by our original definition of behaviour?

Would-be competitors rarely fight; they somehow divide the resources in such a way as to minimize direct competition. This is particularly evident between species for which the evidence presented is spatial or temporal segregation (e.g., Darwin's finches and MacArthur's warblers). In more recent revivals (e.g., Diamond 1978) the very striking evidence presented for competition between congeneric warblers (*Crateroscelis robusta* and *murina*) is an altitudinal segregation which has its intraspecific equivalents in territorial behaviour; the species, like the individuals, keep apart. When real interspecific competition can be demonstrated, it is rarely evenly balanced. (Lawton and Hassell 1981) which suggests that, at different times and places, the victor and the vanquished may not be the same. Our view of species' relationships is partial and ephemeral. Diamond's criteria (1975) are criticized by Connor and Simberloff (1979) on grounds of inadequately established deviations of competitive mechanisms from chance; but spatial segregation is an equally valid argument for the *avoidance of competition by movement* as for either chance or competition. Competitive exclusion effectively states that ecological species avoid competition. The means by which this is achieved in insects is largely the movement by parents between their birthplace and their place of reproduction and, to the field entomologist, the careful choice of oviposition site is a familiar species character-istic. Offspring that themselves successfully reproduce are the measure of its success, i.e. grandchildren, and not merely maximized reproduction in the first generation. In a homogeneous closed system, movement is minimized and no internal circulatory system equivalent to that seen in the locust population can develop even on a small scale. Behavioural mechanisms to avoid competition then break down and a real struggle for resources ensues because the alternative evolutionary strategy, survival by escape, is lost and the associated behaviour frustrated. As the condition persists, the familiar behavioural abnormalities of confinement increase in frequency, (the zoo syndrome) and these can be experimentally reduced by permitting emigration (e.g. Huffaker 1958; Tamarin 1978a) unless the condition has reached a point of no return.

In the environments in which animals have evolved, where no restrictions are imposed on mobility, such laboratory psychoses are less common. The social and sexual events that are so well described by Wynne-Edwards (1962) can be equally well expressed as the movements resulting from the Δ-model, conflict of attraction and repulsion, with spatial exclusion from mating or terri-tory resulting in a displacement activity or ritualized migration. In longer-lived animals it may be more profitable to wait on the sidelines for a second oppor-tunity to achieve fulfilment, rather than to migrate so far as to be unable to take advantage of a change in the competitive conspecific environment if and when it occurs. Thus migration, ritualized or real, should be preferred by the smaller and less experienced individuals of a species (e.g. by young territorial birds). In species in which ritualized migration has not evolved, the periodic abnormal exoduses have been attributed to population stress mediated by the adrenal glands (Selye 1950; Christian 1950). Evidence for Selye's (1950) General Adap-tation Syndrome is not unequivocal but Andrews (1968) has shown that the rate of ACTH production by lemmings does increase in high-abundance years. The behaviour of the unsuccessful contenders in the pseudo-competitive games such as lecking is similar to, though less extreme than, that observed in animals suffering adrenal shock induced by confinement. It seems possible that visible evidence for competition is to some extent an artefact of perspective, for when fighting does occur, it attracts more attention than avoidance. That is not to say that competition does not occur under these conditions, e.g. mate-selection and defence, but what appears to be altruistic social behaviour is perhaps more likely to be a compromise between competition and migration.

In the wild, a would-be competitor faces a choice, to fight or to avoid compe-tition by real or ritualized movement, a choice influenced by the risks involved. Where competition is for limited resources, the greater risk is taken by the late starter and may end in diminished size, fertility, or death if the other competitor has reduced the available resource to below the essential threshold. In compe-tition for a mate, size or strength may be advantageous and again these will often be determined by the same early developmental experience. In insects individual size is greatly affected by environmental factors and may in turn lead to advantage in mating, in swarms for example (Thornhill 1980) where the competitive aggregations effectively advertise successful emergence in environmentally acceptable sites for oviposition. However, the would-be competitor may elect not to compete but to accept lower status and poorer resources, waiting until an opening appears; or it may leave and look for a better home. The second alternative becomes more attractive as the expected waiting time increases, or when the life expectancy is less. Hence in insects it is more obvious than in vertebrates.

We have written as if the choices are faced and made by the competitor itself; however there are good reasons to suppose that the choice is determined in most species by the parents or grandparents, on the basis of conditions they faced earlier, or to compulsions more directly genetically controlled. The most obvious factors are selection of an oviposition site and date of oviposition. These determine to a large degree the larval prospects and may, in survivors, determine the migrant condition (L. R. Taylor and Taylor 1978).

The status ascribed to competition as a potent force in evolution, and the dependent doubts about why increasingly more competitive animals have not evolved, may be resolved by considering migration as an alternative solution and hence as a brake on ever-increasing competition. If so, it is not to be expected that the mechanisms controlling migration, its occurrence, compulsion, timing, and persistence, will be any less complex than those involved in competition. The solution of this balanced choice will be specific and, only when the mobility of the environment is seen to include population density, do the τ/H and $r-K$ spectra become meaningful. Predation is also an active component of population dynamics, both practically and theoretically (e.g. Huffaker 1958; Holling 1959; Hassell 1969). One of the most powerful messages to come from these studies is that predation is a game (Maynard Smith 1976*b*) in which the prey try to avoid the predator while the predator learns ways to improve its ability to catch the prey. There comes a point where collaboration between the most fit prey and the predator is a profitable solution, at the expense of the weakest prey. Here again, strength and weakness frequently refer to motion. For both species the problem is the same, to search for the most benign place, although the places differ with the role played. The most benign place for the prey is likely to involve other individuals as decoys, as well as being the most productive in resources.

Kennedy's conspecific cues, or even the interspecific cues of Kiester (1979), operate to inform both predator and prey, and affect the game itself by attracting or repelling additional players. So the environment becomes more or less benign as a constant response to the movements of predators and prey. Thus an optimum rate of movement exists for both, at which the benefits are maximized and the liabilities minimized. The ability of the prey to hide from predators also decreases as the ability of the environment to support prey increases and prey become more abundant. An optimum prey density therefore exists: the same argument applies to predator density and both of these are achieved locally by movement, not only by mortality as in the purely temporal model. In such a 'game', density is manipulated by movement towards a supra-optimal site or away from a sub-optimal one; selection will be for the appropriate behaviour. The motion is both specific and individual. 'Prey' is not just any individual of the prey species. It is the one in the right place at the right time, in the right density, and with the right intrinsic motion that allows it to be captured. Survival depends on behaviour and behaviour on motion.

A simple mathematical model suffices to demonstrate that movement is at least as powerful a force directing a population's fortunes as reproduction, competition, or even predation. The model generalizes the Lotka–Volterra equations for competition at a point to a universe of u points using matrices.

The density of a species at a defined point in space may change in any of four ways: birth and immigration increase the density in time and space respectively, while death and emigration reduce it in the same two dimensions. If we consider the entire species population, migration to or from the population is all internal and is not a factor. In practice, the data available from experiment or observation deal, not with 'populations', but with local population density which is susceptible to change in both space and time.

If we consider the simplest model for population density at a defined location or, for simplicity, a segment of the population of number N in a unit area,

$$N_{t+1} = N_t + B_t + I_t - D_t - E_t \tag{4}$$

where $B, I, D,$ and E represent the numbers added to or subtracted from the population segment in the interval $t \to t + 1$. Generalizing to an assemblage of locations (perhaps comprising the total population)

$$\mathbf{N}(t+1) = \mathbf{N}(t) + \mathbf{B}(t) + \mathbf{I}(t) - \mathbf{D}(t) - \mathbf{E}(t), \tag{5}$$

where $\mathbf{N}, \mathbf{B}, \mathbf{I}, \mathbf{D},$ and $\mathbf{E}$ are vectors containing all locations inhabited by the population. Since the immigrants (I_i) must have been emigrants from another location (E_j) we can express the number of movements from j to i as the product of a transition matrix of the set of probabilities of all such movements $\mathbf{A}$ and the vector $\mathbf{N}(t)$. Thus the net change in number between t and $(t+1)$ is

$$N(t+1) = \mathbf{A} \left[\mathbf{N}(t) + \mathbf{B}(t) - \mathbf{D}(t) \right]. \tag{6}$$

Vectors $\mathbf{B}(t)$ and $\mathbf{D}(t)$ can be made dependent on density:

$$\mathbf{N}(t+1) = \mathbf{A} \left[\mathbf{N}(t) + \mathbf{R} \, \mathbf{N}(t) \right], \tag{7}$$

where **R** is the matrix containing reproductive rates. Defining a second competing species with population matrix **M**(t), we have

$$M(t+1) = A^1 \left[\dot{M}(t) + R^1 \dot{M}(t) \right],\qquad (8)$$

and competition matrices a and β containing the competition coefficients which
are proportional to the population-depressing effects of one species on the
other at each location. (If a species' competition coefficents are the same at all
locations, the matrices a and β may be replaced by scalars a and β). Hence

$$N(t+1) = A \left[N(t) + R\,N(t) + a\,M(t) \right]\qquad (9)$$

and,

$$M(t+1) = A^1 \left[M(t) + R^1 M(t) + \beta\,N(t) \right].\qquad (10)$$

By comparison with the Lotka–Volterra equations, in which the only way to
compensate for interspecific competition is to modify fecundity, the partici-
pants in this competitive struggle have an alternative parameter to manipulate,
the coefficients in **A** and $\mathbf{A}^1$. By analogy, the same applies to intraspecific
competition in which movements can take individuals away from locations of
high competition.

So far we have not defined the transfer function used to generate the movement transition matrix **A**; possibilities include the island and stepping-stone
models familiar to geneticists. Functions like R. A. J. Taylor's (1978, 1980)
equations relating the decline of density to distance moved have the additional
feature of incorporating distance between locations. In addition the transfer
function could be explicitly density-dependent (e.g. L. R. Taylor and Taylor's
(1977) Δ-model), thus permitting negative feedback on the proportion of
migrants, which will tend to stabilize any perturbation due to competition.
Feedback would be especially effective if the transfer function included distance
and both population densities.

Interestingly, this approach can also be applied to predation theory with
the competition coefficients a and β replaced by predation coefficients.
Williamson (1957) suggested that predation could be regarded as a limiting
condition of competition. If the transition matrix **A** evolves in response to
changes in α, β and **R**, it is easily seen that, in the model at least, the contestants
would be engaged in an evolutionary game in which survival is the perquisite
of the most mobile.

Extrinsic population control by Malthusian killing agencies is partly an
illusion borne of man's obsession with mortality. The evidence presented by
Wynne-Edwards (1962) is real and known to most naturalists. The mechanism

he proposed demands altruism that is partly an illusion of an ethical concept, but also it is the misinterpretation of the avoidance of conflict. This is not altruistic nor quasi-altruistic (D. S. Wilson 1980) but is the purely selfish selection of the lesser of two evils: aggregation resulting in competition which, pushed to its ultimate limit must lead to a fight to the death, or the more individualistic migration with its corresponding risk of failure to find an alternative unoccupied environment; or for vertebrates, the compromise of a wait-and-see indecision (displacement activity). The positive, active *survival* solution is migration and with it the possibility of adaptation to a slightly different environment with continued prospects for change in future generations.

11
Universal correlates of colonizing ability

URIEL N. SAFRIEL and UZI RITTE

Introduction

It can safely be said that, were it not for their movements, the ranges of distribution of most species would be very different from those observed today. Movements related to an expansion of range are usually dispersal movements that carry individuals away from their site of birth. Such movements can be quite common, but most dispersers will either not find a suitable area for settling or end up in an area already occupied by members of their own species. Expansion of a species' geographic distribution will be accomplished only if the dispersers succeed in finding and colonizing a new area. Thus, *colonization* is the establishment of a permanent population in an area previously not occupied by that species.

Most colonization attempts fail because the probability of reaching an unoccupied appropriate niche is low, and because even when such an area is reached success is not guaranteed. A *successful colonization* is one which results in a long-lasting population, independent of arrivals of additional dispersers. Since both permanence and degree of independence are relative, there can be different degrees of successful colonization, and hence it should be possible to associate colonizations with probability estimates. The ability to estimate these probabilities should be of relevance in a number of contexts. In a retroactive sense, the ability to reconstruct events of colonization could be of much help for biogeographers (Pielou 1979). In a prospective sense, the ability to predict successful colonizations can be of value for applied biologists interested in biological control of pests (Beddington, Free, and Lawton 1978), for planners of nature reserves (Diamond and May 1976), and, in general, for everyone interested in natural events or environmental manipulations which may create opportunities for desirable or undesirable colonizations (Safriel and Ritte 1980).

A complete analysis of the determinants of successful colonizations must take into account the geographical and ecological relations between the *colonized area* and the *source* of colonizers, the biological attributes of these species, and the mechanisms by which they are expected to arrive at the colonized area.

* This chapter is contribution No. 14 in the series 'Colonization of the eastern Mediterranean by Red Sea species immigrating through the Suez Canal'.

Colonized areas can be devoid of species, on account of recent origin (volcanic islands) or recent catastrophes (earthquakes, floods, volcanic eruptions). They can be cultivated areas in which the species composition is very different from what it was prior to cultivation, or they can be occupied by natural communities. Colonized areas can be as small as the tiniest oceanic island or as large as whole continets, far away from the source or adjacent to it. Colonists can arrive at the colonized area as a result of active movements, or passively. They can arrive following no change in conditions, following a removal of a barrier or a creation of a passageway between the previously isolated source and colonized areas, or, perhaps, following a genetic change which adapted them to the conditions in the colonized area.

All these factors have to be considered when an attempt is made to estimate the probability of colonization. For example, remote small areas are less colonizable than larger and closer ones; ability to travel long distances is not required when the transfer is passive; areas with a few species are more easily penetrated than areas densely packed with species; and so on. But under each set of circumstances not all species in a given source will succeed in colonizing a propspective colonized area, because the success or failure of a species in colonization is also a function of its biology. The present chapter is devoted to those aspects of the biology of a species which contribute to its colonizing ability. In the first part we review the literature in an attempt to define possible universal attributes of colonizing ability, and in the second part we compare these attributes against available data.

The discussion will be limited to cases of colonization which occur rapidly after the opportunity for it is created, i.e. colonizations which are based on pre-existing adaptations. The degree of success in colonization is measured as the estimated persistence time of the colonist population when established by one colonizing group, although other measures have also been suggested (Stearns and Crandall 1981).

The size of the colonizing population

The use of persistence as a measure of success in colonization implies that in successful colonists the loss of individuals due to death should at least be balanced by addition of new individuals. However, a knowledge of the expected birth and death rates in the colonized area is not sufficient for determining whether colonization will be successful or not. Random events can lead to the extinction of a population even under conditions in which its net growth rate (birth rate minus death rate) is positive. The influence of these events, called *demographic stochasticity* (e.g., May 1973) increases as population size decreases. In this section we review the relationship between the size of a population and its expected persistence time.

Consider an ideal case of colonization, in which N_0, the size of the colonizing group, is so large that the colonized area immediately becomes saturated. If we denote the over-all size of the population when the colonized area is saturated as N^*, then in our example $N_0 = N^*$. Let us also assume that in the colonist population birth rate equals death rate. Can we expect this population to persist as long as the environment remains unchanged? Except for the imaginary case of no deaths, there can be a certainty that N (population size) will stabilize at N^* only if each death is immediately balanced by birth, i.e., if $\mu(N^*) = \lambda(N^*)$ where $\mu(N^*)$ is the instantaneous per capita death rate and $\lambda(N^*)$ the instantaneous birth rate at $N = N^*$. In reality such a high degree of synchronization between demographic events never occurs. Birth and death events are randomly distributed over time, and μ and λ are actually the weighted means of Poisson variables, or the expected values of death and birth rates, respectively. For this reason the values of N also undergo fluctuations and N^* is only the expected mean of the different values it takes during these fluctuations.

Methods for calculating the probability distribution of the different stochastic equilibrium levels of N, and of the variance of these values around N^*, are given by Pielou (1969) and May (1973). In the course of its fluctuations N may reach zero and extinction cannot be ruled out. The period of time until this occurs can be used as a measure of success in colonization (MacArthur and Wilson 1967; Richter-Dyn and Goel 1972; Leigh 1975).

Like birth and death rates, persistence time is also a variable for which only an expected value, different for each value of N^*, can be given. Although the exact relationships between $T(N^*)$, the expected persistence time of a population of size N^*, and N^* depend on the mode of population regulation, the general trend is always the same – $T(N^*)$ increases exponentially with N^*. For example, a population of 40 individuals, in which $\lambda = 1/\text{year}$, $\mu(N<N^*) = 0.91/\text{year}$, and $\mu(N>N^*) \to \infty$, occupying an area with $N^* = 40$, is expected to persist about 160 years. If N^* is increased to 55, persistence time will increase to about 500 years. In spite of the relatively small difference between the two areas, colonization in the second should be much more successful than in the first.

How do our estimates of the expected persistence time change if we relax the assumption of $N_0 = N^*$? If the size of the colonizing group equals 1 ($N_0 = 1$), the expected time of persistence of the population ($T(1)$) also increases with N^*, but $T(1) \ll T(N^*)$, and the difference between the two estimates increases with an increase in N^*.

Because events of colonization generally start with a small colonizing group (not necessarily with $N_0 = 1$, but with $N_0 \ll N^*$), the value of $T(N_0)$ should be much more relevant for our problem than $T(N^*)$. The next section is devoted to the estimation of $T(N_0)$ and its significance in selecting criteria for colonizing ability.

The growth prospects of the colonizing group and the persistence of the colonist population

The problem of the expected persistence time of populations founded by a small number of individuals has been dealt with by a number of models (MacArthur and Wilson 1967; Richter-Dyn and Goel 1972). The basic approach is to divide the period between establishment and extinction into many time intervals and estimate the duration of each. The time intervals are selected in such a way that only a single demographic change (either addition or subtraction of one individual) is expected to occur in each. Thus, the first interval is the time between the moment of establishment and the first birth or death event, the second interval is the time between the first and the second event, and so on until the time interval in which the event (necessarily death) that leads to extinction takes place.

For calculation of the expected duration of each interval one has to know N, $\lambda(N)$, and $\mu(N)$, which are, respectively, population size at the end of the previous interval, and the birth and death rates associated with that size. The expected persistence time of a population founded by one individual is then given by

$$T(1) = \sum_{N=1}^{N^*} \frac{1}{\mu(N)} \prod_{M=1}^{N-1} \frac{\lambda(M)}{(\mu(M))}$$

(Richter-Dyn and Goel 1972; Leigh 1975). Thus, persistence time is a function of the per capita death and birth rates and of saturation size. But the relative importance of these parameters can only be evaluated when the mode by which birth and death rates are affected by the population's own density is determined. If λ and μ are constant and independent of density except at $N=N^*+1$, when $\mu=\infty$, the $T(1)$ equation simplifies to $T(1) \simeq (\lambda/\mu)^{N^*}(1/N^*\lambda)$, which means that relatively small increases in N^* lead to sharp increases in $T(1)$. This prediction also remains valid for the more common cases in which λ and μ remain independent of density only as long as density is low, although the increase of $T(1)$ with N^* is slower than when growth is independent of density for all values of N below N^*.

Note that although N^* is always significant for the expected persistence time of a population founded by a small number of individuals, the size of the colonist population at saturation is by and large a function of the size of the colonized area or its species composition. N^* can thus serve as a species-specific correlate of colonizing ability only if the colonized area is very small, or otherwise limited in the space available to the colonist.

Attributes that can serve as more general species-specific correlates of colonizing ability can be derived from Richter-Dyn and Goel's (1972) demonstration that, in many cases of colonization, long persistence can be guaranteed at a population size which is much lower than N^*. The critical value of N which guarantees long persistence (called N_c) is evident from the existence of a plateau in the curve describing the dependence of $P_s(N,m)$ (the probability of a population founded by m individuals to reach size N) on N. In the case of $m = 1$, for populations in which $\lambda/\mu{>}1$, the value of $P_s(N,1)$ drops from 1 at $N=1$ to $(1-\mu/\lambda)$, and then remains practically constant with increasing values of N (Fig. 11.1(a)). N_c is the value of N at which the probability curve bends for the plateau.

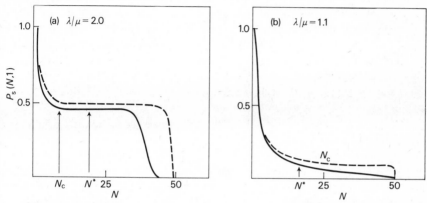

Fig. 11.1. The probability of attaining size N for a colonist population founded by one reproducing individual ($N_0=1$). Solid curves: populations with logistic density-dependent birth and death rates (N^*=saturation size). Dashed curves: populations with expected exponential growth and density-dependent death rate at saturation (N^*=50). λ and μ are, respectively, birth and death rates at founding. N_c ('critical N') is the value of N above which long-term persistence is guaranteed. (After Richter-Dyn and Goel (1972).)

The existence of N_c, beyond which the probability of persistence is the same as when the population is at N^*, suggests that for successful colonization it is sufficient to increase the probability of reaching N_c without going extinct. For a colonizing group of size N_0 this probability is ${\sim}1-(\mu/\lambda)^{N_0}$, so that to increase the probability of success in colonization the ratio λ/μ should be higher than 1, preferably because of a reduced μ.

Two additional population characteristics that may be relevant to the problem of successful colonization and which have not been considered so far are

1. Discrete rather than continuous growth;
2. Time delays in the response of the birth and death rates to the size of the population.

The effect of demographic stochasticity on the persistence time of populations with discrete generations has not been treated as thoroughly as in populations with continuous growth. May (1973) has shown that in populations with discrete growth the expected persistence time increases with population size. More specifically, for a population with non-overlapping generations and with density-independent birth and death rates, founded by a single individual, it was shown (Christiansen and Fenchel 1977) that while extinction is inevitable if the expected number of offspring produced by a colonizing individual is $\leqslant 1$, the probability of persistence increases as the probability of giving birth and size of the colonizing group increases.

Time delays in the response of λ and μ to population size lead to fluctuations in population size both during growth phase and after saturation level (N^*) is reached. As a result population size may oscillate between values above N^* and below N_c, so that long-term persistence cannot be guaranteed. The conditions for these oscillations have been worked out by several authors (e.g. Pielou 1976; May 1976; May and Oster 1976; Royama 1977), who showed that they depend on the relations between N^*, r (the instantaneous rate of increase, or λ-μ), and the length of the delay. Prevention of these oscillations should increase the probability of success in colonization.

The ecological correlates of colonizing ability reviewed so far can be summarized as follows

1. High efficiency of resource utilization, or, in other words, a high carrying capacity per unit area (K). This moderates the potential for high r, which is a determinant of the undesirable oscillations of populations with time delays. An additional outcome of high K, crucial for populations of all types, is the attainment of a large population within the colonized area (high N^*).

2. Low levels of intraspecific competition during the first stages of colonization, so that the birth and death rates are to a large extent density-independent. This escape from density-dependence prolongs persistence time in general, and is specifically important if the colonist population tends to oscillate.

3. When growth is strongly density-dependent, as is usually the case when N^* cannot be high, a short time delay in the response of the demographic parameters is essential. This can best be achieved by reducing the delay in reproduction, for example by shortening the developmental time and by increasing birth rate, so that generations overlap smoothly and the growth rate can be immediately adjusted to density (Barclay 1975).

4. At the early stages of all cases of colonization, a high ratio of the per capita instantaneous rate of increase r to the per capita birth rate λ, preferably achieved by a reduction of the death rate. When density-dependence and time delays are expected, this high r/λ ratio should be coupled with low r.

5. The ability to send a large colonizing group. Although for species with short $T(1)$ long persistence is still guaranteed provided a colonizing group (of one individual) arrives every $1/T(1)$ time units, a high dispersal rate is not a universal attribute of colonizing ability in this context. Most interesting cases of colonization were described from well isolated areas, unlikely to be hit repeatedly by dispersers. On the other hand, for a colon-

izing species with no time delays and $\lambda/\mu \geqslant 1.5$, one colonizing group of 15–20 individuals is sufficient to guarantee success, irrespective of all other requirements for successful colonization. If $\lambda/\mu \simeq 1$ and the growth rate is density-dependent, the probability curve (solid line in Fig. 11.1(b)) does not have a plateau, and the population can go extinct even at N^*. For such populations the best strategy for successful colonization is also an increase in N_0. If this is not possible, and the ratio λ/μ cannot be changed, a smaller value of λ (and, of course, also of μ) will increase the probability of success.

Environmental stability and success in colonization

An implicit assumption in all models discussed so far is that colonization takes place in a uniform and stable environment. Even in the models dealing with fluctuating populations, the assumption is that the fluctuations are caused by random changes in one or more demographic parameters and not by changes in the environment. In the present section we relax the assumption about environmental stability, and examine to what extent the conclusions about prerequisites for successful colonization are modified.

When the environment fluctuates regularly through time so that its carrying capacity is undergoing regular cycles, the time delay of the response of the population to the environmental conditons relative to τ, the period of the environmental cycle, becomes crucial for the ability of the population to persist for a long time. When the response is relatively fast, population size will track environmental variation (May 1976) and periodically approach low levels at which extinction due to demographic stochasticity becomes likely. When the response is slow, population size may stay more or less stable and, although it will never reach the maximum value permitted by the environment, it will also never decline to the precariously low levels expected when the environment is at its worst. Thus, even if the carrying capacity of the colonized area periodically approaches a critical value, a slow responder will manage to stay above these low points of carrying capacity.

The variable which determines whether the response of the population to environmental changes is fast or slow is the product $r\tau$. The response of a species with a given r will be fast if $r\tau \gg 1$, and slow if $r\tau \ll 1$ (May 1976). This dependence of the probability of successful colonization on the product $r\tau$ means that r, a species-specific attribute, can serve as a correlate of colonizing ability only in the context of τ, an attribute of the colonized area. Yet, high r may still serve as an independent, important universal correlate of colonizing ability.

One way to reduce the effects of the environmental fluctuations is to prolong the time it takes the population to respond to the environmental changes, i.e. to invest in environmental resistance. Such resistance is achieved by 'escape responses' such as dormancy, by means of which the population removes itself from the influence of the less favourable parts of the environmental cycle. The reproductive effort is limited to the favourable parts of the cycle when the full potential for growth can be realized (Fretwell 1972).

In cases of colonization it is very unlikely that the fluctuations in the source, to which the colonist population is adapted, will be the same as in the colonized area. It can be assumed that for almost every colonizer the new environment fluctuates more than the environment to which it is adapted.

Whereas in a stable environment a colonist with a high λ/μ is bound to reach and stay at a 'safe' population size, irrespective of the time it takes to reach that size, in a fluctuating environment a population may easily drop to critically low levels and the length of time during which the population remains small become crucial. A speedy emergence from the vulnerable state is guaranteed only by a high r. Indeed, simulations of a genetic model (C. E. King and Anderson (1971) show that when only a small number of reproductive cycles can be fitted into the breeding portion of the environmental cycle, selection for high r prevails. Thus, if the adaptation to the old environment was achieved by means of a high r during the favourable parts of the environmental cycle, the species may succeed also in the new environment even if the favourable parts of the cycle are shorter there. If the colonizer was adapted to its former environment by means of it being resistant to environmental fluctuations, which is usually achieved at the expense of the potential for high r, it may fail in the new environment. As a rule, the species which have a higher probability to succeed are those adapted to the more extreme conditions in their original environments. When colonizing, these species will on the one hand benefit from their high r during the relatively short times available for reproduction. On the other hand they will not be affected by the detrimental influence of high r near saturation, since saturation in a fluctuating environment is achieved only after a very long period, if ever.

The second type of environmental fluctuations to be considered is that of random fluctuations ('*environmental stochasticity*', e.g. May 1973). This factor is introduced into population models, either assuming exponential growth (Lewontin and Cohen 1969) or logistic growth (Levins 1969), by means of diffusion statistics. Whatever the demographic parameter which responds to the environmental fluctuations, the effect of the environment on the population can be predicted through the mean and variance of that parameter. The variance constitutes a measure of the sensitivity of the population to its environment. All models come out with the conclusion that persistence is negatively correlated with the sensitivity of the demographic parameters in question to the environment. Thus, unlike fluctuations in population size which are due to demographic stochasticity, where the probability of persistence depends on the actual size of the population at a given moment, in fluctuations which are due to environmental stochasticity the outcome depends on the mean size of the population (May 1973; Leigh 1975). A population which is currently at a high level is no safer from extinction than a population at a lower level, provided both have the same mean size and the same variance around this mean through time.

The models show that high mean value of r, and avoidance of density-depen-dence, guarantee long persistence in randomly fluctuating environments. Thus, since the relevant demographic parameters are expected to be more sensitive to random fluctuations in the colonized area than in the source (Safriel and Ritte 1980), species with high mean values of r will cope well with the new environ-ment. The function of high r is to shorten the time during which a colonizing population, which is threatened by demographic stochasticity, is exposed to environmental stochasticity. Note, however, that the prerequisites for with-standing both types of stochasticity may not be jointly met, since the maxi-mization of the ratio r/λ could reduce r. Yet, a high r/λ ratio is achieved more easily by reducing μ than by increasing λ only when an equal investment is involved in both responses. In high-r species a further increase in λ should be cheaper than a reduction in μ, and it can be said that these species can with-stand both types of stochasticity.

A joint consideration of the effects of both demographic and environmental stochasticity on the probability of persistence of a population can best be expressed by the modified logistic equation (Leigh 1975)

$$\frac{dn}{dt} = [\, (\bar{r} + \sigma_r^2)N - aN^2 \,] \pm \sqrt{N}$$

where $(\bar{r} + \sigma_r^2)$ accounts for environmental stochasticity, $\sqrt{N}$ for demographic stochasticity (May 1973), and a represents density-dependent effects. This equation makes it clear that during the initial stages of colonization when N is very low, demographic stochasticity is the major factor determining the expected population size, whereas at later stages environmental stochasticity takes over. Under such conditions the $T(1)$ equation becomes quite complicated, but its shows that persistence time increases as the variation in population size relative to its expected mean size decreases. Since the variance in r in the colon-ized area is expected to be high, the importance of a high mean value of r for success in colonization is obvious.

The last aspect of environmental heterogeneity we wish to consider in the context of colonization is spatial heterogeneity. For the same reasons put forward when the temporal heterogeneity was considered, environmental patch-iness in colonized areas is likely to be perceived by a colonist population as greater than in the source area. Because during the initial phase of colonization only a very small proportion of the available patches can be occupied, the colonist population should rapidly increase the number of occupied patches, to reduce the risk that all independent populations will go extinct simultan-eously.

A theoretical treatment of the effects of environmental patchiness on persistence is given by Levins (1970), Agur (1982), and Hanski (1982). In general, the prerequisites for successful colonization of a temporally stable, spatially heterogeneous environment are not different from those for colonizing spatially uniform environments. When temporal variation is superimposed on the spatial variation, the existence of an escape response becomes advantageous. But unlike in uniform environments where the escape is in time (e.g., dormancy), in the spatially heterogeneous environments the escape can also be in space, namely, by dispersal. Although, in many cases of colonization, dispersal ability is not a prerequisite for the initial introduction of the colonist into the colonized area, in heterogeneous environments it should be added to high r as a prerequisite for long-term persistence.

Roff (1974a,b), in a series of simulations, studied the relationships between local population size, r, dispersal rate, and the probability of persistence in a spatially heterogeneous, temporally fluctuating environment. His results can be summarized as follows: With no dispersal, the population becomes rapidly extinct; when dispersal is possible the population persists, but its size fluctuates dangerously when density-dependence is high, and remains relatively constant when density-dependence is low. Persistence time increases with r, especially when coupled with a high rate of dispersal. Temporal stochasticity affects population sizes more strongly when dispersal is low than when it is high, and species with high r are less sensitive to the degree of patchiness than species with low r.

To summarize, high r (and/or high mean r), an avoidance of density-dependence, and high dispersability all guarantee successful clonization in environments expected to fluctuate either systematically, randomly, or spatially.

The challenge from other species

So far our discussion of correlates of colonizing ability has been based on models in which other species in the colonized area have appeared, implicitly, only as potential food for the colonist. In this section we consider the possibility that a suitable niche exists in the colonized area, but is occupied by resident species, so that the success of the colonist depends also on its competitive ability. The importance of competition in determining the success of colonization can be seen from the following example. A species in which $\lambda=2$ and $\mu=1$, colonizing an area in which $N^*=20$, is expected to persist, if the environment is stable and there is no competition, for 25 000 years. If mortality rate is increased because of competition (competition coefficient equals 0.5), the expected persistence time drops to one (!) year (MacArthur 1972).

One way of reducing the effect of competition is to increase the competitive ability of the colonist, so that the effect of the competitor is reduced. An alternative response, which should be more realistic for species with high r, is to reduce the amount of resource overlap between the colonist and its competitor. A species with the potential to change its position along the resource spectrum is likely, therefore, to be a good colonizer (May 1976). Since a high degree of variability for the position along the resource spectrum is generally a characteristic of species with high r, high r may constitute a prerequisite for colonization even in the face of competition.

When the environment is spatially heterogeneous, dispersal ability gains a great deal of significance in competitive situations. Regardless of the mode of competition ('migration' or 'extinction' competition, Christiansen and Fenchel 1977), a high rate of dispersal in the colonist will increase its probability to succeed. Again, a high rate of dispersal is expected in species with high r (Lewontin 1965; Lavie and Ritte 1978), and the predictions regarding universal prerequisites for successful colonization without competitive interactions seem to hold also for colonizations with competition.

The same conclusion can be reached with regard to predation. In general, resident predators will switch to the colonist population only after it has become relatively well established. At the early stages of colonization predation does not constitute more than an additional density-independent mortality factor, and a high r and/or a high rate of dispersal serve well as a preadaptation (Levin 1974).

Genetic correlates of colonizing ability

An additional field of research that should help define species-specific attributes of colonizing ability is the field of genetics. Unfortunately, although much has been written about this subject (e.g. H. G. Baker and Stebbins 1965), most suggestions were based on an analysis of data, and no theoretical foundation has yet been laid on which arguments, similar to those derived from ecological models, can be built.

Obviously, all the universal demographic species-specific attributes of colonizing ability, if they exist, must have a genetic basis. An appropriate genetic constitution at the loci responsible for these attributes can thus be a genetic prerequisite for colonizing ability. However, this requirement is impractical, since the characterization of different species at the genetic level has to be based on their characterization at the phenotypic level and when phenotypic characterization is available the genetic studies are superfluous.

The only cases in which a genetic analysis can help identify potential colonizers are those where a specific genetic difference between species leads to a difference between them with regard to their ability to colonize. Carson and Ohta (1981) have shown that among the picture-winged species group of Hawaiian *Drosophila*, the difference between colonizers and non-colonizers is due to the fact that the former are generalists in their choice of sites for oviposition, while the latter are specialists. A single two-allele locus is responsible for this difference, so that the presence in a population of the allele for generality may serve as an indication of its ability to succeed in colonization. It should be reiterated that in this case the ability to colonize is due to a specific, rather than universal, preadaptation.

Two other aspects of the genetic system that have been considered in the context of universal prerequisites for successful colonization, and which can be studied independently of the demography of the population, are the breeding system of the species and the extent of general genetic variability. As far as breeding systems are concerned, most of the suggestions made about their relevance to success in colonization were based on studies of plants (H. G. Baker 1965; Allard 1965). In animals it has been suggested that hermaphroditism and parthenogenesis may contribute to the probability of successful colonization (Tomlinson 1966). The relative rarity of these phenomena among most animal groups should mean that although they may have led to success in some cases of colonization they cannot be included in the list of possible universal correlates of colonizing ability.

The second aspect is that of genetic variability in general. The possibility that a relation exists between the over-all level of genetic variability to colonize has been raised repeatedly, especially since the application of the technique of electrophoresis to the genetic screening of natural populations, but no general conclusion can yet be suggested.

On the theoretical level, it was assumed intuitively that colonizing ability should be accompanied by a high level of genetic variability. Thus, for example, it was suggested by Fincham (1972) that 'the heterozygous combination of co-dominant alleles could confer a selective advantage – under the changed nutritional conditions such as might accompany colonization'. A similar approach was taken by F. Wilson (1965), and even the findings by Carson (1965) and Dobzhansky (1965) about reduced levels of inversion polymorphism in colonizing species of *Drosophila* led Mayr (1965) to remark that 'one has to remember that chromosomal polymorphism is only one form of polymorphism, and that genic polymorphism may have replaced chromosomal polymorphism in these colonist *Drosophila* species'.

However, the few cases of colonization for which relevant genetic information is available do not unequivocally support the assumption about a correlation between the level of genetic variability and colonization.

In favour of this idea one can quote Wium-Andersen (1970) and Vuilleumier and Matteo (1972), who found higher levels of electrophoretic variability in colonizing species of the marine gastropod *Littorina*, compared to non-colonizing species of the same genus, and Templeton, Carson, and Sing (1976) who found that in *Drosophila mercatorum* the level of variability is a good predictor of success in colonization.

On the other hand, several cases are known in which no correlation exists between the level of genetic variability and colonization. Selander and Kaufman

(1973), for example, found the land snail *Rumina decolata* to be a very success-ful colonizer with no observable electrophoretic variability. Pashtan and Ritte (1978) found extremely high levels of variability in two species of the Red Sea intertidal gastropod genus *Cerithium*, only one of which colonized the Mediterranean after the opening of the Suez Canal (see below). Templeton (1979) concluded, for several species of *Drosophila* from Hawaii, that electro-phorectic loci do not play a role in colonization.

Since no method is available for the indentification of loci which may take a direct part in the genetic changes associated with colonization, it seems that no genetic factor can yet be suggested as a universal prerequisite for successful colonization. We therefore end our digression into the field of genetics, and return to our discussion of demographic parameters.

Testing the validity of the postulated correlates of colonizing ability

A summary of the conclusions of the models reviewed above leads to the follow-ing theoretical prescription for success in colonization. Potential colonizers should be recruited from among the species that perceive their environments as fluctuating, and in which the environmental heterogeneity is expressed mainly by fluctuations in adult mortality. Adaptations to this type of environ-ment include attributes of the so-called *r*-selection (Pianka 1970), such as rapid development, early maturity, large reproductive effort, short life span, small body size, and dispersal ability. If success in colonization is indeed correlated with one or several of these characters, then characterization of species in a given assemblage (a taxon or a community) according to them can enable the species to be ranked according to colonizing ability. This section is devoted to short reviews of several cases of colonization, in an attempt to see to what extent the assumptions about the existence and nature of these correlates are supported by the data. The examples are divided into several categories, and are presented in the order of increasing difficulty of establishing a permanent population in the colonized area.

Colonization of newly-created, 'virgin', and empty habitats

The best examples of habitats of this type are oceanic islands of volcanic origin. But although considerable biogeographical insight can be gained by studying the present distribution and dynamics of species on such islands (e.g., Diamond 1975), evolutionary history must have masked features that contributed to the success of their initial colonizers (e.g., Ricklefs and Cox 1972). For this reason the investigation of recently created volcanic islands should be of value.

An island of recent origin is Surtsey, which appeared in 1965 off Iceland at the end of major volcanic activity. Surtsey is about 1.5 km² and 25 km from the nearest colonization source (Fridriksson 1975). The colonists of the first years included representatives of several groups of arthropods, but the arthropod groups in the source (only groups with 9 species or more were included in the survey) were not represented evenly among them. The relative representation of the different groups, which varied between 4 per cent (Hemiptera) and 50 per cent (Diptera), may be associated with differences in ecological generality and/or in rates of reproduction, but it is difficult to see a consistent trend regarding any of these attributes when all groups are taken into account.

Colonization of 'recreated' empty habitats

The most famous example in this category is that of the six tiny (11–25 m in diameter) experimentally defaunated mangrove islets in the Florida Keys (Wilson and Simberloff 1969; Simberloff and Wilson 1969; Simberloff 1976a, b,c, 1978). The islets were monitored closely throughout the first three years after defaunation. As in Surtsey, morphological adaptations for reaching the islets (e.g., wings of Orthoptera) did not guarantee high rates of invasion. Furthermore, species that arrived quickly did not necessarily become permanent residents, as was evident from several species of spiders that arrived frequently but went extinct almost as fast.

Of the 130 species recorded on the islets throughout the first 3 years, only 12 became unequivocally extinct during that period (Simberloff 1976a), but no information is available about relevant ecological differences between these species and closely related species that did not go extinct. Predators and parasites had higher extinction probabilities on the smaller islets and the failure of several other species could easily be attributed to their being highly specialized (Simberloff 1976b). The importance of competitive interactions (Simberloff 1978) and of 'assembly rules' based on trophic relationships (Simberloff 1976c) for the success of individual species could not be assessed, but its seems that they do not have to be invoked to explain the observed patterns of distribution and abundance.

Another example of colonization of defaunated areas is that of the recolonization by birds of the islands Long and Ritter and in several adjacent islets in the Bismarck Archipelago (New Guinea), following two volcanic eruptions about 200 and 100 years ago. According to Diamond (1974), many of the colonists were 'supertramps', or species with a presumed history of successful colonizations. Furthermore, most of the bird species on the islands did not come from the large 'mainland' of New Guinea, but from the equally distant but smaller island of New Britain, suggesting that the species that manage to persist on Long and Ritter are those that have already distinguished themselves as good colonizers on New Britain. Diamond (1975) observed among these species of the 'supertramp' category a higher incidence of juvenile movement between islands, temporary high local abundance, large number of clutches per year, and diet generalization, but noted that the same species are eventually excluded from most islands. Diamond attributed this to their low competitive ability, but this need not necessarily be the case (Connor and Simberloff 1979).

Colonization through experimental introductions

Introductions are experiments where the physical barrier for colonization is overcome, and success or failure in colonization cannot be attributed to differences in the ability to cross that barrier. Experiments of this kind were performed with rodents in islands of Penobscot Bay, Maine (Crowell 1973; Crowell and Pimm 1976).

Within an 8-km radius of a large (72 km²) island inhabited by one representative of each of the three rodent genera *Microtus, Peromyscus,* and *Clethrionomys,* there are 50 small (<50 ha) islands. These contain habitats suitable for all three species, but are inhabited only by *Microtus.* Crowell introduced propagules of *Peromyscus* or *Clethrionomys* to nine of these islands and monitored the success of each, as well as the response of the native *Microtus.* He concluded that although the inability to reach the islands cannot be responsible for the absence of *Peromyscus* and *Clethrionomys,* neither of the two can persist there for a long time.

In addition, Crowell collected demographic data, in order to find out whether the difference in colonizing ability between the three species (which declined in the order *Microtus>Peromyscus>Clethrionomys*) could be predicted without reference to their patterns of distribution. A recalculation of Crowell's estimates of the values of r_s, the intrinsic growth rate in the colonized area (based on the growth curve at the exponential phase, usually during the first year, and on the larger islands) shows r_s for *Peromyscus* to be 13 per cent larger than that for *Clethrionomys* (Table 11.1). No test for statistical significance could be performed because of insufficient sample size. The recalculated r_s value for *Microtus,* which is based on the growth curves following the low points of the normal cycles on the larger islands, is also given in Table 11.1, which also includes estimates for the values of λ and μ for the three species. In general the relations between the different parameters of the three species fit the difference in relative colonizing ability between them. *Peromyscus* has both a higher *r*-value and a higher λ/μ ratio than *Clethrionomys,* while *r* of the really successful

TABLE 11.1 *Demographic parameters for rodents colonizing islands in Penobscot Bay, Maine. (Recalculated from Crowell 1973)*

	λ	μ†	r	λ/μ	r/λ
Clethrionomys gapperi	4.39	2.11	2.28	2.08	0.52
Peromyscus maniculatus	3.71	1.13	2.58	3.28	0.70
Microtus pennsylvanicus‡	7.81	2.79	4.84	2.63	0.62

† Assuming that most deaths during the early stages of colonization were of first-year individuals, $\mu = \ln(N_{x=1}/N_{x=0})$.

‡ Crowell's values were corrected by a factor obtained by comparing his and our values for the other species.

colonizer among the three (*Microtus*) is much higher than that of the other two, but its λ/μ ratio has an intermediate value. One can conclude that in this system environmental rather than demographic stochasticity determines the success of colonization.

In the course of another experiment, both *Peromyscus* and *Clethrionomys* were introduced together to the same island, and their habitat utilization and degrees of overlap and displacement were estimated (Crowell and Pimm 1976). The various estimates (Table 11.2) show that ranking according to *r* can predict

TABLE 11.2. *Several ecological characteristics of the rodent species in islands of Penobscot Bay, Maine. (Data from Crowell (1973) and Crowell and Pimm (1976))*

	r†	Index of specialization‡	Competition coefficient††	Invasion rate*	Extinction rate*
Clethrionomys gapperi	2.28	69	0.65;0.44	0	0.100
Peromyscus maniculatus	2.58	60	0.66;0.20	0.028	0.065
Microtus pennsylvanicus	4.84	37	0.17;0.14	0.043	0.057

† From Table 11.1.
‡ (Variability in habitat usage divided by total habitat variability) x 100.
†† Each value stands for the effect on one of the two other species.
* Yearly rates.

the ranking according to specialization, competitive ability, and colonizing ability. Unlike the arthropods of the mangrove islets, invasion rate is here positively correlated with persistence. Furthermore, as is generally maintained, specialization is positively correlated with competitive ability and negatively correlated with *r*. Thus, both *r* and the degree of specialization are reliable correlates of colonizing ability in this system. The fact that the worst competitor is the best colonizer can be explained provided that low competitive ability is associated with a high rate of dispersal. *Microtus*, which can actively cross water barriers up to 1 km wide, is thus expected to reach vacant habitats first. As summarized by Crowell and Pimm (1976), 'competitive ability is not required for a species to eliminate competitors, provided that the lesser competitor is present first in sufficient numbers. This is more likely to be achieved by an opportunist (a generalist) even though it is competitively subordinate'.

To conclude, *Clethrionomys,* the species with the lowest *r*, is a poor colonizer in all the islands, and the difference between *Microtus* and *Peromyscus* apparently depends on the size of the island and its degree of isolation. On a large island both *Microtus* and *Peromyscus* will persist, although *Microtus*, with the much higher *r*, might have to rely on repeated invasions. If the island is well isolated, *Peromyscus*, once introduced, will persist longer than *Microtus*.

Colonization through a man-made connection between two previously isolated areas

The case we wish to discuss under this title, in somewhat greater detail, is that of the large-scale colonization that took place after the opening of the Suez Canal, connecting the Red Sea (the Gulf of Suez) with the eastern Mediterranean.

The Red Sea and the eastern Mediterranean are adjacent to one another, and indeed share many genera, but differ markedly in their communities: the biota of the Red Sea is unmistakably tropical Indo-Pacific, whereas that of the eastern Mediterranean is more Atlantic in character. The climatic conditions in the two seas are sufficiently similar (although far from identical) to provide suitable habitats for many of the species of the other basin. Nevertheless, the degree of isolation between the two basins is still large, mostly because of the nature of the canal, which is long, narrow, poor in habitat diversity, and/or with harsh environmental conditions (Por 1978). Moreover, a definite south-to-north current regime in the Canal defines the Gulf of Suez as the source for most of the colonizing species, and the Mediterranean as the colonized area (Agur and Safriel 1981).

At least 128 Red Sea species have successfully colonized the Mediterranean, whereas only 3 Mediterranean species are known for certain to have colonized the Red Sea. Whatever the reason for this 'division of roles' between the two basins, the system provides a unique opportunity for studying the prerequisites for successful colonization, by means of looking for answers to the following two questions

1. What do the 128 colonists have in common?
2. What is it that distinguishes these species from the many hundreds of Red Sea species that have not (yet?) used the opportunity to colonize the Mediterranean?

Safriel and Lipkin (1975), Por (1978), and Vermeij (1978) addressed themselves to the first question, and came out with a tentative profile of the colonist:

(a) *Dispersability*. The colonists are either active swimmers, or can be passively transported along relatively short distances; if benthic, they are not permanently attached to the substrate. When they have planktonic larvae, most are lecithotrophic and short-lived; holoplanktonic species hardly ever take part in colonization (Por 1978).

(b) *Reproduction*. The colonists are said to have long reproductive seasons (Por 1978), although this has not yet been quantitatively substantiated; short larval stages or direct development are commoner than indirect development; and at least in one case (a seastar) asexual reproduction was associated with success in colonization (Por 1978).

(c) *Ecological plasticity*. Many of the colonists are euryhaline and eurythermal, with non-selective substrate requirements and non-selective diets (Por 1978), i.e., they are generalists.

(d) *Competitive ability*. Unlike the example of *Microtus* (p. 230), in the Mediterranean the colonizers are good competitors, as can be inferred either from their relative abundance in high-diversity communities (Safriel and Lipkin 1975; see also MacArthur 1972) or from what seems to be an actual case of exclusion of an indigenous species (possibly one case among the seastarts), or partial displacement (several cases among fishes and crustaceans; Por 1978).

(e) *Predation*. Most colonists are rather poor with respect to the possession of anti-predatory morphological adaptations (Vermeij 1978).

(f) *Distribution*. Most colonists of the Mediterranean tend to avoid the very deep waters (Por 1978) and, if intertidal, the highest levels (Safriel and Lipkin 1975). Species associated with colonizers, e.g. parasites of colonizers, are likely to be colonizers themselves, but this was not always the case (Por 1978).

All the above conclusions are at best gross generalizations, mostly because the biology of virtually all colonists is poorly known. A step toward solving this difficulty has been taken while attempting to answer the second question. The idea was to select one Red Sea colonizer and another closely related, Red Sea non-colonizer species, and compare them with regard to several potential attributes of colonizing ability. As a control, each comparison involved a native, closely-related Mediterranean species, in an attempt to distinguish, in the colonizer, between attributes that are specific for successful colonization in the Mediterranean, and the general attributes that are independent of the specific conditions in the colonized area (Safriel and Ritte 1980). Two groups of intertidal molluscs of the rocky shores — mytilid bivalves and cerithid gastropods — were used by ourselves and our students for these comparisons. The research is still underway, but some conclusions can already be drawn.

Of the four intertidal mytilid species indigenous to the Mediterranean coast of Israel only one, *Mytilaster minimus*, has colonized the Suez Canal (but not the Gulf of Suez). The smallest species in this group, *Musculus costulatus*, is locally and regionally rare. The largest, *Modiolus barbatus*, is mostly subtidal, which probably means that it is adapted to relative environmental stability. The fourth species, *Gregariella petagnae*, is small, and similar in size to *Mytilaster minimus*. Both may be regionally rare but very common locally. The difference between them is that the non-colonizer is mainly a low-midlittoral species, while the colonizer is a mid-midlittoral species (Safriel, Gilboa, and Felsenburg 1980). Thus, an opportunistic life-style, as is evident from the occupation of a physically variable environment, coupled with a potential for a very high local abundance is a possible correlate of colonizing ability among the eastern Mediterranean mytilids.

In contrast to the Mediterranean colonist, which colonized only the Suez

Canal, one Red Sea mytilid, *Brachidontes variabilis*, became a successful colonizer in both the Suez Canal and the Mediterranean. The only other mytilid species with which *Branchidontes* can be compared is *Modiolus auriculatus*, which also reached the Mediterranean, as is evident from some washed up valves, but never managed to establish itself (Safriel *et al*. 1980). *Brachidontes* and *Modiolus* differ from one another in size, habitat, and distribution. The latter is twice as large as the former, and also matures at a size which is more than twice the size at maturity of the other (Felsenburg 1974; D. Lavee, personal communication). *Modiolus* lives in a lower intertidal level than *Brachidontes*, in areas which are highly accessible to predatory fish. The rapid growth of its individuals, as well as their larger sizes, seem to be antipredatory adaptations, achieved at the expense of the ability to reproduce at an early age. *Brachidontes*, on the other hand, lives in areas which in themselves make it less exposed to predation. Even when predation pressure is high it does not invest much in trying to avoid it, and becomes common only where the pressure is low. As a result, *Modiolus* is widespread but relatively uncommon locally, while *Brachidontes* is regionally very rare but may be exceedingly common locally.

 Brachidontes reaches its highest densities in the Suez Canal (Safriel *et al*. 1980), and this may be attributed to the harsher conditions there, which eliminate competitors and predators and favour opportunistic species, as *Brachidontes* may indeed be, at least relative to *Modiolus*. *Brachidontes* is also common in the large lagoon of Bardawil (Northern Sinai) in the eastern Mediterranean, where again the conditions are rather variable. Elsewhere in the Mediterranean it is rare locally (about 70 times as rare as in the Suez Canal), usually living side by side with *Mytilaster*, which can be 250 times commoner.

 Although it is not surprising that *Brachidontes*, rather than *Modiolus*, colonized the Mediterranean, the answer to the question why *Brachidontes* did not become more abundant in the Mediterranean is less certain. In the Suez Canal, where the two species are common, *Brachidontes* does not seem to be affected by *Mytilaster*. Can it be that the competitive interactions between them are different in the Mediterranean? Comparisons between the two species (Gilboa 1976) reveal that *Brachidontes* is less *r*-selected than *Mytilaster* (Fig. 11.2), suggesting that when competing over a vacant patch, under equal conditions, *Mytilaster* will have the advantage. The numerical advantage of *Mytilaster* over *Brachidontes* also contributes to its advantage, and, indeed, when a dense population of *Brachidontes* was experimentally created in the Mediterranean, it survived well (Yiftakh, Gilboa, and Safriel 1978). Thus, when a patch in the *Mytilaster* bed becomes vacant, it is more likely to be colonized by the much commoner larvae of *Mytilaster*. In the western Mediterranean and in the Atlantic *Mytilaster* is not as common, probably because of competition with larger mytilids, which are absent in the eastern Mediterranean.

 Because both *Brachidontes* and *Mytilaster* tend toward the *r* end of the

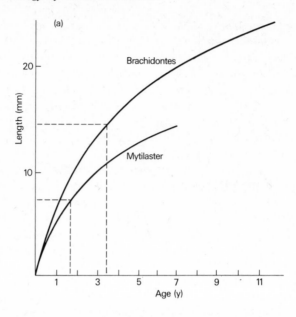

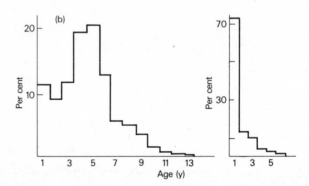

Fig. 11.2. Comparisons between the Red Sea-colonizing mytilid species *Brachidontes variabilis* (data from Ras Muhammad, Sinai), and the Mediterranean species *Mytilaster Minimus* (data from Palmachim, the Mediterranean coast of Israel) (from Gilboa 1976). (a) Individual growth curves. Dashed lines mark the age at which the animals attain half final size. (b) Age structures. Left: *B. variabilis* (n=263, April 1974); right: *M. minimus* (n=2592, March 1975).

r–K continuum, the possibility that the success of *Brachidontes* is due to its being adapted specifically to the conditions of the eastern Mediterranean cannot be ruled out. A demonstration that *r*-attributes are not a common characteristic of all species in this region came from the studies of the cerithid gastropods of the system.

Out of the nine cerithid species that live in the intertidal rocks of the Gulf of Suez (Ayal and Safriel 1981), only one, *Cerithium scabridum*, has successfully colonized the eastern Mediterranean. Using a technique for scaling the most relevant environmental variables (exposure to waves, retainment of moisture at low tide, and depth and rate of movement of sand on the rocky sediment) (Ayal and Safriel 1980), it was possible to scale the other *Cerithium* species with respect to their ecological similarity to *scabridum* (Ayal 1978). *C. caeruleum* was found to be most similar, and was therefore selected for the comparison with *scabridum* regarding the correlates of colonizing ability.

After comparing the patterns of distribution of *scabridum* and *caeruleum* and the habitats occupied by them, it was concluded that the Suez Canal could not have served as a barrier for immigration by *caeruleum* and that the eastern Mediterranean actually offers more suitable habitats and resources to *caeruleum* than to *scabridum*. What, then, is missing in *caeruleum* but possessed by *scabridum* that is responsible for the difference between them in colonizing ability? Among the various *Cerithium* species of the Gulf of Suez, *scabridum* is not the smallest and *caeruleum*, being twice as large as *scabridum*, is not the largest. Both are generalists in terms of habitat preferences and resistance to the abiotic environment. Both are opportunistic and their distributions are patchy. The most significant differrences between the two species are their life histories, as follows:

(a) *Dispersability*. Both species have planktonic larvae, but the adults of *caeruleum* produce 2–3 times more larvae per year than adults of *scabridum*, and the larvae of the former remain planktonic for twice as long as those of the later.

(b) *Individual growth rate and age at maturity*. *Scabridum* grows fast until it starts reproducing, at the age of 15–22 months. The growth rate of *caeruleum* rapidly declines with time, and it starts reproducing at the age of 4 years (Fig. 11.3).

(c) *Birth and death rates*. The death rate of *scabridum* is much higher than that of *caeruleum*, and its life span is much shorter. An adult reproduces at most during 2 seasons, compared to 7–10 seasons in *caeruleum*, but the eggs of both are similar in size, and there is no difference between the larvae in size at hatching and at settlement.

In order to define the relative position of these two species, both of which exhibit *r*-attributes, along the $r–K$ continuum, it was necessary to estimate their innate capacity for increase (Andrewartha and Birch 1954). Ayal (1978) measured age-specific fecundities and mortalities for cohorts of both species from the Red Sea under optimal conditions. Since larval survivorship could not be measured, *r* was calculated for a wide range of hypothetical values of larval survivorship. For each such value, an upper bound for *r* (labelled r_m in

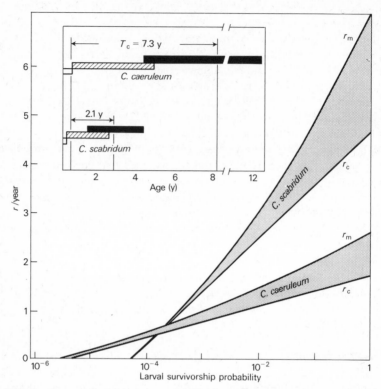

Fig. 11.3. Innate capacity for increase *r* and life-history parameters of a cohort of the Red Sea snail *Cerithium scabridum*, a colonizer of the Mediterranean, and a cohort of the non-colonizer *C. caeruleum*. T_c is cohort generation time; solid bars: life span of adults; striped bars: length of period during which juveniles can be found; blank bars: length of time between onset of egg laying and onset of recruitment. For further details see text. (After Ayal (1978); Ayal and Safriel (unpublished).)

Fig. 11.3) was obtained using Lotka's equation for an exponentially growing population, and a lower bound (r_c) was obtained using a simpler equation for growth of a population with discrete generations (Ayal and Safriel, unpublished). The results (Fig. 11.3) show that the r_m value is generally higher in *scabridum* than in *caeruleum*, in spite of the fact that the total number of eggs laid by a female throughout her lifetime is much greater in *caeruleum*.

Thus, although both species can grow in the eastern Mediterranean, *scabridum* is expected to grow there faster at the early stages of colonization. In reference to the possibility that this high value of *r* is a specific adaptation for occupying the eastern Mediterranean, it is worthwhile to consider the only indigenous intertidal cerithid of this area, *C. rupestre*. Relative to *scabridum*, *rupestre* has typical *K*-attributes (Ayal 1978). It is twice as large, has a thicker shell, its eggs are heavier and 30 times as few; it starts reproducing at the age of 3 years (compared to the age of 1 year in *scabridum* in the Mediterranean), and can live up to the age of 5 years (compared to 1–2 years in Mediterranean *scabridum*). In addition, its daily rate of egg laying is about 100 times, and its embryonic development is about 5 times slower than in *scabridum*. High r, at least in this case, proves to be general, rather than a specific, correlate of colonizing ability.

By studying the habitats occupied by *scabridum* and *rupestre* in the Mediterranean, Ayal concluded that the present-day distribution of *rupestre* is largely affected by the presence of *scabridum*. The ability of *scabridum* to invade a system with a potentially efficient resident competitor can be ascribed to the fact that the distribution of *rupestre* is patchy, with only a certain proportion of suitable habitats occupied at any given time. The larvae of *rupestre* do not have a planktonic stage, and *scabridum* was able to establish itself first in those parts of its habitat which were outside the range of *rupestre* (the highest intertidal levels), as well as in the empty patches within the range of *rupestre*. It then could rapidly swamp the resident species by means of its faster reproduction and its higher immigration rate between patches. It can be said that the quantitative advantage of *scabridum* outweighted the qualitative advantage of *rupestre*, which would have gone extinct were it not for its ability to evade its competitor in the lower intertidal and subtidal levels (Ayal 1978).

As *caeruleum* is also characterized by high fecundity and high dispersability, it is worth speculating why it did not join *scabridum* in colonizing the Mediterranean. A possible answer to this question may lie in the fact that *scabridum* and *caeruleum* adapted in two different ways to their common environment. *scabridum* is a typical 'r-strategist', adapted to cope primarily with spatial unpredictability by means of a potential for rapid utilization of local opportunities. *caeruleum*, on the other hand, is a 'bet-hedger' (Stearns 1976), adapted to cope with both spatial and temporal unpredictabilities. In addition to spreading the risks spatially, like *scabridum*, it also invests in spreading the risks in time, by improving its ability to respond fast to short-lived opportunities. In the case of colonizing the Mediterranean, the 'r-strategist' proved to have the upper hand, by means of being first in establishing 'stepping stones' in the Canal, first in reproducing, and first in sending the larvae further on. Later arrivals of *caeruleum* could not be established, presumably due to the 'migration and extinction' competitive advantage of *scabridum*.

It should be pointed out, however, that the advantage of the 'r-strategy' is due to the larvae being the only dispersing stage. If adults disperse as well (as the case may be with many fish), a species with a resistant propagule of high reproductive value – as is expected in 'bet-hedgers' – may succeed more than a species with non-resistant propagules.

To conclude, the ideal colonizer is the species that can both *arrive* fast in the colonized area and *grow* fast once it has arrived.

Other cases of colonization

Artificial introductions can also be viewed as colonization attempts through a man-made connection between previously isolated areas. Whereas the rate of arrival and the size of the colonizing group in these cases are controlled by man, the introduced species are usually confronted with communities which are very hard to penetrate. It is not our intention to give a thorough review of the immense amount of literature dealing with the subject of introductions, especially since no rigorous analysis of successful introductions *versus* non-successful ones is available.

Similarly, the subject of introductions for biological control has so far only been briefly examined in the context of colonization theory (Remmington 1968; Beddington *et al.* 1978) and an extensive review on this subject is desirable. We will only note that pests themselves are good colonizers, by virtue of being generalists (Conway 1976). The species introduced to combat pests are naturally derived from a higher trophic position, are likely to be specialists, and are usually expected to encounter difficulties in colonization. The applied biologist prefers that they will be highly specific in their diet requirements, but at the same time they have to be good colonizers, i.e. good in dispersal and superb in persistence (DeBach 1974). An understanding of the population dynamics and the life-history tactics of candidates for introduction should be of great value, and the emphasis in these studies should shift from an inquiry into their preadaptations to the specific community into which they have to be introduced to an inquiry into their universal correlates of colonizing ability.

Finally, we want to point out that the ability to identify a potential colonizer from any assemblage of species declines as the ease of getting into the colonized area increases. When a given habitat is transformed, it is immediately exposed to frequent invasions and the eventual composition of the community may depend more on 'assembly rules' (Diamond 1975) than on preadaptations of individual species. In general, fast arrival and rapid growth are of prime importance for the early colonists, while the success of later arrivals depends more on their specific adaptations relative to the existing community, and on their invasion rates (Cody 1975; Diamond 1975).

Conclusions

The number of species that can be expected to colonize in any given system is a function of the accessibility of the colonized area, the degree to which its community is already packed with species, and the adaptations of the potential colonizers for overcoming the physical and biological limitations to successful colonization. Our review indicates that under most circumstances attributes of the 'r-strategy' are associated with these adaptations, so that in any consideration of the possibility of colonizations a knowledge of the relative positions of the various candidates along the $r-K$ continuum of life-history tactics should be of great help. It should be remembered, however, that the collection of data for estimating the relevant demographic parameters is a very tedious process, so that in practice these parameters will have to be inferred from morphological, ecological, or distributional characters that are easier to study. The relevant morphological characters are body size, or the existence of 'escape' stages and dispersal adaptations. The ecological characters may be multivoltinism, early maturity, and general ecological plasticity. As far as distribution is concerned, one should look for occurrence in fluctuating environments, wide geographical distribution, particular 'incidence curves' (Diamond 1975), intimate associations with known colonizing species, or a history of successful colonization. It should be pointed out, however, that even the best adaptations for successful colonization cannot guarantee that a colonization will actually take place. It may perhaps be more practical to suggest that, after species have been ranked according to their colonizing ability, rather than going to the group of species with high ability and pointing out the species that are likely to colonize, one should go to the other side of the range, and identify those species that regardless of the specific circumstances cannot be expected to be included in the group of potential colonists.

12
The function of distance movements in vertebrates

A.R.E. SINCLAIR

There are no aseasonal habitats: whatever the environment, certain periods of the year are relatively worse for animals than other periods (Sinclair 1975). The bad periods are usually caused by climatic changes, which indirectly affect populations through lack of resources such as food and shelter from climatic extremes or predators. The size of the populations relative to resources (density) is the other factor determining the severity of the bad period.

The time window for this period and the severity of the resource bottleneck vary with latitude. In lowland tropical forests, the bottleneck is only slightly more severe than the good period but it may last eight months (Ward 1969; Fogden 1972). In tropical savannah, there is a marked difference between dry-season paucity of resources lasting 2–4 months and wet-season super-abundance (Sinclair 1975, 1978). In temperate and subpolar regions, the differences between poor and good seasons becomes even greater, with the time period of low resources varying between two months in lower latitudes to nine months at higher latitudes.

This universal resource fluctuation of high and low abundance with its variable time period, severity, and predictability has acted as an agent of natural selection to produce a number of different life-history adaptations. Amongst these are a group of movement strategies.

Although food abundance is probably the most common selection factor producing movements in animal populations, there are a few cases where some other requirement of the animals has selected for specific movements. One such factor is the need in some colonial birds and sea mammals for breeding sites which prove to be limited in number and distribution so requiring the population to converge at a certain time of year; another is the need to find mates that occur in a few local concentrations, requiring the opposite sex to converge upon them.

In this chapter, I shall consider some strategies that animals have evolved to counter these selection agents. Such a review cannot fail to mention the unequalled comprehensive work of R. R. Baker (1978) who covers the literature up to about 1973. There are some difficulties with definitions, however. Baker includes all forms of movement (apart from trivial ones) under the name of 'migration' to the extent that a boulder moving downhill performs migration. In this sense, the term is synonymous with 'movement' and is redundant. By including everything, the word becomes meaningless, and Baker then invents a whole new jargon that is both cumbersome and confusing.

There is a gradation in movement types from irregular one-way to regular seasonal round-trip; the best treatment of such gradation is arbitrarily to divide it into sections in order to compare the strategies and the selecting agents producing them. Thus, I define 'migration' as a regular round trip within a life-span of the individual, 'emigration' as a directional one-way movement, and 'dispersal' as one-way movement with no predetermined direction in population. I shall be concerned with the first two as adaptations to changing environments. Dispersal, which can occur in stable saturated environments as, for example, an adaptation to avoid inbreeding and to reduce competition between relatives, is treated elsewhere in this volume (see Chapters 7 and 8).

Migration

Evolution of migration

If one accepts the theory of natural selection, then it follows that migration (and other movement patterns) are adaptations to improve long-term reproductive success (R. R. Baker 1978). Such an obvious conclusion, however, does not help in elucidating the actual selection factors that have lead to these adaptations.

The conventional view concerning the evolution of migration, at least in birds, is that it evolved to avoid the resource bottleneck in temperate regions (Lack 1954; Newton 1972; R. R. Baker 1978). Implicit in this idea is the suggestion that ancestral members of a species were originally nonmigratory, and that they survived less well during the winter period, so creating strong selection in favour of those that moved out at this period. But this idea remains incomplete for its begs the question of why they were in temperate regions in the first place. It seems to me more complete understanding is achieved if one approaches the evolution of migration from the other end, namely that it is an adaptation to find better breeding habitats.

Many studies have now indicated that, at least in mammals and birds, the critical factor determining the success of breeding is the amount of food available to parents for building up body reserves prior to birth or egg laying (Perrins 1970; Schifferli 1973; Sinclair 1977, 1978, 1982). The better this pre-birth period, the more successful is the mammalian mother at providing milk and conceiving again, and the avian mother at laying more and larger eggs and at feeding the juveniles.

Thus, there must be strong selection for animals to place themselves in environments where there is a superabundant food supply prior to breeding. Organisms have always been able to breed and then remain in tropical regions, despite there being a food restriction at certain times of year. The favourable season in the tropics, determined by monsoon rains, is both less predictable and less food-abundant relative to the summer season in temperate latitudes. Therefore, individuals from a tropical population that moved to temperate regions to breed, would have an initial advantage over the resident individuals. After breeding, they would have to return to low latitudes because the northern winter is now a worse environment than the tropics.

This hypothesis that migration is an adaptation to find greater food resources for breeding is both more complete and more general than current suggestions that migration is merely an adaptation to avoid the resource bottleneck: the latter hypothesis does not account for why species should not remain in the tropics to breed. Clearly, breeding in the tropics is possible, and many migratory bird species have close relatives that do so.

Migration to find food resources for breeding

Birds The most conspicuous examples of migration for breeding in superabundant environments occur in insectivorous birds. The wheatear (*Oenanthe oenanthe*), a member of the thrush family, spends the winter in tropical savannah Africa in close association with the congeneric capped wheatear (*O. pileata*). The latter is breeding during this period and has been seen to chase and exclude the migrants from its territory (Sinclair 1978). In April, at the height of the tropical breeding season, the migrant wheatear travels to the northern Palaearctic, crossing the Atlantic to Greenland in the west, while in the east spreading across the whole of Asia to Siberia and even to Alaska.

The white stork (*Ciconia ciconia*) also spends the winter in tropical and southern Africa and adults fly north to Europe to breed. Immature birds, one year old, remain in the tropical areas or move only part way north following the monsoon, showing that tropical environments remain habitable all year; and in South Africa, white storks have occasionally been resident there and produced young (Moreau 1966, p. 123). Black storks (*Ciconia nigra*) are resident in Malawi and further south, but some merely winter in Malawi and migrate to Europe to breed. Other examples of closely related bird species, some of which breed in the tropics, others migrating to Europe to breed, include the migratory European roller (*Coracius garrulus*) and the sedentary African Lilac-breasted Roller (*C. caudata*), the migratory European Spotted flycatcher (*Muscicapa striata*) and many resident flycatcher species in Africa (e.g., *Alseonax*, *Bradornis* spp.).

Perhaps a good example of how migration may have evolved, is seen amongst the bee-eaters. There are a number of forest and savannah species of the genus *Melittophagus* that are sedentary in the tropical zone. Other species such as the white-throated bee-eater (*Aerops albicollis*) are inter-tropical migrants, breeding just south of the sahara in the June–September wet period, and then moving south of the equator to southern Tanzania and Zaire for their rainy season. There are two species (perhaps only subspecies) of carmine bee-eaters, one of which, *Merops nubicus*, behaves like the previous species, but the other *M. nubicoides*, shows the opposite strategy, breeding in South Africa September–March and then migrating north to the equator (Fig. 12.1). This suggests how bird migration could have evolved within the tropics.

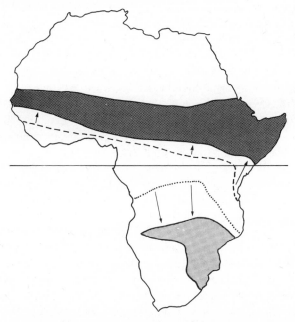

Fig. 12.1. Intra-African migration of southern (*Merops nubicoides*) and northern (*M. nubicus*) carmine bee-eaters. Each spends the nonbreeding season closer to the equator (broken lines), then migrates to higher latitude to breed (stipled area). (After Moreau (1966).) (With permission, Academic Press.)

Finally, there is one species, the blue-cheeked bee-eater *M. supercillosus*) which has two subspecies, one breeding in Turkey and the Middle East, the other breeding in Madagascar and both migrate to the equator in the nonbreeding season. The European bee-eater *M. apiaster*) behaves in a way similar to the northern race of the former species (occasionally it has also bred in South Africa during the northern winter; Moreau 1966). It is suggestive that many of these northern species would have bred in the Sahara region before it was degraded a few thousand years ago by man. As the Sahara became more inhospitable, birds were forced to fly further to pass it.

The migration patterns of the white-throated and carmine bee-eaters are typical of most intra-African bird migrants. The best documentation of this is given by Elgood, Fry, and Dowsett (1973) for movements in Nigeria. In that area, there is a simple north–south stratification of vegetation from dry savannah in the north to moist forest in the south. The rains move north with

244 The ecology of animal movement

the sun from January to June, then south again. Consequently, the wet period is longest in the south.

Some general principles emerge from the Nigeria study

1. Almost no forest species migrate, indicating that the cost of movement is greater than the difficulty of remaining sedentary through the relatively mild dry season.

2. Of the non-forest species, migration is uniformly a strategy to follow the rains north in spring and south in autumn.

3. Seed-eating birds tend to be resident, while insectivores are usually migratory. Clearly, the former breed in the rains, whenever they occur, and feed their young on the abundant insects at that time. Then after fledging the whole population switches to a graminivorous diet coinciding with seed-set of the vegetation at the end of the rains and throughout the dry season. That is, by switching diet they can remain sedentary. The insectivores adopt the opposite strategy of following the insect superabundance north then south.

4. Fewer species of vertebrate feeders are migratory compared to the insectivores, and Elgood *et al.* (1973) suggest that this is because the higher trophic levels in the consumer chain are more buffered against seasonal changes in food abundance compared to lower trophic levels.

5. Since migrant species follow the rains throughout the year, most times of year should be potentially suitable for breeding, and we should expect the species to spread their breeding throughout the year; in fact this does occur to a large extent. Moreau (1966) shows there are species which breed in the south and move north for the remaining part of the year (19 of 67 species), while others do the reverse (31 species). The preponderance of north Nigerian breeders suggests that there is a greater superabundance of food there in conjunction with the greater seasonality.

There are, however, other species such as the Grey-headed Kingfisher (*Halcyon leucocephala*), which stop to breed in Nigeria on their way north, then continue north after breeding (Fig. 12.2). Even more versatile is the sunbird (*Nectarinia erythroceria*) which M. P. L. Fogden (in Elgood *et al.* 1973) has found breeding twice a year in the same territories on the equator in Uganda, but migrating alternately north and south after each breeding period. He considers they may also breed during both migratory periods, making four breeding seasons in the year.

These examples illustrate the general conclusion of Elgood *et al.* (1973) that migration originates in a species independently of other life-history events such as breeding or moulting.

6. Palaearctic migrants, like those migrating within Africa, move slowly north with the rains, resting and feeding as they go. They differ from intra-African migrants only in that they break away from the seasonality

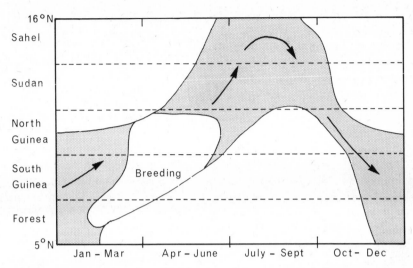

Fig. 12.2. Diagram of breeding and migration of the grey-headed kingfisher (*Halcyon leucocephala*) in Nigeria. They migrate north in January, stop to breed about April, then continue north again. Vegetation zones are not of equal width. (After Elgood, *et al.* (1973).) (With permission, *Ibis*, journal of the British Ornithologist's Union.)

of the wet Intertropical Front and make use of the seasonality in temperate regions determined by temperature and day-length. Temperate summers are both more predictable and longer-lasting than tropical rain seasons.

Elgood *et al.* (1973) point out that Palaearctic migrants are usually members of families that are migratory in general, and include some intra-African migrants. It is worth considering here why Palaearctic migrants do not also breed during their stay in Africa, especially since many congeneric residents breed at that time. It appears that most species cannot stay for long enough in any one locality to complete a breeding cycle successfully (see below). However, Moreau (1966, p. 123) reports instances of house martins (*Delichon urbica*) and European bee-eaters nesting in South Africa during their stay, and the common sandpiper (*Tringa hypoleucos*) may do so in the tropics.

7. In Africa, about 180 Palaearctic migrant bird species spend their non-breeding season in savannah habitats — very few use the forest habitats. They share the savannah with about 800 species of African resident birds (Moreau 1972). One particular question posed by Moreau was how did the migrants compete successfully with the resident species since the former appeared to arrive during the tropical dry season when resources were scarce. Most migrant species are either insectivores or vertebrate feeders, and their food supply is determined by rainfall. Recent

work (Sinclair 1978) has suggested that the migrants are nomadic for much of the time they are in Africa, following the rainstorms as they move south and north with the Intertropical Front. As a result of this strategy, they are usually following a local insect superabundance which itself is often composed of insects following the rain. Thus, competition between migrants and residents is likely to be minimal.

I have concentrated on the African migrants because they are in some ways better documented than those of other continents. However, the studies of Nix (1976), Braithwaite (1975, 1976), and Frith (1967) show that in eastern Australia with predictable seasons, the same principles apply. In North America, MacArthur (1959) has pointed out that bird species leave summer breeding areas if food supply is predictably scarce in winter but abundant in summer. What happens to these species in Central and South America is poorly documented. There is little savannah available to them and many species appear to use the forest habitats (N. G. Smith, personal communcation). However, many of the Nearctic migrants are forest species (unlike those in Europe and Asia) so tropical-forest habitats may be more suitable for them.

Mammals I have discussed birds as the most obvious examples of species moving to find optimal resources for breeding, but most large-mammal species also perform movements for the same purpose. Their migrations are usually not so extensive, except in the case of whales.

The great Mysticete whales either filter microplankton or eat small fish, copepods, and euphausid crustacea. Typically they spend the winter in warm tropical waters and move to polar regions in the summer (Norris 1967) in order to feed. It is the polar waters which have the highest densities of food, some 10–20 times that of tropical waters (R. R. Baker 1978). When the ice forms, the whales return to tropical waters where they are usually found in the few high-density food areas feeding to some extent. However, it is generally considered that throughout their tropical sojourn there is little or even no feeding. This pattern is typical of the humpback (*Megaptera novaenglia*) and blue whales (*Balaenoptera musculus*).

While in tropical waters and during the period of minimal feeding, whales give birth and suckle their young. Suckling is very frequent (every half-hour in some dolphins) and growth is rapid – the young blue whale, for example, doubles its birth weight in seven days (R. R. Baker 1978). It is clear that the mother can only do this by using her fat supplies built up during the previous summer: blubber, therefore, is a food reserve in preparation for birth and its function as an insulator is of secondary importance (Norris 1967). Indeed heat load may be more of a problem for large whales than heat loss.

The return to tropical waters, therefore, is not to avoid cold climates, nor to find food, but more likely to provide a suitable temperature regime for the small new-born whale which may have an insulation problem. In the primitive grey whale (*Eschrichtius gibbosus*), mothers give birth in the shallow bays of Baja California, a behaviour designed to avoid predation by killer whales (*Orcinus orca*). The movements of Mysticete whales, therefore, support the thesis that migration is an adaptation to provide the best habitat in preparation for breeding. In the case of the whales, a whole summer's feeding is required to build up body condition in preparation for birth, because the animals are so large. As a corollary, the actual environmental conditions for these large mammals during birth and lactation may be of lesser importance. It follows that smaller animals would need a shorter time to build up body condition, and this is in fact seen in the smaller Odontocete whales that eat fish and cephalopods. The small (50 kg) arctic porpoise (*Phocoena phocoena*) builds up a blubber supply amounting to 40 per cent of body weight in the summer, makes a short migration away from the ice edge, gives birth, and returns. The new-born is relatively large, being 30 per cent of the mother's weight. Thus, the small size allows a rapid build-up of body supplies but precludes long migrations to warm waters.

Because Odontocetes eat fish, they are high in the trophic chain and therefore are less subject to extreme seasonal changes in food supply experienced by Mysticetes; and as in birds, show less extreme migration patterns. Many Odontocetes simply follow the concentrations of fish in the ocean and are, therefore, far more nomadic. And the killer whale, a top carnivore, is effectively a resident. Pinniped movements are in many ways similar to those of the Odontocetes: the Pacific Walrus (*Odobenus rosmarus*) performs continuous movements following the advance and retreat of the ice edge, and harp seals (*Pagophilus groenlandicus*) show similar migrations in time with their main fish stocks (Norris 1967).

All ungulates perform seasonal migrations from breeding to nonbreeding ranges. In almost every case, the animals can remain year round and even breed in their normal nonbreeding ranges, and there are several examples where this has occurred in aberrant years. On the other hand, very few species can remain year round in their normal breeding range, which suggests that migration places animals in a temporary but good habitat for breeding, and then returns them to the more permanent nonbreeding habitat.

Examples of ungulate migration are numerous: North American deer (*Odocoileus* spp.) have a winter range usually in some protected valley bottom with woodland, and a higher summer range. Moose (*Alces alces*) show a similar pattern except they give birth half-way between the two ranges, and in late summer they move down again to make use of temporarily available aquatic vegetation. African bovids all show seasonal shifts in distribution with, in general, a wet-season dispersal into lush but temporary habitats followed by a retreat to dry-season ranges closer to water supplies (Sinclair 1982). For example, impala (*Aepyseros melampus*) use open *Acacia* woodland during the wet season but migrate in the dry season a few hundred metres to habitats close to rivers Jarman and Sinclair 1979). African buffalo (*Syncerus caffer*) prefer riverine grasslands during the dry season for these are the only remaining green long-grass habitats, but in the rains they spread out, using most habitats equally (Sinclair 1977).

Births take place in African ungulates during the wet season with smaller species giving birth earlier in the season than larger species, because the former can build up body condition quicker. In very large ungulates, such as elephant (*Loxodonta africana*) and white rhinoceros (*Ceratotherium simum*), like the large whales, the whole wet season is required to build up condition and births occur at the end of the rains and into the dry season (Sinclair 1980).

Although most African ungulates migrate only a short distance to their wet-season range, those using the extensive semi-arid savannahs often move a hundred kilometres or more: such species include several gazelles in the Sahara and springbuck (*Antidorcas marsupialis*) in the Kalahari region; and wildebeest (*Connochaetes taurinus*), zebra (*Equus burchelli*), and eland (*Taurotragus oryx*) in southern and eastern Africa. Wildebeest in the Serengeti region of Tanzania regularly return to the semi-arid plains to produce their young, if the plains are green. But if the rains fail, and they did in late 1966 and 1979, then wildebeest give birth in their dry-season habitats.

Large carnivores, in contrast to the ungulates, are much more sedentary. In the Serengeti, the main predators of wildebeest, lion (*Panthera leo*) and hyaena (*Crocuta crocuta*), are confined to territories and cannot follow the migrants. This has an important consequence for the wildebeest; all predators experience some part of the year when there are no migrant ungulates present, and this is the time when predator numbers are restricted by relatively scarce resident prey (Hanby and Bygott 1979). As a result, the wildebeest population cannot be regulated by its predators – the migratory strategy allows the wildebeest to escape its predators. This is, therefore, a benefit additional to the primary one of finding green-food supplies (Sinclair 1979).

Some large carnivores, however, do show migrations in response to similar movements of their ungulate prey. Wolves (*Canis lupus*), for example, follow the movements of the caribou (*Rangifer tarandus*) in northern Canada. In spring, they move ahead of their prey, find a den, and give birth to coincide with the arrival of the herds and the birth of young, easily caught caribou.

Seasonal migrations for the specific purpose of finding improved food supplies for breeding is less well documented in other groups of vertebrates such as reptiles, amphibians, and fishes though no doubt they occur. The large populations of pelagic fish such as herring (*Clupea harengus*) and cod (*Gadus morhua*) appear to move seasonally in response to changes in their food supplies (Harden Jones 1968; R. R. Baker 1978).

Migration to find breeding sites

Some migrations clearly have the function of finding suitable areas in which to breed. Many seabirds nest on rocky cliffs or isolated islands and these are relatively scarce. Consequently, the whole population of these species converges from all over the oceans to these tiny areas. Gannets, shearwaters, and albatrosses are examples. Their colonies are situated near waters with high food stocks so that they do not have to move far to feed their young. After breeding, the populations then disperse and become nomadic across the oceans. An example of this is seen in the slender-billed shearwater or mutton-bird (*Puffinus tenuirostris*), which breeds on a few islands between Australia and Tasmania. After breeding, adults and young migrate north into the Pacific, and the young do not return to breed until at least 5 years old. Because recaptures of marked birds were necessarily along the Pacific coastal rim, it appeared originally as if the birds were performing a clockwise circuit around the Pacific before returning to breed (D. L. Serventy in Dorst 1962) but current opinion suggests this may be an artefact of where observers are found (i.e. along the coast) and the birds probably wander over the whole Pacific. Another example is the wandering albatross (*Diomedea exulans*) which breeds on sub-Antarctic islands such as South Georgia. Breeding lasts a year and following this the animals stay in the southern ocean, apparently circumnavigating the Antarctic by moving with the prevailing westerly winds.

Amongst the mammals, pinnipeds migrate to particular rocky shores to give birth and breed. The northern fur seal (*Callorhinus ursinus*) breeds on a few islands around the Pacific rim. Young born on the Pribilof islands, for example, wander throughout the Pacific but always return to their natal shore to breed. Adult females, however, migrate directly to the California coast (R. R. Baker 1978). A similar example is seen in the California sea lion (*Zalophus californianus*) which breeds along that coast line. After breeding, the sexes go in different directions, females south and males much further north (Norris 1967).

Movement towards breeding sites also accounts for the migrations of the green turtle (*Chelonia mydas*). This species lays its eggs on particular sandy beaches throughout the tropical oceans. There are several apparently discrete populations. Females return to the same beach to lay every 2–3 years and in the meantime wander over a certain area of ocean to feed. Some movements are particularly long: turtles feeding off the Brazilian coast migrate 2000 km to the small mid-Atlantic island of Ascension to lay their eggs (Carr and Coleman 1974).

Many of the anadromous fish, such as Salmonids, perform migrations to particular breeding sites; indeed the adults always return to their natal stream for breeding. They feed little, if at all, on their way upstream and die after breeding. The young feed in the stream for a certain period but eventually they migrate to the ocean where they spend 2–4 years, depending on the species.

In general, the main characteristics of these species are a generally nomadic existence in the oceans and a migration to highly specific breeding sites. Where species have many breeding sites available to them, migration usually disappears;

the Weddell seal (*Heptonychotes wedelli*), for example, remains in a home range on the Antarctic sea ice throughout the year, feeding under the ice by using holes which it keeps open (Stirling 1969).

Migration to find mates

The need to find mates has resulted in some species migrating to traditional mating grounds. This is seen in some widely dispersed species where family groups aggregate in the breeding season. African elephant family groups migrate and aggregate in the rains, sometimes forming groups of several hundred animals in the Serengeti region, and similar groupings have been noted in Kenya and Uganda (Leuthold and Sale 1973). At this time, the large, usually solitary bulls, associate with the females and conceptions take place (Croze 1974). These aggregations do not appear necessary to obtain food or other resources and seem more to have a social function. In a similar way, many of the large whale family groups aggregate in tropical waters, and although this may be due to patchy food resources, the timing coincides with breeding and may, like elephants, help females find the dominant males.

In species which form leks, it is necessary for females to move to the lek site to find the males; such movements are usually not great. Lekking has been reported in two antelope species, the kob (*Kobus kob*) and the Lechwe (*K. leche*) (Buechner and Roth 1974; Schuster 1976), and in several bird species, particularly in the grouse family. These movements have the same function as those mentioned above, namely to find and select the best male for mating.

Emigration

Evolution of nomadism

It is no accident that most of the world's large populations are of migrant species. Within the same environments, resident species are at much lower density. A nonmigratory existence is of advantage in stable environments, such as tropical forests, and this strategy is often associated with other strategies such as co-operative breeding in birds (Grimes 1976; Woolfenden 1976).

Within fluctuating environments, resident-species populations are regulated by resources available at the worst time of year. Some species can reduce the resource bottleneck by sequestering resources; deermice (*Peromyscus* spp.) make seed caches for winter use as do many squirrel species in North America. The small lagomorph, the pika (*Ochotona princeps*), collects hay piles for the same purpose, while the acorn woodpecker (*Melanerpes formicivorus*) stores acorns in specially made holes in dead trees (MacRoberts and MacRoberts 1974). In northern Canada, the Canada jay (*Perisoreus canadensis*) is one of the few species that can remain in the boreal forest year round and it does so by storing food in bark crevices.

Interesting though these alternate strategies are for dealing with the resource restriction, the majority of species adopt the strategy of migration by following temporarily good conditions. Most bird species, and many mammals, such as carnivores and pinnipeds, however, are constrained at some time of year for the purpose of breeding: they are forced to be sedentary while their young are developing and this places a constraint on where and when they can breed, for they must find areas where resources are abundant for long enough to allow successful reproduction. Resource superabundances that last for shorter periods of time cannot be used by these species. Therefore, there should be selection for individuals that can reduce the period of breeding residency or young dependence, in environments which are changing rapidly. Such environments are also characterized by another feature — that of unpredictability in both the timing and location of the high-resource areas. Thus, selection would favour those that adopted a nomadic existence, moving in unpredictable ways rather than following a prescribed regular migration from nonbreeding range to breeding area and back.

This progression towards nomadism is evident in ungulates. Within the antelope group Alcelaphinae, the topi (*Damaliscus korrigum*) has new-born that hide in long grass, and the females are confined to relatively permanent habitats until the young can follow her. A related species, the wildebeest, has precocial young that follow the mother within 5 minutes of birth and can run as fast as her by one day old. These animals are effectively nomadic, following local rainstorms throughout the year in a predictable pattern on a gross scale (Fig. 12.3) but entirely unpredictable on a fine scale (Pennycuick 1975; Maddock 1979). The success of this strategy is seen in the comparison of population numbers — one and a half million wildebeest compared to forty thousand topi in the same area. Nomadism in the wildebeest has the added advantage of reducing the risk of predation, a point I will return to later. Caribou in northern Canada also show the nomadic movement typical of wildebeest; although caribou move towards the treeline in winter and away from it in summer, there is great variation in where they occur from year to year (e.g., G. R. Parker 1972). In desert environments, most of the large ungulates in Africa are nomadic as are kangaroos in Australia. In the oceans, Odontocete whales are clearly nomadic because their young can stay with the adults and pinnipeds are nomadic in the nonbreeding season.

Amongst birds, most of the typical palaearctic migrants that perform predictable seasonal movements for breeding are, within Africa, quite nomadic (Sinclair 1978). Although some species are known to return to the same winter site, they probably have several such areas which they visit as they follow the rains south across the equator and then north again.

Emigration in unpredictable environments

Ungulates and whales are not constrained by a sedentary breeding period,

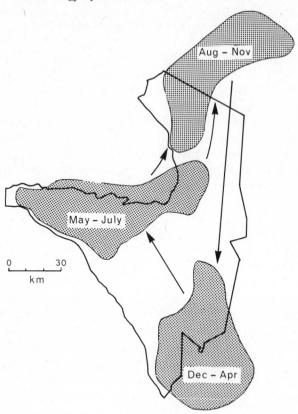

Fig. 12.3. Distribution and migration of the wildebeest in the Serengeti Park, Tanzania. They spend the wet season (December–April) on the plains, then move north-west in the early dry season, and end up in the far north at the end of the dry season.

while most birds and pinnipeds can only become nomadic at certain times of year. However, in highly unpredictable environments, birds must abandon the regular migratory pattern and adopt a strategy whereby they emigrate from their natal site and search for new breeding sites. Thus, nomadism is extended to include the breeding locations as well.

Examples of this strategy can be found in the Australian waterfowl (Frith 1967; Braithwaite 1975, 1976). In general, most species that breed in southern Australia (e.g., *Aythya australis* hardheads; *Anas superciliosa* black duck) with predictable breeding habitats, perform regular migrations. Those that breed in the arid interior (*Anas gibberifrons* grey teal; *Malacorhynchus membranaceus* pink- eared duck) do so whenever appropriate habitat is created by the unpredictable flooding of rivers, and the birds find the floods by long-distance emigration from their birthplace. The pink-eared duck has the added adaptation of gonads divided into parts, some of which are always ready for breeding at any time.

Other birds, such as the budgerigar (*Melopsittacus undulatus*) are found in very large flocks. The budgerigar is an opportunistic breeder and, when conditions are good, can produce several broods in quick succession. When conditions are bad, they cease breeding and emigrate in search of better conditions (Serventy 1971). S. J. J. F. Davies (1976) has stressed that the Australian arid-land birds often have regular breeding seasons; small amounts of rain fall in isolated and unpredictable locations but predictably at certain times of year. The rain drains into sumps which for a short period, produces green vegetation. Birds, such as the budgerigar, zebra finch (*Taeniopygia castanotis*), and emu (*Dromaius novaehollandiae*) can maintain their regularity by efficient searching for these locally abundant areas. Hence, a strategy of emigration is advantageous because the natal site is unlikely to offer good opportunities for much longer and there is a reasonable opportunity of finding new areas. Nix (1976) shows that emigration in central Australia can be in any direction but there is a bias towards north or south in accordance with rainfall patterns.

Unlike true migrant birds which are mainly insectivorous, many arid-adapted emigrant species are seed-eaters (probably because arid-adapted plants produce many seeds as a dormant phase to withstand drought). Some of the small birds have been able to condense the breeding time as an adaptation to the temporary nature of their habitats, and a classic example of this is the red-billed quelea (*Quelea quelea*), a small seed-eater in arid savannahs of Africa (Ward 1971). In eastern Africa, this species follows the rain-front south from Somalia across the equator to Tanzania and then back north again. This is unlike other seed-eaters which perform only local migrations between wet- and dry-season habitats. The quelea also differs from the insectivorous birds which follow the rain, because it is not a true migrant, but an emigrant that goes through several generations as it follows this cycle. The birds remain a few weeks behind the rain-front coinciding with areas that have a new growth of maturing grass. They breed in exceptionally large communal nests (up to a million birds); the time from egg laying to fledging is only 30 days, and the whole breeding cycle lasts six weeks. After fledging their young, the flocks move on to new areas in which grass has grown up ahead of them. They may either bypass other flocks or join them to produce another brood. The quelea is unusual in that it has used a breeding strategy associated with temporary habitats to follow a predicatable weather phenomenon, unlike other species.

The European finches provide an interesting comparison of the different strategies of migration and emigration (Newton 1970, 1972). Within Europe, there is a group of finches that eat herbaceous seeds (e.g., goldfinch, *Carduelis carduelis*; and linnet, *C. cannabina*) and perform a migration from southwestern Europe in winter to northeastern Europe in summer and back again. Many of the southern populations are only partial migrants with some birds remaining to breed in their winter areas while others move north – the double strategy indicating that migration is beneficial in some years but not in others (there may be two genotypes or one genotype with two responses to the environment, as suggested by R. R. Baker 1978).

Another group of finches feed on tree seeds and concentrate wherever their food is plentiful at the time (e.g., redpoll, *C. flammea;* siskin, *C. spinus*; pine grosbeak, *Pinicola enucleator*). The food supply of conifer seeds is patchily distributed; in northern climates, trees need 4–5 years to build up reserves before producing a seed crop, and in any one area most trees (even of different species) crop in phase because they come under the same weather, but in widely separated areas seed crops are out of phase. Conifer seed-eating birds in an area with a poor crop, will emigrate until they find a suitable area. In the boreal forests of the Palaearctic (and presumably of the Nearctic), there is a constant but unpredictable movement of these birds – a situation similar to the seed-eating species in arid environments.

The three species of crossbill finches in Europe are also conifer seed-eaters and perform annual emigrations similar to the above species. However, they are noted for their large-scale emigrations every few years to southwestern Europe (irruptions). The common crossbill (*Loxia curvirostra*) feeds on spruce seeds and populations are high in areas with a good spruce crop (Fig. 12.4). Newton considers that irruptions occur when a high population of birds (following a good spruce crop) coincides with a crop failure over a large area. The birds

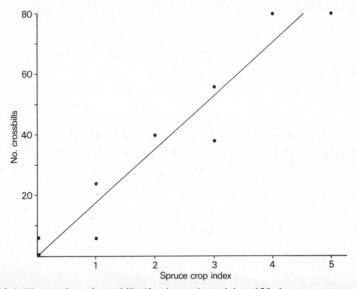

Fig. 12.4. The number of crossbills (*Loxia curvirostra*) in a 120–km transect was directly related to an arbitrary index of spruce cone crop, northern Finland. Seed production determines whether birds invade or evacuate an area. (From Reinikainen in Newton (1972).) (With permission, Collins Publishers.)

then perform long-distance movements. Captures of birds at this time show they have desposited fat supplies, suggesting that emigration is an adaptive strategy for which they have prepared rather than the last-gasp movements of starving birds (Newton 1972). Presumably, they adopt the irruption strategy, as R. R. Baker (1978) suggests, when a threshold of density per food is passed. Birds performing these irruptions, breed in southern Europe, and a few recaptures have shown that some at least return north in later years. In general, emigration is the extreme form of the nomadic strategy adapted to track unpredictable food resources.

Migration as an evolutionarily stable strategy

To this point, I have considered migration and emigration as being adaptations which are uniformly beneficial for a population. However, there are some instances where only a portion of the population migrates, the rest being resident, suggesting these animals are adopting a mixed evolutionarily stable strategy (Maynard Smith and Price 1973). A mixed strategy occurs when an individual can adopt one of two or more alternative strategies with equal benefit, the choice being made solely on what other members of the population are doing. The wildebeest in the Serengeti region of Tanzania appears to be a good example of this; there is a large migratory population and several resident populations. One resident population occurs in the adjacent Ngorongoro Crater and does not overlap in area with the migrants. The Crater population suffers severe predation pressure, as mentioned earlier, with some 14 per cent of the population being killed annually (J. P. Elliott and Cowan 1978). Predation almost certainly holds the population constant and promotes a high recruitment and turnover of the population; that is, an average female has a relatively short lifespan of about 7 years. But she has high fertility, producing her first calf at 2 years old and then one each year thereafter, so that on average, she gives birth to 6 in her lifetime. Two-thirds of these will die before maturity so leaving two offspring as a lifetime contribution to the next generation – as should be expected. By comparison, the migrant animals have an advantage in very much reduced predation (1 per cent annually). Against this must be considered the cost of migration which appears to be quite small; the present population has excess food and little mortality from starvation – recruitment is 14 per cent and the population is increasing at 10 per cent annually, so that mortality is only 4 per cent (Sinclair 1979). Since 1 per cent mortality is due to predation, 3 per cent can be considered as due to migration (drowning when crossing rivers, broken legs, and indirect effects of energy expenditure). At first sight, it appears migrants have an over-all advantage, but one must first consider what happens when the migrant population reaches a stable state. Evidence strongly suggests that this is effected when the population reaches the limits of its food supply (Sinclair 1979). Under these conditions, recruitment of yearlings drops to about 10 per cent, which means that the average lifespan of a female in this stable population is 10 years. It is probable (though not yet substantiated) that the effects of undernutrition reduce the fertility of the female by delaying puberty an extra year and inhibiting conception in older females, so that she produces 7–8 offspring in a lifetime. Food stress imposes heavy mortality on new-born animals so that only 25–30 per cent survive to one year old. This means that again only two offspring survive to the next generation. In other words, either strategy (migrant or resident) is equally suitable when the populations are stable. Furthermore, it would not be advantageous for an individual to switch strategies when both populations are stable, for that animal would suffer two sets of costs with no compensating benefit.

This result explains the even more curious situation of a second resident population of wildebeest in the western Serengeti which actually overlaps with the migrants in the early dry season. The residents have the double disadvantage of increased predation and increased food competition when the migrants arrive and literally remove all the food. When this happens, the migrants move on (an example of a paradoxical strategy where the intruder has an advantage over a resident). The residents can only remain because there is enough soil moisture to allow a small amount of regrowth of food after the migrants have gone to support a small population (in this case, 10 000 residents compared to 1.5 million migrants). This also explains why residents are not found where the migrants occur at the worst time of year – the end of the dry season in the northern Serengeti; here they would have both predation and severe competition. The western residents and the migrants do not appear to switch strategies; both have synchronized breeding seasons which are at different times of year – any substantial mixing would have destroyed these differences.

Why, then, does an individual decide to be a resident or a migrant? The decision depends on the sizes of the two populations relative to their stable sizes. If all animals were migrants, then any individual who stays behind in an area capable of supporting him year round, would have an initial advantage of superabundant food and hence no competition, and probably little predation. This advantage, of course, is only temporary, lasting until numbers have increased to the stable level; but for the first individuals it is a real advantage. Similarly, if all animals were resident, there would be an initial advantage to becoming migrant. Hence, the advantage depends only on what the other members of the populations are doing.

I have detailed this example because most of the population dynamics is known. However, there are a number of cases where it seems a similar mixed strategy could be employed. A recent example is that of the Aldabra tortoise (*Geochelone gigantea*) reported by Swingland and Lessells (1979). They found that some tortoises, but not all, would migrate from the interior of the island to the coast (a few hundred metres) at the beginning of the rainy season. The advantage of migrating lay in the extra food that could be found on the coast at this season, particularly for females thus insuring that they laid that year. The disadvantage of migration lay in the hazards of being caught in the open, too far from shade to prevent death from heat stress on unpredictable clear days: Swingland and Lessells found a much higher mortality in the migrant population and they argued that this counteracted the advantages, so that the two strategies of migrant and resident produced equal returns in the long term. These strategies can also be considered as conditional ones; they are conditional on the sex of the individual since it might pay a female to migrate, but a male to stay at home.

A similar situation may explain the dichotomy in the sockeye salmon (*Oncorhynchus nerka*) of British Columbia and Alaska. McCart (1970) has described two morphs, the sockeye and the Kokanee. The sockeye shows the classical anadromous strategy in which eggs are laid in freshwater streams, and the young migrate to the sea where they grow to large adults. After four years, they return to the home stream to spawn once and die. The kokanee, on the other hand, spawns in the same streams, but remains in adjoining lakes throughout its life, growing to a small size before spawning. Apparently the two morphs are genetically identical and are produced from the same eggs. The decision whether to migrate or not may even depend on the food supply for the fry in the lake – if food is abundant, then more fry remain behind.

The advantage of migration in the salmon lies in the greater food supply in the ocean which leads to large size, and for females, this means more eggs. As in the case of the tortoises, more females than males migrate. The disadvantage again lies in the hazards of the journey, mainly through predation. For males, a more complicated situation arises. Large males have greater success at courting large females, so there is an advantage for males to migrate. McCart also found that on average, a pair of courting sockeye had associated with them 2–4 small satellite kokanee males; they acted as 'sneakers' by creeping up to the large courting pair (for being small, the kokanee remained unnoticed) and at the time of egg release, darted in and released their sperm before or at the same time as the large males. Clearly, only a certain proportion of these 'sneakers' can exist for too many would interfere with each other.

There is, however, another advantage in being small, applying to both sexes. In dry years, some spawning streams are too shallow to allow large fish to travel up them, and small fish have a decided advantage (McCart 1970); in fact, kokanee tend to be found in areas far from the sea where access is more difficult. Both in the case of the 'sneakers' and in the shallow streams, the proportion of the population adopting a strategy depends on what the others are doing; if all are migrants, then there is an advantage in some males becoming resident 'sneakers' and, in either sex, becoming resident to use shallow streams.

Conclusion

In general, one should expect a proportion of a population of migrants to become resident whenever year-round conditions allow it, as can be seen in the European goldfinch and linnet, the wildebeest, Aldabra tortoise, and sockeye salmon. The size of the resident and migrant portions of this mixed strategy, depends on the relative abundances of resources available to them.

If, as a result of adopting a mixed resident—migration strategy, the two populations change their breeding seasons, then the initial conditions for sympatric speciation through disruptive selection are set. Thus, one can see the first stage of this in the black stork which has resident and migrating populations that breed at the same time; the wildebeest has slightly different breeding seasons, while the white stork that was resident in South Africa bred six months out of phase with the migrant population wintering there. Moreau (1966) mentioned that of some 32 bird species having resident African and Palaearctic migrant populations, 24 of them show subspecific differences in their populations. This may point to why so many Palaearctic species overlap with closely related species in Africa — resident species from which they could have evolved sympatrically.

A universal resident strategy occurs only where environmental conditions are so uniformly good that all areas have exploited resources. Therefore, movement does not result in the discovery of unexploited resources but merely imposes an additional cost due to the movement. Such conditions are found in tropical forests for bird populations.

Populations with no resident members occur in areas where no year-round favourable conditions are found. I have suggested that the different movement patterns of migration and emigration have evolved in response to different environmental patterns.

Migration, which is a regular seasonal movement, has evolved in response to predictably changing food sources; animals move in order to place themselves in optimum conditions for as long as possible. A few migrations have evolved in response to limited resources, such as breeding sites or to find mates.

Emigration, which is a one-way movement from natal site to breeding site with a strong directional bias in the population, has evolved in fluctuating but unpredictable environments. This strategy is the extreme form of nomadism, involving associated adaptations for rapid breeding. In its mild form, nomadism is seen as local movements to make use of clumped food supplies.

References

Abegglon, J. J. (1976). On socialization in hamadryas baboons. Ph.D. dissertation. University of Zurich, Zurich, Switzerland.

Abramsky, Z. and Tracy, C. R. (1979). Population biology of a 'noncycling' population of prairie voles and an hypothesis on the rate of migration in regulating microtine cycles. *Ecol.* **60**, 349–61.

—— —— (1980). Relation between home range size and regulation of population size in *Microtus ochrogaster*. *Oikos* **34**, 347–55.

—— Van Dyne, G. M. (1980). Field studies and a simulation model of small mammals inhabiting a patchy environment. *Oikos* **35**, 80–92.

Agur, Z. (1982). Unpublished Ph.D. thesis, Hebrew University, Jerusalem, Israel.

—— Safriel, U. N. (1981). Why is the Mediterranean more readily colonized than the Red Sea by Organisms using the Suez Canal as a passageway? *Oecologia* **49**, 359–61.

Albrecht, F. O. (1967). *Polymorphism phasaire et biologie des acridiens migrateurs*. Masson, Paris.

Alcock, J. (1975). *Animal behavior*. Sinauer, Sunderland, Mass.

—— (1980). Natural selection and the mating systems of solitary bees. *Amer. Sci.* **68**, 146–53.

—— Jones, C. E., and Buchmann, S. C. (1977). Mating strategies in the bee *Centris pallida* Fox (Hymenoptera: Anthrophoridae). *Amer. Naturalist* **111**, 143–55.

Alerstam, T. and Enckell, P. H. (1979). Unpredictable habitats and evolution of bird migration. *Oikos* **33**, 228–32.

Alexander, R. D. (1968). Life cycle origins, speciation and related phenomena in crickets *Quart. Rev. Biol.* **43**, 1–41.

—— (1974). The evolution of social behavior. *Ann. Rev. Ecol. Systemat.* **5**, 325–83.

—— (1977). Natural selection and the analysis of human sociality. In *The changing scenes in the natural sciences, 1776–1976* (ed. C. E. Goulden), pp. 283–337. Academy of Natural Sciences, Philadelphia.

—— (1979). *Darwinism and human affairs*. Pitman, London.

Alison, R. M. (1975). Breeding biology and behaviour of the Oldsquaw (*Clangula hyemalis* L.). *Ornithol. Monogr.* **18**, 1–52.

Allard, R. W. (1965). Genetic systems associated with colonizing ability in predominantly self-pollinated species. In *The genetics of colonizing species* (ed. H. G. Baker and G. L. Stebbins), pp. 49–75. Academic Press, New York.

—— (1975). Mating system and microevolution. *Genetics* **79**, 115–26.

Allee, W. C., Emerson, A. E., Park, O., Park, T., and Schmidt, K. P. (1949). *Principles of animal ecology*. W. B. Saunders, Philadelphia.

Allen, G. (1965). Random and non-random inbreeding. *Eugenics Quart.* **12**, 181–98.

Ambrose, H. W. (1972). Effect of habitat familiarity and toeclipping on rate of owl predation in *Microtus pennsylvanicus*. *J. Mammal.* **53**, 909–12.

Anderson, J. H. (1975). Phenotypic correlations among relatives and variability in reproductive performance in populations of the vole *Microtus townsendii*. Unpublished Ph.D. thesis. University of British Columbia.

Anderson, P. K. (1970). Ecological structure and gene flow in small mammals.

Symp. zool. Soc. Lond. **26**, 299–325.

Andersson, M. (1980). Nomadism and site tenacity as alternative reproductive tactics in birds. *J. animal Ecol.* **49**, 175–84.

Andrejewski, R., Petrusewicz, K., and Walkova, W. (1963). Absorption of newcomers by a population of white mice. *Ekol. Polska Ser. A* **11**, 223–40.

Andrewartha, H. G. and Birch, L. C. (1954). *The distribution and abundance of animals*. University of Chicago Press.

Andrews, R. V. (1968). Daily and seasonal variation in adrenal metabolism of the brown lemming. *Physiol. Zool.* **41**, 86–94.

Antonovics, J., Bradshaw, A. D., and Turner, R. G. (1971). Heavy metal tolerance in plants. *Advan. ecolog. Res.* **7**, 2–58.

Arata, A. A. (1967). Muroid, gliroid, and dipodoid rodents. In *Recent mammals of the world* (ed. S. Anderson, and J. K. Jones), pp. 226–53. Ronald Press, New York.

Armitage, K. B. and Downhower, J. F. (1974). Demography of yellow-bellied marmot populations. *Ecology* **55**, 1233–45.

Armstrong, J. T. (1965). Breeding home range in the nighthawk and other birds; it's evolutionary significance. *Ecology* **46**, 619–29.

Arnold, E. N. (1979). Indian Ocean giant tortoises : their systematics and island adaptations. *Phil. Trans. R. Soc. Lond.* **B286**, 163–76.

Austin, O. L., Sr. (1949). Site tenacity, a behavior trait of the common tern. *Bird Banding* **20**, 1–39.

Ayal, Y. (1978). Ph.D. dissertation. Department of Zoology, Hebrew University, Jerusalem, Israel.

— — Safriel, U. N. (1980). Intertidal zonation and key-species associations of the flat rocky shores of Sinai used for scaling environmental variables affecting cerithid gastropods. *Isr. J. Zool.* **29**, 110–24.

— — — — (1981). Species composition, geographical distribution and habitat characteristics of rocky intertidal Cerithiidae (Gastropoda:Prosobranchia) along the Red Sea shores of Sinai. *Argamon, Isr. J. Malac.* **7**, 53–72.

— — — — (unpublished). *r*-curves and the cost of the planktonic stage.

Ayala, F. J. and Campbell, C. A. (1974). Frequency-dependent selection. *Ann. Rev. Ecol. Systemat* **5**, 115–38.

Bailey, N. T. J. (1975). *The mathematical theory of infectious disease and its application*. Griffin, London.

Bairlein, F. (1978). Uber, die biologie einer sudwest deutschen popluation der Monchsgrasmuche (Sylvia atricapilla). *J. Ornithologie* **119**, 14–51.

Baker, H. G. (1965). Characteristics and modes of origin of weeds. In *The genetics of colonizing species* (ed. H. G. Baker and G. L. Stebbins), pp. 147–68. Academic Press, New York.

— — Stebbins, G. L. (Eds.) (1965). *The genetics of colonizing species*. Academic Press, New York.

Baker, M. C. and Mewaldt, L. R. (1978). Song dialects as barriers to dispersal in white-crowned sparrows, *Zonotrichia leucophrys* Nuttali. *Evolution* **32**, 712–22.

Baker, R. R. (1978). *The evolutionary ecology of animal migration*. Hodder and Stoughton, London.

Balat, F. (1976). Dispersion prozesse und brutortstreue beim feldsperling *Passer montanus*. *Zoologica Listy* **25**, 39–49.

Baltensweiler, W. (1971). The relevance of changes in the composition of larch bud moth populations for the dynamics of its numbers. *Proc. Advan. Study Inst.* Dynamics numbers populations, pp. 208–219. (Oosterbeck, 1970).

—— Fischlin, A. (1979). The role of migration for the population dynamics of the larch bud moth, *Zeiraphera diniana* Gn. (Lep. Tortricidae). *Bull. Société Entomol. Suisse,* **52**, 259—71.

Barclay, H. (1975). Population strategies and random environments. *Can. J. Zool.* **53**, 160—5.

Bartholomew, G. A. and Tucker, V. A. (1964). Size, body temperature, thermal conductance, oxygen consumption and heart rate in Australian varanid lizards. *Physiol. Zool.* **37**, 341—54.

Bateson, P. P. G. (1978). Sexual imprinting and optimal outbreeding. *Nature, Lond.* **273**, 659—60.

Batzli, G. O. (1973). Population determination and nutrient flux through lemmings. *US IBP Tundra Biome Data Rep.* 73—119.

—— (1975). The role of small mammals in arctic ecosystems. In *Small mammals: their productivity and population dynamics* (ed. F. B. Golley, K. Petrusewicz, and L. Ryszkowski), pp. 243—68. Cambridge University Press.

—— (1981). Populations and energetics of small mammals in tundra ecosystems. In *Tundra ecosystems; a comparative analysis* (ed. J. J. More) pp. 377—97. Cambridge University Press, Cambridge.

—— Getz, L. L., and Hurley, S. S. (1977). Suppression of growth and reproduction of microtine rodents by social factors. *J. Mammalogy* **58**, 583—91.

—— White, R. G., MacLean, S. F., Pitelka, F., and Collier, B. D. (1980). The herbivore-based trophic system. In *An arctic ecosystem: the coastal tundra at Barrow* (ed. J. Brown *et al.*) Chapter 10, pp. 335—410. Dowden, Hutchinson, and Ross, Stroudsburg, Pennsylvania.

Beacham, T. D. (1979a). Size and growth characteristics of dispersing voles, *Microtus townsendii. Oecologia.* **42**, 1—10.

—— (1979b). Dispersal tendency and duration of life of littermates during population fluctuations of the vole *Microtus townsendii. Oecologia.* **42**, 11—22.

—— (1979c). Selectivity of avian predation in declining populations of the vole *Microtus townsendii. Can. J. Zool.* **57**, 1767—72.

—— (1980a). Growth rates of the vole *Microtus townsendii* during a population cycle. *Oikos* **35**, 99—106.

—— (1980b). Demography of declining populations of the vole *Microtus townsendii. J. animal Ecol.* **49**, 453—64.

—— (1981). Some demographic aspects of dispersers in fluctuating populations of the vole *Microtus townsendii. Oikos* **36**, 273—80.

Beddington, J. R., Free, C. A. and Lawton, J. H. (1978). Characters of successful natural enemies in models of biological control of insect pests. *Nature, Lond.* **273**, 513—19.

Beebe, W. (1949). Insect migration at Rancho Grande in North-central Venezuela. General account. *Zoologica* **34**, 107—10.

Begon, M. (1976). Dispersal, density and microdistribution in *Drosophila subobscura* Collin. *J. animal Ecol.* **45**, 441—56.

—— (1978). Population densities in *Drosophila obscura* Fallen and *D. Subobscura* Collin. *Ecol. Entomol.* **3**, 1—12.

Bekoff, M. (1977). Mammalian dispersal and the ontogeny of individual behavioral phenotypes. *Amer. Naturalist* **111**, 715—32.

Bell, M. A. (1976). Evolution of phenotypic diversity in *Gasterosteus aculeatus* on the Pacific coast of North America. *Systemat. Zool.* **25**, 211—27.

Belovsky, G. E. and Slade, J. B. (In preparation.) Body size—home range area and an energy maximising explanation.

Bengtsson, B. O. (1977). Evolution of the sex ratio in the wood lemming,

Myopus schisticolor. In *Measuring selection in natural populations* (ed. F. Christiansen and T. Fenchel), pp. 333–43. Springer-Verlag, Berlin.

— — (1978). Avoiding inbreeding: at what cost? *J. theoret. Biol.* **73**, 439–44.

Bent, A. C. (1962). *Life histories of North American wild fowl*, Parts I and II. Dover, New York.

— — (1968). *Life histories of North American cardinals, grosbeaks, buntings, towhees, finches, sparrows and allies*, Part I. Dover, New York.

Berger, E. (1976). Heterosis and the maintenance of enzyme polymorphism. *Amer. Naturalist* **110**, 823–39.

Berndt, R. and Sternberg, H. (1968). Terms, studies and experiments on the problems of bird dispersion. *Ibis* **110**, 256–69.

Berry, R. E. and Taylor, L. R. (1968). High-altitude migration of aphids in maritime and continental climates. *J. animal Ecol.* **37**, 713–22.

Berry, R. J. (1979). Genetical factors in animal population dynamics. In *Population dynamics* (ed. R. M. Anderson, B. D. Turner, and L. R. Taylor), pp. 53–80. Blackwell Scientific Publications, Oxford.

— — Jakobson, M. E. (1974). Vagility in an island population of the house mouse. *J. Zool. Lond.* **173**, 341–54.

Bertram, B. C. (1975). Social factors influencing reproduction in wild lions. *J. Zool.* **177**, 463–82.

— — (1976). Kin selection in lions and in evolution. In *Growing points in ethology* (ed. P. P. G. Bateson and R. A. Hinde), pp. 281–301. Cambridge University Press, Cambridge.

Birch, L. C. (1957). The meanings of competition. *Amer. Naturalist,* **91**, 5–18.

Blackwell, T. L. and Ramsey, P. R. (1972). Exploratory activity and lack of genotypic correlates in *Peromyscus polionotus. J. Mammal.* **53**, 401–3.

Blair, W. F. (1953). Population dynamics of rodents and other small mammals. *Advan. Genet.* **5**, 41.

— — (1960). *The rusty lizard: a population study.* University of Texas Press, Austin.

Blaustein, A. R. and O'Hara, R. K. (1981). Genetic control for sibling recognition? *Nature* **290**, 246–8.

Boonstra, R. (1978). Effect of adult Townsend voles (*Microtus townsendii*) on survival of young. *Ecology* **59**, 242–8.

— — Krebs, C. J. (1977). A fencing experiment on a high-density population of *Microtus townsendii. Can. J. Zool.* **55**, 1166–75.

Bourlière, F. (1975). Mammals, small and large; the ecological implications of size. In *Small mammals: their productivity and population dynamics* (ed. F. B. Golley, K. Petrusewicz, and L. Ryszkowski), pp. 1–8. Cambridge University Press, Cambridge.

Bradbury, J. W. and Vehrencamp, S. L. (1976). Social organization and foraging in Emballonurid bats. I. Field studies. *Behav. Ecol. Sociobiol.* **1**, 337–81.

— — — — (1977). Social organization and foraging in Emballonurid bats. III Mating systems. *Behav. Ecol. Sociobiol.* **2**, 1–17.

Brady, R. H. (1979). Natural selection and the criteria by which a theory is judged. *Systemat. Zool.* **28**, 600–21.

Braithwaite, L. W. (1975). Waterfowl on a dry continent. *Nat. Hist.* **84**, 60–9.

— — (1976). Breeding seasons of waterfowl in Australia. *Proc. 16th Int. ornithol. Cong.,* Canberra, pp. 235–47.

Brannon, E. L. (1967). Genetic control of migrating behavior of newly hatched sockeye salmon fry. *Int. Pac. Salmon Comm. Progr. Rep.* no. 16.

Brett, J. R. (1971). Energetic response of salmon to temperature. A study of some thermal relations in the physiology and freshwater ecology of sockeye salmon (*Oncorhynchus nerka*). *Amer. Zoologist* **11**, 99–114.

Brinkhurst, R. O. (1959). Alary polymorphism in the Gerroidea (Hemiptera-Heteroptera). *J. animal Ecol.* **28**, 211–30.

Brown, E. S. (1951). The relation between migration-rate and type of habitat in aquatic insects, with special reference to certain species of Corixidae. *Proc. zool. Soc. Lond.* **121**, 539–45.

—— (1970). Nocturnal insect flight direction in relation to the wind. *Proc. R. entomol. Soc.Lond.* [A], **45**, 39–43.

—— (1943). Territoriality and home range concepts as applied to mammals. *J. Mammal.* **24**, 346–52.

—— Betts, E., and Rainey, R. C. (1969). Seasonal changes in distribution of the African armyworm, *Spodoptera exempta* (Wlk.) (Lep. Noctuidae), with special reference to Eastern Africa. *Bull. entomol. Res.* **58**, 661–728.

Brown, J. L. (1963). Aggressiveness, dominance and social organization in the Steller's jay. *Condor* **65**, 126–53.

—— (1964). The evolution of diversity in avian territorial systems. *Wilson Bull.* **76**, 160–9.

—— (1975). *The evolution of behaviour.* W. W. Norton, New York.

—— Orians, G. H. (1970). Spacing patterns in mobile animals. *Ann. Rev. Ecol. Systemat.* **1**, 239–62.

Brown, L. E. (1966). Home range in small mammal communities. In *Survey of biological progress* vol. 4 (ed. B. Glass). Academic Press, London.

Bruce, H. M. (1959). An exteroreceptive block to pregnancy in the mouse. *Nature* **184**, 105.

Buechner, H. K. and Roth, H. D. (1974). The lek system in Uganda kob antelope. *Amer. Zool.* **14**, 145–62.

Bulmer, M. C. (1972). Multiple niche polymorphism. *Amer. Naturalist* **106**, 254–7.

—— (1973). Inbreeding in the great tit. *Heredity* **30**, 313–25.

—— Taylor, P. D. (1980a). Dispersal and the sex ratio. *Nature* **284**, 448–9.

—— —— (1980b). Sex ratio under the haystack model. *J. theoret. Biol.* **86**, 83–9.

Burger, J. (1972). Dispersal and post-fledging survival of Franklin's gulls. *Bird Banding* **43**, 267–75.

Burt, W. H. (1940). Territorial behavior and populations of some small mammals in southern Michigan. *Miscell. Publ. Museum Zool., Univ. Michigan* **45**, 1–58.

Busse, C. and Hamilton, W. J. (1981). Infant carrying by male chacma baboons. *Science* **212**, 1281–3.

Bygott, G. D., Bertram, B. C. R., and Hanby, J. P. (1979). Male lions in large coalitions gain reproductive advantages. *Nature* **282**, 839–41.

Caldwell, R. L. and Rankin, M. A. (1974). Separation of migratory from feeding and reproductive behaviour in *Oncopeltus fasciatus*. *J. comp. Physiol.* **88**, 383–94.

Campbell, J. S. (1972). A comparative study of the anadromous and freshwater forms of brown trout (*Salmo trutta* L.) in the River Tweed. Ph.D. thesis. University of Edinburgh.

Capranica, R. R., Frishkopf, L. S., and Nevo, E. (1973). Encoding of geographic dialects in the auditory system of the cricket frog. *Science* **182**, 1272–5.

Carl, E. (1971). Population control in arctic ground squirrels. *Ecology* **52**, 395–413.

Carpenter, F. L. and MacMillen, R. E. (1976a). Threshold model of feeding

264   References

territoriaļity and test with a Hawaiian honeycreeper. *Science* **194**, 639–42.
— — (1976*b*). Energetic cost of feeding territories in an Hawaiian honey-creeper. *Oecologia* **26**, 213–24.
Carpenter, F. M. (1977). Geological history and evolution of the insects. *Proc. 15th int Congr. Entomol.,* Washington D. C. 1976, pp. 63–70.
Carr, A. and Carr, M. H. (1972). Site fixity in the Caribbean green turtle. *Ecology* **43**, 425–9.
— — Coleman, P. J. (1974). Seafloor spreading theory and the odyssey of the green turtle. *Nature, Lond.* **249**, 128–30.
Carson, H. L. (1965). Chromosomal morphism in geographically widespread species of *Drosophila*. In *The genetics of colonizing species* (ed. H. G. Baker and G. L. Stebbins), pp. 503–31. Academic Press, New York.
— — (1967). Inbreeding and gene fixation in natural populations. In *Heritage from Mendel* (ed. R. A. Brink), pp. 281–308. University of Wisconsin Press, Madison.
— — Ohta, A. T. (1981). Origin of the genetic basis of colonizing ability. In *Evolution Today: Proceedings of the International Congress of Systematics and Evolutionary Biology* (Vancouver, July 1980), (ed. G. G. E. Scudder and J. L. Reveal). pp. 365–70.
Cavalli-Sforza, L. (1962). The distribution of migration distances: models, and applications to genetics. In *Les déplacements humains* (ed. J. Sutter), pp. 139–58. Union Européene d'Editions, Monaco.
Charlesworth, B. (1974). Inversion polymorphism in two-locus genetic system. *Genet. Res. (Cambridge)* **23**, 259–80.
Charnov, E. L. (1976). Optimal foraging: the marginal value theorem. *Theoret. pop. Biol.* **9**, 129–36.
— — Finerty, J. (1980). Vole population cycles; a case for kin-selection? *Oecologia* **45**, 1–2.
— — Orians, G. H., and Hyatt K. (1976). The ecological implications of resource depression. *Amer. Naturalist* **110**, 247–59.
Chepko-Sade, B. D. and Olivier, T. J. 1979. Coefficient of genetic relationship and the probability of intragenealogical fission in Macaca mulatta. *Behav. Ecol. Sociobiol.* **5**, 263–78.
Chesser, R. K. and Ryman, N. (1980). The optimization of inclusive fitness in subdivided populations. Manuscript.
Chewning, W. C. (1975). Migratory effects in predator–prey models. *Math. Biosci.* **23**, 253–62.
Chitty, D. (1955). Adverse effects of population density upon the viability of later generations. In *The numbers of man and animals* (ed. J. B. Cragg and N. W. Pirie), pp. 57–67. Edinburgh University Press.
— — (1960). Population processes in the vole and their relevance to general theory. *Can. J. Zool.* **38**, 99–113.
— — (1967). The natural selection of self-regulatory behaviour in animal populations. *Proc. ecol. Soc. Austral.* **2**, 51–78, 313–31.
— — (1970). Variation and population density. *Symp. zool. Soc. Lond.* **26**, 327–34.
— — (1977). Natural selection and the regulation of density in cyclic and non-cyclic populations. In *Evolutionary ecology* (ed. B. Stonehouse and C. Perrins), pp. 27–32. MacMillan Press, London.
— — Pimentel, D., and Krebs, C. J. (1968). Food supply of overwintered voles. *J. animal Ecol.* **37**, 113–20.
Christian, J. J. (1950). The adreno-pituitary system and population cycles in

mammals. *J. Mammalogy,* **31**, 247–59.

— — (1970). Social subordination, population density and mammalian evolution. *Science* **168**, 84–90.

— — (1971). Fighting, maturity and population density in *Microtus pennsylvanicus. J. Mammal.* **52**, 556–67.

Christiansen, F. B. and Fenchel, T. M. (1977). *Theories of populations in biological communities. Ecological studies* 20. Springer-Verlag, Berlin.

Clark, A. B. (1978). Sex ratio and local resource competition in a prosimian primate *Science* **201**, 163–5.

Clarke, B. (1975). Frequency–dependent and density–dependent natural selection. In *The role of natural selection in human evolution* (ed. F. M. Salzano). North-Holland, Amsterdam.

— — (1979). The evolution of genetic diversity. *Proc. R. Soc. Lond.* **B205**, 453–74.

Clough, G. C. (1965). Lemmings and population problems. *Amer. Scientist* **53**, 199.

— — (1968). Social behaviour and ecology of Norwegian lemmings during a population peak and crash. *Papers of the Norwegian State Game Reserve Institute, Ser. 2,* **28**, 1–50.

Clutton-Brock, T. H. (1974). Primate social organisation and ecology. *Nature* **250**, 539–42.

— — Harvey, P. H. (1976). Evolutionary rules and primate societies. In *Growing points in ethology* (ed. P. P. G. Bateson and R. A. Hinde), pp. 195–237. Cambridge University Press.

— — — — (1977a). Primate ecology and social organization. *J. Zool.* **183**, 1–39.

— — — — (1977b). Species differences in feeding and ranging behaviour in primates. In *Primate ecology: studies of feeding and ranging behaviour in lemurs, monkeys and apes* (ed. T. H. Clutton-Brock). Academic Press, London.

— — — — (1979). Home range size, population density and phylogeny in primates. In *Primate ecology and human origins* (ed. I. S. Bernstein and E. O. Smith). Garland Press, New York.

— — — — (In press). The functional significance of variation in body size among mammals. In *Mammalian Behaviour* (ed. J. F. Eisenberg). Special publication of the American Society of Mammalogists.

Cody, M. L. (1971). Finch flocks in the Mohave Desert. *Theoret. pop. Biol.* **2**, 142–8.

— — (1974). Optimization in ecology. *Science* **183**, 1156–64.

— — (1975). Towards a theory of continental species diversities. In *Ecology and evolution of communities* (ed. M. L. Cody and J. M. Diamond), pp. 214–50. Belknap Press, Cambridge, Massachusetts.

Coelho, A. M. (1974). Socio-bioenergetics and sexual dimorphism in primates. *Primates* **15**, 263–9.

Cohen, D. (1967). Optimization of seasonal migratory behavior. *Amer. Naturalist* **101**, 5–17.

Cole, F. and Batzli, G. O. (1978). Influence of supplemental feeding on a vole population. *J. Mammal.* **59**, 809–19.

Cole, L. C. (1960). Competitive exclusion. *Science* **132**, 348–9.

Collett, R. (1911–12). *Norges pattedyr.* A. Aschehoug., Kristiania (Oslo). (In Norwegian).

Collins, M. D., Ward, S. A., and Dixon, A. F. G. (1981). Handling time and the functional response of *Aphelinus thompsoni,* a predator and parasite of

the aphid *Drepanosiphum platanoidis. J. animal Ecol.* 50, 479–87.
Colwell, R. K. (1981). Group selection is implicated in the evolution of female-biased sex ratios. *Nature* 290, 401–4.
Comins, H. N., Hamilton, W. D., and May, R. M. (1980). Evolutionarily stable disperal strategies. *J. theoret. Biol.* 82, 205–30.
Common, I. F. B. (1954). A study of the ecology of the adult Bogung moth, *Agrotis infusa* (Boisd.) (Lepidoptera:Noctuidae) with special reference to its behaviour during migration and aestivation. *Austral. J. Zool.* 2, 223–63.
Connell, J. H. (1961). Effects of competition, predation by *Thais lapillus,* and other factors on natural populations of the barnacle *Balanus balanoides. Ecol. Monogr.* 31, 61–104.
Connor, E. F. and Simberloff, D. (1979). The assembly of species communities: chance or competition? *Ecology* 60, 1132–40.
Conway, G. (1976). Man versus pests. In *Theoretical ecology* (ed. R. M. May), pp. 257–81. Saunders, Philadelphia.
Cook, R. M. and Hubbard, S. F. (1977). Adaptive searching strategies in insect parasites. *J. animal Ecol.* 46, 115–26.
Cook, S. B. and Cook, C. B. (1978). Tidal amplitude and activity in the pulmonate limpets *Siphonaria normalis* (Gould) and *S. alternata* (Say). *J. exp. mar. Biol. Ecol.* 35, 119–36.
Cooke, F., MacInnes, C. D., and Prevett, J. P. (1975). Gene flow between breeding populations of the lesser snow goose. *Auk* 92, 493–510.
Coope, G. R. (1978). Constancy of insect species versus inconstancy of Quarternary environments. *Diversity of insect faunas* (ed. L. A. Mound and N. Waloff), pp. 176–87. *R. Entomol. Soc. Symp.* 9. Blackwells, Oxford.
—— (1979). Late Cenozoic fossil Coleoptera: evolution, biogeography and ecology. *Ann. Rev. Ecol. Systemat.* 10, 247–67.
Cooper, K. W. (1939). The nuclear cytology of the grass mite, *Pediculopsis graminum* (Reut.), with special reference to karyomerokinesis. *Chromosoma* 1, 51–103.
Coulson, J. C. (1971). Competition for breeding sites causing segregation and reduced young population in colonial animals. *Proc. Advan. Study Inst.*, pp. 248–57. Dynamics numbers populations. (Oosterbeck, 1970).
Cowan, D. P. (1979). Sibling matings in a hunting wasp: Adaptive inbreeding? *Science* 205, 1403–5.
Cronin, T. W. and Forward, Jr. R. B. (1979). Tidal vertical migration: an endogenous rhythm in estuarine crab larvae. *Science* 205, 1020–2.
Crook, J. H. (1964). The evolution of social organisation and visual communication in the weaver birds (Ploceinae). *Behav. Suppl.* 10, 1–178.
—— (1965). The adaptive significance of avian social organisation. *Symp. zool. Soc. Lond.* 14, 181–218.
—— (1970). The socio-ecology of primates. In *Social behaviour in birds and mammals* (ed. J. H. Crook), pp. 103–166. Academic Press, London.
—— (1972) Sexual selection, dimorphism and social organization in the primates. In *Sexual selection and the descent of man* (ed. B. Campbell). Aldine, Chicago.
—— Gartlan, J. S. (1966). Evolution of primate societies. *Nature* 210, 1200–3.
Crow, J. F. and Kimura, M. (1970). *An introduction to population genĕtics theory.* Harper Row, New York.
Crowell, K. L. (1973). Experimental zoogeography: introduction of mice to small islands. *Amer. Naturalist* 107, 535–58.
—— Pimm, S. L. (1976). Competition and niche shifts of mice introduced onto

small islands. *Oikos* **27**, 251–8.

Croze, H. (1974). The Seronera bull problem. *East African Wildlife J.* **12**, 1–27.

Curry-Lindahl, K. (1962). The irruption of the Norway lemming in Sweden during 1960. *J. Mammal.* **43**, 171–84.

—— (1975). *Fjäll-lämmel, en artmonografi.* Sesam, Stockholm. (In Swedish).

Darlington, C. D. (1958). *The evolution of genetic systems.* Basic Books, New York.

Davies, N. B. (1977). Prey selection and the search strategy of the spotted flycatcher (Muscicapa striata): a field study on optimal foraging. *Animal Behav.* **25**, 1016–33.

—— Houston, A. I. (1981). Owners and satellites: the economics of territory defence in the pied wagtail, *Motacilla alba. J. animal Ecol.* **50**, 157–80.

—— Krebs, J. R. (1978). Introduction: ecology, natural selection and social behaviour. In *Behavioural ecology an evolutionary approach* (ed. J. R. Krebs and N. B. Davies), Chapter 1, pp. 1–22. Blackwells, Oxford.

Davies, S. J. J. F. (1976). Environmental variables and the biology of Australian arid zone birds. *Proc. 16th int. Ornithol. Cong.,* Canberra, pp. 481–8.

Dawkins, R. (1980). Good strategy or evolutionarily stable strategy? In *Sociobiology: beyond nature/nurture* (ed. G. W. Barlow and J. Silverberg) Westview Press, Boulder, Colorado.

Dawson, W. R. (1974). Appendix: conversion factors for units used in the symposium. In *Avian energetics* (ed. R. A. Paynter). Publ. Nuttal Ornithol. Club no. 15.

DeBach, P. (1974). *Biological control by natural enemies.* Cambridge University Press.

Defries, J. C. and McClearn, G. E. (1970). Social dominance and Darwinian fitness in the laboratory mouse. *Amer. Naturalist* **104**, 408–11.

DeLong, K. T. (1967). Population ecology of feral house mice. *Ecology* **48**, 611–34.

Dempster, J. P. and Lakhani, K. H. (1979). A population model for cinnabar moth and its food plant, ragwort. *J. animal Ecol.* **48**, 143–63.

Diamond, J. M. (1974). Colonization of exploded volcanic islands by birds: the supertramp strategy. *Science* **184**, 803–6.

—— (1975). Assembly of species communities. In *Ecology and evolution of communities* (ed. M. L. Cody and J. M. Diamond), pp. 342–444. Belknap Press, Cambridge, Massachusetts.

—— (1978). Niche shifts and the rediscovery of interspecific competition. *Amer. Scientist* **66**, 322–31.

—— May, R. M. (1976). Island biogeography and the design of natural reserves. In *Theoretical ecology* (ed. R. M. May), pp. 163–86. Saunders, Philadelphia.

Dice, L. R. and Howard, W. E. (1951). Distances of dispersal by prairie deer mice from birthplaces to breeding sites. *Contribut. Lab. vertebrate Biol., Univ. Mich.* **50**, 1–15.

Dingle, H. (1972). Migration strategies of insects. *Science* **175**, 1327–35.

—— (ed.) (1978). *Evolution of insect migration and diapause.* Springer-Verlag, New York.

—— (1979). Adaptive variation in the evolution of insect migration. In *Movement of highly mobile insects* (ed. R. L. Rabb and G. G. Kennedy), pp. 64–87. North Carolina State University.

Dixon, A. F. G. (1959). An experimental study of the searching behaviour of the predatory coccinelid beetle *Adalia decempunctata. J. animal Ecol.* **28**,

259–81.

Dobson, F. S. (1979). An experimental study of dispersal in California ground squirrel. *Ecology* **60**, 1103–9.

Dobzhansky, T. (1950). Heredity, environment and evolution. *Science* **111**, 161–3.

—— (1965). "Wild" and "domestic" species of *Drosophila*. In *The genetics of colonizing species* (ed. H. G. Baker and G. L. Stebbins), pp. 533–46. Academic Press, New York.

—— (1970). *Genetics of the evolutionary process*. Columbia University Press, New York.

—— Ayala, F. J., Stebbins, G. L., and Valentine, J. W. (1977). *Evolution*. W. H. Freeman, San Francisco.

—— Wright, S. (1943). Genetics of natural populations X. Dispersion rates in *Drosophila pseudoobscura*. *Genetics* **28**, 304–40.

Dorst, J. (1962). *The migrations of birds*. Heinemann, London.

Doyle, R. W. (1975). Settlement of planktonic larvae, a theory of habitat selection in varying environments. *Amer. Naturalist* **109**, 113–26.

Draper, J. (1980). The direction of desert locust migration. *J. animal Ecol.* **49**, 959–74.

Dreyfus, A. and Breuer, M. E. (1944). Chromosomes and sex determination in the parasitic hymenopteran *Telenomus fariai* (Lima). *Genetics* **29**, 75–82.

Dry, W. W. and Taylor, L. R. (1970). Light and temperature thresholds for take-off by aphids. *J. animal Ecol.* **39**, 493–504.

Eanes, W. F. (1979). The Monarch butterfly as a paradigm of genetic structure in a highly dispersive species. *Movement of highly mobile insects* (ed. R. L. Rabb and G. G. Kennedy), pp. 88–102. North Carolina State University Press.

Egli, O. (1950). Centre d'observation pour les migrations de papillons, Zurich. *Circulaire* **24**, 3.

Ehrlich, P. R. and Raven, P. H. (1969). Differentiation of populations. *Science* **165**, 1228–32.

—— White, R. R., Singer, M. C., McKechnie, S. W., and Gilbert, L. E. (1975). Checkerspot butterflies: a historical perspective. *Science* **188**, 221–8.

Ehrman, L. (1970). The mating advantage of rare males in *Drosophila*. *Proc. nat. Acad. Sci. US* **65**, 345–8.

Eisenberg, J. F. (1966). The social organization of mammals. *Handb. Zool.* **10**, 1–92.

Ekblom, T. (1941). Untersuchungen über den Flügeldimorphismus bei *Gerris asper* L. *Nottingham Entomol.* **21**, 49–64.

Elgood, J. H., Fry, C. H., and Dowsett, R. J. (1973). African migrants in Nigeria. *Ibis* **115**, 1–45, 375–411.

Elliott, J. P. and Cowan, I. McT. (1978). Territoriality, density and prey of the lion in Ngorongoro Crater, Tanzania. *Can. J. Zool.* **56**, 1726–34.

Elliott, L. (1978). Social behavior and foraging ecology of the eastern chipmunk (*Tamias striatus*) in the Adirondack Mountains. *Smithsonian Contrib. Zool.*, no. 265, 1–107.

Ellis, P. E. and Ashall, C. (1957). Field studies on diurnal behaviour, movement and aggregation in the desert locust. *Anti-Locust Bull*, **25**, pp. 94.

Elton, C. S. (1925). The dispersal of insects to Spitzbergen. *Trans. entomol. Soc. Lond.* **1925**, 289–99.

—— (1942). *Voles, mice and lemmings: problems of population dynamics*. Oxford University Press, London.

Emlen, J. M. (1966). The role of time and energy in food preference. *Amer. Naturalist* **100**, 611–17.

—— (1973). *Ecology: an evolutionary approach*. Addison-Wesley, Reading, Massachusetts.

Emlen, S. T. and Oring, L. W. (1977). Ecology, sexual selection, and the evolution of mating systems. *Science* **197**, 215–23.

Endler, J. A. (1977). *Geographic variation, speciation and clines*. Princeton University Press, New Jersey.

—— (1979). Gene flow and life history patterns. *Genetics* **93**, 263–84.

Errington, P. L. (1946). Predation and vertebrate populations. *Quart. Rev. Biol.* **21**, 144–77, and 221–45.

—— (1963). *Muskrat populations*. Iowa State University Press, Ames, Iowa.

Estes, R. D. (1969). Territorial behaviour of the wildebeest (*Connochaetes taurinus* Burchell 1823). *Z. Tierpsychol.* **26**, 284–370.

—— (1976). The significance of breeding synchrony in the wildebeest. *East African Wildlife J.* **14**, 135–52.

Fairbairn, D. J. (1977). The spring decline in deer mice: death or dispersal? *Can. J. Zool.* **55**, 84–92.

—— (1978*a*). Dispersal of deer mice. *Peromyscus maniculatus:* Proximal causes and effects on fitness. *Oecologia* **32**, 171–93.

—— (1978*b*). Behaviour of dispersing deer mice (*Peromyscus maniculatus*). *Behav. Ecol. Sociobiol.* **3**, 265–82.

Falconer, D. S. (1960). *Introduction to quantitative genetics*. Oliver and Boyd, London.

Farrow, R. A. (1974). Comparative plague dynamics of tropical *Locusta* (Orthoptera: Acrididae). *Bull. entomol. Res.* **64**, 401–11.

Felsenburg, T. (1974). M.Sc. dissertation. The Hebrew University of Jerusalem, Jerusalem, Israel.

Felt, E. P. (1925). Dispersal of butterflies and other insects. *Nature, Lond.* **116**, 365–8.

—— (1926). The physical basis of insect drift. *Nature, Lond.* **117**, 754–5.

—— (1928). Dispersal of insects by air currents. *NY State Museum Bull.* **274**, 59–129.

Fincham, J. R. S. (1972). Heterozygous advantage as a likely general basis for enzyme polymorphism. *Heredity* **28**, 387–91.

Findley, J. S. (1951). Habitat preferences of four species of *Microtus* in Jackson Hole, Wyoming. *J. Mammal.* **32**, 118–20.

—— (1954). Competition as a possible limiting factor in the distribution of *Microtus*. *Ecology* **35**, 418–20.

Fisher, H. I. (1975). Mortality and survival in the Laysan Albatross *Diomedea immutabilis*. *Pacific Sci* **29**, 279–300.

—— (1976). Some dynamics of a breeding colony of Laysan Albatrosses. *Wilson Bull.* **88**, 121–42.

Fisher, R. A. (1922). On the dominance ratio. *Proc. R. Soc. Edinburgh* **42**, 321–41.

—— (1930). *The genetical theory of natural selection*. Clarendon Press, Oxford.

—— (1931). The evolution of dominance. *Biol. Rev.* **6**, 345–68.

Fitch, H. S. and Shirer, H. W. (1970). A radiotelemetric study of spatial relationships in the oppossum. *Amer. Midland Naturalist* **84**, 170–86.

Fitzgerald, B. M. (1977). Weasel predation on a cyclic population of the montane vole (*Microtus montanus*) in California. *J. animal Ecol.* **46**, 367–97.

Flanders, S. E. (1947). Elements of host discovery exemplified by parasitic hymenoptera. *Ecology* 28, 299–309.

Fleming, T. H. (1979). Life-history strategies. In *Ecology of small mammals* (ed. D. M. Stoddart). Chapman and Hall, London.

Fletcher, T. B. (1925). Migration as a factor in pest-outbreaks. *Bull. entomol. Res.* 16, 177–81.

Fogden, M. P. L. (1972). The seasonality and population dynamics of equatorial forest birds in Sarawak. *Ibis* 114, 307–43.

Ford, E. B. (1964), *Ecological genetics*. Methuen, London.

Forester, D. C. (1977). Comments on the female reproductive cycle and philopatry by *Desmognathus ochrophaeus* (Amphibia, Urodela, Plethodontidae). *J. Herpetol.* 11, 60–80.

Frame, L. H., Malcolm, J. R., Frame, G. W., and Lawick, H. van (1979). Social organisation of African wild dogs (Lycaon pictus) on the Seregeti Plains, Tanzania 1967–1978. *Z. Tierpsychol.* 50, 225–49.

Frank, F. (1956). Grundlagen, Möglichkeiten und Metoden der Sanieurung von Feldmansplagegebieten. *Nachrichtenbl. Deutsch. Pflantzenschutzd.* 8, 147–58.

Fredga, K., Gropp, A., Winking, H., and Frank, F. (1976). Fertile XX- and XY-type females in the wood lemming, *Myopus schisticolor*. *Nature* 261, 225–7.

—— Gropp, A., Winking, H., and Frank, F. (1977). A hypothesis explaining the exceptional sex ratio on the wood lemming (*Myopus schisticolor*). *Hereditas* 85, 102–4.

Freeman, J. A. (1945). Studies in the distribution of insects by aerial currents. The insect population of the air from ground level to 300 feet. *J. animal Ecol.* 14, 128–54.

French, N. R., Grant, W. R., Grodzinski, W., and Swift, D. M. (1976). Small mammal energetics in grassland ecosystems. *Ecol. Monogr.* 46, 201–20.

French, R. A. (1969). Migration of *Laphygma exigua* Hübner (Lepidoptera: Noctuidae) to the British Isles in relation to large-scale weather systems. *J. animal Ecol.* 38, 199–210.

Fretwell, S. D. (1972). *Populations in seasonal environment*. Princeton University Press, New Jersey.

—— (1978). The evolution of migration in relation to the factors regulating bird numbers. In *Migratory birds in the tropics* (ed. E. S. Morgan and A. Keast). Smithsonian Press, Washington, DC.

—— Lucas, H. L. (1970). On territorial behavior and other factors influencing habitat distribution in birds. I. Theoretical development. *Acta Biotheoret.* 19 16–36.

Fredriksson, S. (1975). *Surtsey. Evolution of life on a volcanic island*. Butterworths, London.

Frith, H. J. (1967). *Waterfowl in Australia*. Angus and Robertson, Sydney.

Fritts, T. H. (Unpublished) Morphometrics of Galapagos tortoises: evolutionary implications.

Furuya, Y. (1969). On the fission of troops of Japanese monkeys: 2. General view of troop fission in Japanese monkeys. *Primates* 10, 47–69.

Fuzeau-Braesch, S. (1961). Variations dans la longeur des ailes en fonction de l'effect de groupe chez quelques espéces de Gryllides. *Bull. Soc. Zoologique France* 86, 785–8.

Gadgil, M. (1971). Dispersal: population consequences and evolution. *Ecology* 52, 253–61.

Gaines, M. S. (1981). Importance of genetics to population dynamics. In *Mam-*

malian population genetics (ed. M. H. Smith and J. Joule), pp. 1–27. University of Georgia Press.

— — Krebs, C. J. (1971). Genetic changes in fluctuating vole populations. *Evolution* **25**, 702–23.

— — and McClenaghan, Jr., L. R. (1980). Dispersal in small mammals. *Ann. Rev. Ecol. Systemat.* **11**, 163–96.

— — Baker, C. L., and Vivas, A. M. (1979*a*). Demographic attributes of dispersing southern bog lemming (*Synaptomys cooperi*) in Eastern Kansas. Oecologia **40**, 91–101.

— — Vivas, A. M., and Baker, C. L. (1979*b*). An experimental analysis of dispersal in fluctuating vole populations: demographic parameters. *Ecology* **60**, 814–29.

Garten, C. T., Jr. (1976). Relationships between behavior, genetic heterozygosity and population dynamics in the oldfield mouse, *Peromyscus polionotus. Evolution* **30**, 59–72.

— — (1977). Relationships between exploratory behavior and genetic heterozygosity in the oldfield mouse. *Animal Behav.* **25**, 328–32.

— — Smith, M. H. (1974). Movement by old field mice and population regulation. *Acta theriologica* **19**, 513–14.

Gauthreaux, S. A., Jr. (1978). The ecological significance of behavioural dominance. In *Perspectives in ethology* (ed. P. P. G. Bateson and P. H. Klopfer), Vol. 3, pp. 17–54. Plenum Press, London.

Gee, J. M. and Williams, G. B. (1965). Self and cross-fertilization in *Spirorbis borealis* and *S. pagenstecheri. J. marine Biol. Assoc.* **45**, 275–85.

Getz, L. L. (1962). Aggressive behaviour of the meadow and prairie voles. *J. Mammal.* **43**, 351–8.

— — (1978). Speculations on social structure and population cycles of microtine rodents. *The Biologist* **60**, 134–47.

— —Carter, C. S. (1980). Social organization in *Microtus ochrogaster* populations. *The Biologist* **62**, 56–69.

— — Verner, L., Cole, F. R., Hofmann, J. E., and Avalos, D. E. (1979). Comparisons of population demography of *Microtus orchrogaster* and *M. pennyslvanicus. Acta theriologica* **24**, 319–49.

Ghiselin, M. T. (1974). *The economy of nature and the evolution of sex*. University of California Press, Berkeley, California.

Ghouri, A. S. K. and McFarlane, J. E. (1958). Occurrence of a macropterous form of *Gryllodes sigillatus* (Walker) (Orthoptera: Gryllidae) in laboratory culture. *Can. J. Zool.* **36**, 837–8.

Gilboa, A. (1976). M.Sc. dissertation. Department of Zoology, Hebrew University, Jerusalem, Israel.

Gileva, E. A. and Chebotar, N. A. (1979). Fertile XO males and females in the varying lemming, *Dicrostonyx torquatus* Pall. (1779): A unique genetic system of sex determination. *Heredity* **42**, 67–77.

Gill, D. E. (1974). Intrinsic rate of increase, saturation density and competitive ability. II. The evolution of competitive stability. *Amer. Naturalist* **108**, 103–16.

— — (1978*a*). On selection at high population density. *Ecology* **59**, 1289–91.

— — (1978*b*). The metapopulation ecology of the red-spotted newt, *Notophthalmus viridescens* (Rafinesque). *Ecol. Monogr.* **48**, 145–66.

Gill, F. B. and Wolf, L. R. (1975*a*). Economics of feeding territoriality in the golden-winged sunbird. *Ecology* **56**, 333–45.

— — — — (1975*b*). Foraging strategies and energetics of East African sunbirds at mistletoe flowers. *Amer. Naturalist* **109**, 491–510.

Gillett, S. D., Hogarth, P. J., and Noble, F. E. J. (1979). The response of predators to varying densities of *gregaria* locust nympths. *Animal Behav.* **27**, 592–6.

Gilpin, M. E. (1975). *Group selection in predator–prey communities.* Princeton University Press, New Jersey.

Gipps, J. H. W. and Jewell, P. A. (1979). Maintaining populations of bank vole, *Clethrionomys glareolus*, in large outdoor enclosure, and measuring the response of population variables to the castration of males. *J. animal Ecol.* **48**, 535–56.

Glick, P. A. (1939). The distribution of insects, spiders and mites in the air. *Tech. Bull. US Dept, Agricult.,* no. 673, 150 pp.

Gliwicz, J. (1980). Island populations of rodents: their organization and functioning. *Biol. Rev.* **55**, 109–38.

—— (1981). Competitive interactions within a forest rodent community in Central Poland. *Oikos* **37**, 353–62.

Godfrey, J. (1958). The origin of sexual isolation between bank voles. *Proc. R. phys. Soc. Edinburgh* **27**, 47–55.

Golley, F. (1960). Energy dynamics of a food chain of an oldfield community. *Ecol. Monogr.* **30**, 187–206.

Gould, J. L. (1980). The case for magnetic sensitivity in birds and bees (such as it is). *Amer. Scientist* **68**, 256–67.

Grant, P. R. (1969). Experimental studies of competitive interaction in a two-species system. I. *Microtus* and *Clethrionomys* species in enclosures. *Can. J. Zool.* **47**, 1059–82.

—— (1970). Experimental studies of competitive interactions in a two-species system. II. The behaviour of *Microtus. Peromyscus* and *Clethrionomys* species. *Animal Behav.* **18**, 411–26.

—— (1971a). Experimental studies of competitive interactions in a two-species system. III. *Microtus* and *Peromyscus* species in enclosures. *J. animal Ecol.* **40**, 323–50.

—— (1971b). The habitat preference of *Microtus pennsylvanicus*, and its relevance to the distribution of this species on islands. *J. Mammal.* **52**, 351–61.

—— (1972). Interspesific competition among rodents. *Ann. Rev. Ecol. Systemat.* **3**, 79–106.

—— (1975). Population performance of *Microtus pennsylvanicus* confined to woodland habitat occupancy. *Can. J. Zool.* **53**, 1147–65.

—— (1978a). Competition between species of small mammals. In *Populations of small mammals under natural conditions* (ed. D. P. Snyder). Special Publ. Ser. Pymatuning Lab. Ecol., University of Pittsburg.

—— (1978b). Dispersal in relation to carrying capacity. *Proc. Nat. Acad. Sci. USA* **75**, 2854–8.

—— Grant, B. R., Smith, J. N. M., Abbott, I. J., and Abbott, L. K. (1976). Darwin's finches: population variation and natural selection. *Proc. Natl. Acad. Sci. USA* **73**, 257–61.

Grant, V. (1963). *The origin of adaptations.* Columbia University Press, New York.

—— (1971). *Plant speciation.* Columbia University Press, New York.

—— (1975). *Genetics of flowering plants.* Columbia University Press, New York.

Greenbank, D. O. (1957). The role of climate and dispersal in the initiation of outbreaks of the spruce budworm in New Brunswick. II. The role of dispersal. *Can. J. Zool.* **35**, 385–403.

—— (1963). The analysis of moth survival and dispersal in the unsprayed area. In *The dynamics of epidemic spruce budworm populations* (ed. R. F. Morris), pp. 87–99. *Memoirs entomol. Soc. Can.* **31**.

—— Schaeffer, G. W., and Rainey, R. C. (1979). Spruce budworm (Lepidoptera: Tortricidae) moth flight and dispersal: new understanding from canopy observations, radar, and aircraft. *Memoirs entomol. Soc. Can.* **110**, 1–49.

Greenberg, L. (1979). Genetic component of bee odor in kin recognition. *Science* **206**, 1095–7.

Greenwood, J. J. D. (1974). Effective population numbers in the snail *Cepaea nemoralis. Evolution* **28**, 513–26.

Greenwood, P. J. (1980). Mating systems, philopatry and dispersal in birds and mammals. *Animal Behav.* **28**, 1140–62.

—— Harvey, P. H. (1976). The adaptive significance of variation in breeding area fidelity of the blackbird (*Turdus merula* L.). *J. animal Ecol.* **45**, 887–98.

—— —— (1977). Feeding strategies and dispersal of territorial passerines: a comparative study of the blackbird, *Turdus merula*, and the greenfinch, *Carduelis chloris. Ibis* **119**, 528–31.

—— —— Perrins, C. M. (1978). Inbreeding and dispersal in the great tit. *Nature Lond.* **271**, 52–4.

—— —— —— (1979a). The role of dispersal in the great tit (*Parus major*): the causes, consequences and heritability of natal dispersal *J. animal Ecol.* **48**, 123–42.

—— —— —— (1979b). Mate selection in the great tit *Parus major* in relation to age, status and natal dispersal *Ornis Fenn.* **56**, 75–86.

—— —— —— (1979c) Kin selection and territoriality in birds? A test. *Animal Behav.* **27**, 645–51.

Griffing, B. (1976). Selection in reference to biological groups. VI. Use of extreme forms of nonrandom groups to increase selection efficiency. *Genetics* **82**, 723–31.

Grime, J. P. (1979). Competition and the struggle for existence. In *Population dynamics* (ed. R. M. Anderson, B. D. Turner, and L. R. Taylor), pp. 123–39. Blackwell Scientific Publications, Oxford.

Grimes, L. G. (1976). Co-operative breeding in African birds. *Proc. 16th Int. Ornithol. Congr.*, Canberra, pp. 667–73.

Grinnell, J. (1904). The origin and distribution of the chestnut-backed chickadee. *Auk* **21**, 364–82.

—— (1922). The role of the "accidental". *Auk* **34**, 373–80.

Gurney, W. S. C. and Nisbet, R. M. (1978). Single-species population fluctuations in patchy environments. *Amer. Naturalist* **112**, 1075–90.

Hagen, D. W. (1967). Isolating mechanisms in three-spine sticklebacks (Gasterosteus). *J. Fisheries Res. B. Can.* **24**, 1637–92.

—— (1973). Inheritance of numbers of lateral plates and gill rakers in *Gasterosteus aculeatus. Heredity* **30**, 303–12.

—— Gilbertson, L. G. (1973). Selective predation and the intensity of selection acting upon the lateral plates of three-spine sticklebacks. *Heredity* **30**, 273–87.

—— McPhail, J. D. (1970). The species problem within *Gasterosteus aculeatus* on the Pacific coast of North America *J. Fisheries Res B. Can.* **27**, 147–55.

Hairston, N. G., Tinkle, D. W., and Wilbur, H. M. (1970). Natural selection and the parameters of population growth. *J. wildlife Management* **34**, 681–90.

Haldane, J. B. S. (1932). *The causes of evolution*. Longmans Green, London.

Halgren, L. A. and Taylor, L. R. (1968). Factors affecting flight responses of alienicolae of *Aphis fabae* Scop. and *Schizaphis graminum* Rondani (Homoptera:Aphididae). *J. animal Ecol.* **37**, 583–93.

Halliburton, R. and Mewaldt, L. R. (1976). Survival and mobility in a population of Pacific coast song sparrows (*Melospiza melodia gouldii*). *Condor* **78**, 499–504.

Halliday, T. R. (1978). Sexual selection and mate choice. In *Behavioural ecology* (ed. J. R. Krebs and N. B. Davies), pp. 180–213. Blackwells, Oxford.

Hamilton, W. D. (1964). The genetical evolution of social behaviour. I, II. *J. theoret. Biol.* **7**, 1–52.

—— (1967). Extraordinary sex ratios. *Science* **156**, 477–88.

—— (1971). Selection of selfish and altruistic behaviour in some extreme models. In: *Man and beast : comparative social behaviour* (ed. J. P. Eisenberg and W. S. Dillon), pp. 57–91. Smithsonian, Washington.

—— (1972). Altruism and related phenomena, mainly in social insects. *Ann. Rev. Ecol. Systemat.* **3**, 193–232.

—— (1975). Innate social aptitudes of man: an approach from evolutionary genetics. In *Biosocial anthropology* (ed. R. Fox). Malaby Press, London.

—— R. M. May. (1977). Dispersal in stable habitats. *Nature* **269**, 578–81.

Hamrick, J. L., Linhart, K. B., and Mitton, J. B. (1979). Relationships between life history characteristics and electrophorectically detectable genetic variation in plants. *Ann. Rev. Ecol. Systemat.* **10**, 173–200.

Hanby, J. P. and Bygott, J. D. (1979). Population changes in lions and other predators. In *Serengeti: dynamics of an ecosystem* (ed. A. R. E. Sinclair and M. Norton-Griffiths), pp. 249–62. University of Chicago Press.

Hansen, T. A. (1978). Larval dispersal and species longevity in lower tertiary gastropods. *Science* **199**, 885–7.

Hanski, I. (1982). Dynamics of regional distribution: the core and satellite species hypothesis. *Oikos* **38**, 210–21.

Hansson, L. (1977*a*). Spatial dynamics of field voles, *Microtus agrestis*, in heterogeneous landscapes. *Oikos* **29**, 539–44.

—— (1977*b*). Landscape ecology and stability of populations. *Landscape Planning* **4**, 85–93.

—— (1979*a*). On the importance of landscape heterogeneity in northern regions for the breeding population densities of homeotherms: a general hypothesis. *Oikos* **33**, 182–9.

—— (1979*b*). Food as a limiting factor for small rodent numbers. Tests of two hypotheses. *Oecologia* **37**, 297–314.

Harden-Jones, F. R. (1968). *Fish migration*. Edward Arnold, London.

Hardy, A. C. and Milne, P. S. (1938). Studies in the distribution of insects by aerial currents. Experiments in aerial tow-netting from kites. *J. animal Ecol.* **7**, 199–229.

Harestad, A. S. and Bunnell, F. L. (1979). Home range and body weight – a re-evaluation. *Ecology* **60** 389–402.

Harris, M. P. (1970). Abnormal migration and hybridization of *Larus argentatus* and *L. fuscas* after interspecies fostering experiments. *Ibis* **112**, 488–498.

Harrison, R. G. (1979). Flight polymorphisms in the field cricket *Gryllus pennsylvanicus Oecologia* **40**, 125–32.

—— (1980), Dispersal polymorphism in insects. *Ann. rev. ecol. System.* **11**, 95–118.

Hartnoll, R. G. (1976). Reproductive strategy in two British species of *Alcyon-*

ium. In *Biology of benthic organisms* (ed. B. F. Keegan, P. O. Ceidigh, and P. J. S. Boaden), pp. 321—8. Pergamon Press, Oxford.

Harvey, P. H. and Clutton-Brock, T. H. (1981). Primate home-range size and metabolic needs. *Behav. Ecol. Sociobiol.* **8**, 151—5.

—— Greenwood, P. J. and Perrins, C. M. (1979). Breeding area fidelity of the great tit (*Parus major*). *J. animal Ecol.* **48**, 305—13.

—— Mace, G. M. (1982). Comparisons between taxa and adaptive trends: problems of methodology. In *Current Problems in sociobiology* (ed. Kings College Sociobiology Group). Cambridge University Press.

Hasler, D. A., Scholz, A. T., and Horrall, R. M. (1978). Olfactory imprinting and homing in salmon. *Amer. Scientist* **66**, 347—55.

Hasler, J. F. and Banks, E. M. (1975). Reproductive performance and growth in captive collard lemmings (*Dicrostonye groenlandicus*). *Can. J. Zool.* **53**, 777—87.

Hassell, M. P. (1969). A population model for the interaction between *Cyzenis albicans* (Fall.) (Tachinidae) and *Operophtera brumata* (L.) (Geometridae) at Wytham, Berkshire. *J. animal Ecol.* **38**, 567—76.

—— (1971). Mutual interference between searching insect parasites. *J. animal Ecol.* **40** 473—86.

—— (1980). Some consequences of habitat heterogeneity for population dynamics. *Oikos* **35**, 150—60.

—— R. M. May. (1973). Stability in insect host-parasite models. *J. animal Ecol.* **42**, 693—736.

—— —— (1974). Aggregation in predators and insect parasites and its effect on stability. *J. animal Ecol.* **43**, 567—9.

Hastings, A. (1977). Spatial heterogeneity and the stability of predator—prey systems. *Theoret. pop. Biol.* **12**, 37—48.

—— (1978). Spatial heterogeneity and the stability of predator—prey systems: predator-mediated coexistence. *Theoret. pop. Biol.* **14**, 380—95.

Hay, D. E. and McPhail, J. D. (1975). Mate selection in three-spine sticklebacks (*Gasterosteus*). *Can. J. Zool.* **53**, 441—50.

Healey, M. C. (1967). Aggression and self-regulation of population size in deer mice. *Ecology* **48**, 377—92.

Heape, W. (1931). *Emigration, migration and nomadism.* Heffer, Cambridge.

Henson, W. R. (1951). Mass flights of the spruce budworm. *Can. Entomol.* **83**, 240.

Henttonen, H., Kaikusalo, A., Tast, J., and Viitala, J. (1977). Interspecific competition between small mammals in subarctic and boreal ecosystems. *Oikos* **29**, 581—90.

Hilborn, R. (1975). Similarities in dispersal tendency among siblings in four species of voles (*Microtus*). *Ecology* **56**, 1221—5.

—— Krebs, C. J. (1976). Fates of disappearing individuals in fluctuating populations of *Microtus townsendii*. *Can. J. Zool.* **54**, 1507—8.

Hilden, O. (1965). Habitat selection in birds. *Ann. Zool. Fenn.* **2**, 53—75.

Hill, M. G. (1980). Wind dispersal of the coccid *Icerya Seychellarum* (Margarodidae: Homoptera) on Aldabra Atoll. *J. animal Ecol.* **49**, 939-57.

Hill, J. L. (1974). Peromyscus: effect of early pairing on reproduction. *Science* **186**, 1042—4.

Hinde, R. A. (1956). The biological significance of the territories of birds. *Ibis* **98**, 340—69.

Hinton, H. E. (1977). Enabling mechanisms. *Proc. 15 Int. Cong. Entomol.*, Washington DC 1976, pp. 71—83.

Hirth, H. F. (1971). Synopsis of biological data on the green turtle, *Chelonia mydas* (Linnaeus) 1758. *FAO Fisheries Synopsis* **85**.

Hoff, J. M. (1980). Unpublished thesis, University of Oslo (in Norwegian). *Heterogeneitet i tid og rom; implikasjoner for populasjoners utvikling.*

Hoffmeyer, I. (1973). Interaction and habitat selection in the mice *Apodemus flavicollis* and *A. sylvaticus. Oikos* **24**, 108–16.

Holling, C. S. (1959). Some characteristics of simple types of predation and parasitism. *Can. Entomol.* **91**, 385–98.

Horn, H. S. (1978). Optimal tactics of reproduction and life-history. In *Behavioural ecology: an evolutionary approach* (ed. J. R. Krebs and N. B. Davies), pp. 411–29. Blackwells, Oxford.

–– (1981). Sociobiology. In *Theoretical ecology: principles and applications*, (2nd ed). (ed. R. M. May) Blackwells, Oxford.

–– MacArthur, R. H. (1972). Competition among fugitive species in a harlequin environment. *Ecology* **53**, 749–52.

Howard, W. E. (1960). Innate and environmental dispersal of vertebrates. *Amer. Midland Naturalist* **63**, 152–61.

–– (1965). Interaction of behaviour, ecology and genetics of introduced mammals. In *The genetics of colonizing species* (ed. H. G. Baker and G. L. Stebbins). Academic Press, New York.

Hoyt, D. F. and Taylor, C. R. (1981). Gait and the energetics of locomotion in horses. *Nature* **292**, 239–40.

Hrdy, S. B. (1977). *The Langurs of Abu.* Harvard University Press, Cambridge, Massachusetts.

Hubbard, S. F. and Cook, R. M. (1978). Optimal foraging by parasitoid wasps. *J. animal Ecol.* **47**, 593–604.

Huffaker, C. B. (1958). Experimental studies of predation: dispersion factors and predator–prey oscillations. *Hilgardia* **27**, 343–83.

Ibbotson, A. and Kennedy, J. S. (1951). Aggregation in *Aphis fabae* Scop. 1. Aggregation on plants. *Ann. appl. Biol.* **38**, 65–78.

–– –– (1959). Interaction between walking and probing in *Aphis fabae* Scop. *J. exp. Biol.* **36**, 377–90.

Immelmann, K. (1975). Ecological significance of imprinting and early learning *Ann. Rev. Ecol. Systemat.* **6**, 15–37.

Itani, J. (1972). A preliminary essay on the relationship between social organization and incest avoidance in non-human primates. In *Primate socialization* (ed. F. E. Poirer), pp. 165–71. Random House, New York.

Ivlev, V. S. (1960). On the utilization of food by plankton phage fishes. *Bull. math. Biophys.* **22**, 371–85.

Jacquard, A. (1975). Inbreeding: one word, several meanings. *Theoret. pop. Biol.* **7**, 338–63.

Jarman, P. J. and Sinclair, A. R. E. (1979). Feeding strategy and the pattern of resource partitioning in ungulates. In *Serengeti: dynamics of an ecosystem.* (ed. A. R. E. Sinclair and M. Norton-Griffiths), pp. 130–63. University of Chicago Press.

Järvinen, O. and Vēpsälainen, K. (1976). Wing dimorphism as an adaptive strategy in water striders (Gerris). *Hereditas* **84**, 61–8.

Jeannel R. G. (1960). *Introduction to entomology* (English edition translated by H. Oldroyd). Hutchinson, London.

Jenkins, P. F. (1977). Cultural transmission of song patterns and dialect development in a free-living bird population. *Animal Behav.* **25**, 50–78.

Jewell, P. A. (1966). The concept of home range in mammals. *Symp. zool. Soc.*

Lond. **18**, 85–109.

Johnson, B. (1958). Factors affecting the locomotor and settling responses of alate aphids. *Animal Behav.* **6**, 9–26.

Johnson C. G. (1957). The distribution of insects in the air and the empirical relation of density to height. *J. animal Ecol.* **26**, 479–94.

—— (1960). A basis for a general system of insect migration and dispersal by flight. *Nature, Lond.* **186**, 348–50.

—— (1961). Aphid dispersal and its bearing on the general character of insect migration by flight. *Animal Behav.* **9**, 233–4.

—— (1962). A functional approach to insect migration and dispersal and its bearing on future study. XI. *Internationaler Kongress für Entomologie, Wien* 1960, pp. 50–3.

—— (1963). Physiological factors in insect migration by flight. *Nature, Lond.* **198**, 423–7.

—— (1965). Migration. In *The physiology of insecta*, Vol. 2 (ed. K. D. Roeder), pp. 187–226. Academic Press, New York.

—— (1966). A functional system of adaptive dispersal by flight. *Ann. Rev. Entomol.* **11**, 233–60.

—— (1969). *The migration and dispersal of insects by flight.* Methuen, London.

—— (1976). Lability of the flight systems: a context for functional adaptation. *R. Entomol. Soc. Symp.* **7**, 217–34.

—— Taylor, L. R. (1957). Periodism and energy summation with special reference to flight rhythms in aphids. *J. exp. Biol.* **34**, 209–21.

Johnson, W. W. (1974). Coadaptation and recessive lethal content in DDT resistant populations of *Drosophila melanogaster*. *Evolution* **28**, 251–8.

Johnston, R. F. (1961). Population movements of birds. *Condor* **63**, 386–9.

Jones, R. E. (1977). Movement patterns and egg distribution in cabbage butterflies *J. animal Ecol.* **46**, 195–212.

—— Gilbert, N., Guppy, M., and Nealis, V. (1980). Long-distance movement of *Pieris rapae* L. *J. animal Ecol.* **49**, 629–42.

Jonkel, C. J. and McCowan, I. McT. (1971). The black bear in the spruce pine forest. *Wildlife Monogr.* **27**, 5–57.

Joule, J. and Cameron, G. N. (1975). Species removal studies. I. Dispersal strategies of sympatric *Sigmodon hispidus* and *Reithrodontomys fulvescens* populations. *J. Mammal.* **56**, 378–96.

Kaddou, I. K. (1960). The feeding behavior of *Hippodamia quinauesignata* (Kirby) larvae. *University of California Publications in Entomology* 16: 181–228.

Kalela, O. (1961). Seasonal change of habitat in the Norwegian lemming, *Lemmus lemmus* (L.). *Ann. Acad. Sci. Fenn., Ser. A, IV, Biol.* **55**, 1–72.

—— Kilpeläinen, L., Koponen, T., and Tast, J. (1971). Seasonal differences in habitats of the Norwegian lemming, *Lemmus lemmus* (L.), in 1959 and 1960 at Kilpisjärvi, Finnish Lapland. *Ann. Acad. Scient. Fennicae, Ser. A. II Biol.* **178**, 1–22.

—— Koponen, T. (1971). Food consumption and movement of the Norwegian lemming in areas characterized by isolated fells. *Ann. Zool. Fenn.* **8**, 80–4.

—— Oksala, T. (1966). Sex ratio in the wood lemming, *Myopus schisticolor* (Lilljeb.), in nature and captivity. *Ann. Univ. Turk. II. Biol.–Geo.* **37**, 5–24.

Kamil, A. C. (1978). Systematic foraging for nectar by Amakihi, *Loxops virens*. *J. Comp. Physiol. Psychol.* **92**, 388–96.

Kaufman, J. H. (1962). Ecology and behaviour of the coati *Nasua narica* on Barro Colorado Island, Panama. *Univ. Cal. Publ. Zool.* **60**, 95–222.

Keith, L. B. and Windberg, L. A. (1978). A demographic analysis of the snow-shoe hare cycle. *Wildlife Monogr.* **58**, 1–70.

Kempton, R. A., Lowe, H. J. B., and Bintcliffe, E. J. B. (1980). The relationship between fecundity and adult weight in *Myzus persicae. J. animal Ecol.*, **49**, 917–26.

Kendall, M. G. and Buckland, W. R. (1957). *A dictionary of statistical terms* Oliver and Boyd, Edinburgh.

Kennedy, J. S. (1951). The migration of the desert locust (*Schistocerca gregaria* Forsk.) I. The behaviour of swarms. II. A theory of long-range migrations. *Phil. Trans. R. Soc. Lond.* (B), **235**, 163–290.

– – (1956). Phase transformation in locust biology. *Bio. Rev.* **31**, 349–70.

– – (1961*a*) A turning point in the study of insect migration. *Nature, Lond.* **189**, 785–91.

– – (1961*b*). Continuous polymorphism in locusts. In *Insect polymorphisms* (ed. J. S. Kennedy), pp. 11–19 Symp. R. entomol. Soc. Lond. **1**, 11–19.

– – (1966*a*). Some outstanding questions in insect behaviour. In *Insect behaviour* (ed. P. T. Haskell), pp. 97–112. Symp. R. entomol. Soc. London.

– – (1966*b*) The balance between antagonistic induction and depression of flight activity in *Aphis fabae* Scopoli. *J. Exp Biol.* **45**, 215–28.

– – (1969). *The relevance of animal behaviour.* Imperial College, London.

– – (1972). The emergence of behaviour. *J. Austral. entomol Soc.* **11**, 168–76.

– – (1975). Insect dispersal. In *Insects, science and society* (ed. D. Pimentel), pp. 103–19. Academic Press, New York.

– – Crawley, L. (1967). Spaced-out gregariousness in sycamore aphids *Drepanosiphum platanoides* (Schrank) (Hemiptera, Callaphididae). *J. animal Ecol.* **36**, 147–70.

– – Way, M. J. (1979). Summing up the conference. In *Movement of highly mobile insects* (ed. R. L. Rabb and G. G. Kennedy), pp. 446–56. North Carolina State University.

Kenney, A. M., Evans, R. L., and Dewsbury, D. A. (1977). Post-implantation pregnancy disruption in *Microtus ochrogaster, M. pennsylvanicus,* and *Peromyscus maniculatus. J. Reproduct. Fertility* **49**, 365–7.

Kenyon, K. W. (1960). Territorial behavior and homing in the Alaska fur seal. *Mammalia* **24**, 431–44.

Kermack, K. A. and Haldane, J. B. S. (1950). Organic correlation and allometry. *Biometrika* **37**, 30–41.

Kerster, J. W. (1964). Neighborhood size in the rusty lizard, *Sceloporus olivaceus. Evolution* **18**, 445–57.

Ketterson, E. D. and Nolan, V. (1976). Geographic variation and its climatic correlates in the sex ratio of eastern-wintering dark-eyed juncos (Junco hyemalis hyemalis). *Ecology.*

Kettlewell, H. B. D. (1952). A possible genetic explanation and understanding of migration of continuous brooded insects. *Nature, Lond.* **169**, 832–3.

Kiester, A. R. (1979). Conspecifics as cues: a mechanism for habitat selection in the Panamanian grass anole (*Anolis auratus*). *Behav. Ecol. Sociobiol.* **5**, 323–30.

Kimura, M. and Crow, J. F. (1963). The measurement of effective population number. *Evolution* **17**, 279–88.

– – Ohta, T. (1971). *Theoretical aspects of population genetics.* Princeton University Press, Princeton, New Jersey.

King, C. E. and Anderson, W. W. (1971). Age-specific selection. II. The interaction between r and K during population growth. *Amer. Naturalist* **105**,

137–55.

King J. A. (1955). Social behaviour, social organisation and population dynamics in a black tailed prairie dogtown in the Black Hills of South Dakota. *Contri. Lab. vertebrate Biol., Univ. Mich.* **67**, 1–123.

King, J. C. (1955). Evidence for the integration of the gene pool from studies of DDT resistance in *Drosophila. Cold Spring Harbor Symp. quant. Biol.* **20**, 311–17.

King J. R. (1974). Seasonal allocation of time and energy resources in birds. In *Avian energetics* (ed. R. A. Paynter). Publ. Nuttall Ornithological Club, no. 15.

— — Farner, D. S. (1961). Energy metabolism, thermal regulation and body temperature. In *Biology and comparative physiology of birds*, vol. 2 (ed. A. L. Marshall). Academic Press, New York.

Kleiber, M. (1961). *The fire of life.* John Wiley, New York.

Knight-Jones, E. W. (1953). Decreased discrimination during settling after prolonged planktomic larvae of *Sprirobis borealis* (Serpulidae). *J. Marine Biol. Assoc.* UK **32**, 337–45.

Knowlton, N. and Parker, G. A. (1979). An evolutionarily stable strategy approach to indiscriminate spite. *Nature* **279**, 419–21.

Koenig, W. D. and Pitelka, F. A. (1979). Relatedness and inbreeding avoidance: counterplays in the communally nesting acorn woodpecker. *Science* **206**, 1103–5.

Kojima, E. (1971). Is there a constant fitness value? No! *Evolution* **25**, 281–5.

Koponen, T., Kokkonen, A., and Kalela, O. (1961). On a case of spring migration in the Norwegian lemming. *Ann. Acad. Sci. Fenn., Ser. A. IV Biol.* **52**, 1–30.

Koshinka, T. V. (1962). Migrations of *Lemmus lemmus. Zool. Zh.* **41**, 1859–74. (In Russian, cited from Batzli 1975).

Kowalski, K. (1977). Fossil lemmings (*Mammalia, Rodentia*) from the Pliocene and early Pleistocene of Poland. *Acta Zool. Cracov* **22**, 297–317.

Kozakiewicz, M. (1976). Migratory tendencies in a population of bank voles and a description of migrants. *Acta Theriologica* **21**, 321–38.

Krebs, C. J. (1964). The lemming cycle of Baker Lake, North-west Territories, during 1959–62. *Arctic Inst. N. Amer. Tech. Paper*, no. 15.

— — (1966). Demographic changes in fluctuating populations of *Microtus californicus. Ecol. Monogr.* **36**, 239–73.

— — (1978*a*). A review of the Chitty Hypothesis of population regulation. *Can. J. Zool.* **56**, 2463–80.

— — (1978*b*). Aggression, dispersal, and cyclic changes in populations of small mammals. In *Aggression, dominance and individual spacing* (ed. L. Kramer, P. Pliner, and T. Alloway), pp. 49–60. Plenum, New York.

— — (1979). Dispersal, spacing behaviour, and genetics in relation to population fluctuations in the vole *Microtus townsendii. Fortschr. Zool.* **25**, 61–77.

— — Boonstra, R. (1978). Demography of the spring decline in populations of the vole *Microtus townsendii. J. animal Ecol.* **46**, 1007– 15.

— — DeLong, K. T. (1965). A *Microtus* population with supplementary food. *J. Mammal.* **46**, 566–73.

— — Gaines, M. S., Keller, B. L., Myers, J. H., and Tamarin, R. H. (1973). Population cycles in small rodents. *Science* **179**, 35–41.

— — Keller, B. L., and Tamarin, R. H. (1969). *Microtus* population biology: demographic changes in fluctuating populations of *M. ochrogaster* and *M. pennsylvanicus* in southern Indiana. *Ecology* **50**, 587–607.

—— Myers, J. H. (1974). Population cycles in small mammals. *Advan. ecol. Res.* 8, 268–399.

—— Redfield, J. A., and Taitt, M. J. (1978). A pulsed-removal experiment on the vole *Microtus townsendii. Can. J. Zool.* 56, 2253– 62.

—— Wingate, J., LeDuc. J., Redfield, J. A., Taitt, M., and Hilborn, R. (1976). *Microtus* population biology: dispersal in fluctuating populations of M. townsendii. *Can. J. Zool.* 54, 79–95.

Krebs, J. R. (1978). Optimal foraging: decision rules for predators. In *Behavioural ecology: an evolutionary approach* (ed. J. R. Krebs and N. B. Davies. Blackwells, Oxford.

—— (1980). Optimal foraging, predation risk and territory defence. *Ardea* 80, 83–90.

—— Cowie, R. J. (1976). Foraging strategies in birds. *Ardea* 64, 98– 116.

—— Ryan, J. C., and Charnov, E. L. (1974). Hunting by expectation or optimal foraging? A study of patch use by chickadees. *Animal Behav.* 22, 953–64.

Lack, D. (1954). *The natural regulation of animal numbers.* Clarendon Press, Oxford.

—— (1958). The evolution of reproductive rates. In *Evolution as a process* (ed. J. Huxley, A. C. Hardy, and E. B. Ford) pp. 143–156. George Allen and Unwin, London.

—— (1966) *Population studies of birds.* Oxford University Press.

—— (1968) *Ecological adaptations for breeding in birds.* Methuen, London.

—— Lack, E. (1951). Migration of insects and birds through a Pyrenean pass. *J. animal Ecol.* 20, 63–7.

Laing, J. (1937). Host-finding by insect parasites. I. Observations on the finding of hosts by *Alysia manducator, Mormoniella vitripennis* and *Trichogramma evanescens. J. animal Ecol.* 6, 298–317.

—— (1938). Host-finding by insect parasites. II. The chance of *Trichogramma evanescens* finding its hosts. *J. exp. Biol.* 15, 281–302.

Lakhani, K. H. and Dempster, J. P. (1981). Cinnabar moth and its food plant, ragwort: further analysis of a simple interaction model. *J. animal Ecol.* 50 231–49.

Lasieweki, R. C. and Dawson, W. R. (1967). A re-examination of the relation between standard metabolic rate and body weight in birds. *Condor* 69, 13–23.

Lavie, B. and Ritte, U. (1978). The relation between dispersal behavior and reproductive fitness in the flour beetle, *Tribolium castaneum. Can. J. genet. Cytol* 20, 589–95.

—— —— (1980). Correlated effects of the response to conditioned medium in the flour beetle *Tribolium castaneum. Res. popul.* Ecol. 21, 228–32.

Lawlor, L. R. (1978). A comment on randomly constructed model ecosystems. *Amer. Naturalist* 112, 445–7.

—— Maynard Smith, J. (1976). The coevolution and stability of competing species. *Amer. Naturalist* 110, 79–99.

Lawton, J. H. and Hassel, M. P. (1981). Asymmetrical competition in insects. *Nature, Lond,* 289, 793–5.

—— McNeill, S. (1979). Between the devil and the deep blue sea: on the problem of being a herbivore. In *Population dynamics* (ed. R. M. Anderson, B. D. Turner, and L. R. Taylor), pp. 223–44. Blackwell Scientific Publications, Oxford.

Lees, A. D. (1956). The physiology and biochemistry of dispause. *Ann Rev. Entomol.* 1, 1–16.

—— The control of polymorphism in aphids. *Advan. insect Physiol.* **3**, 207—77.

—— (1967). The production of apterous and alate forms in the aphid *Megoura vuciae* Buckton, with special reference to the role of crowding. *J. Insect Physiol.* **13**, 289—318.

—— (1975). Aphid polymorphism and 'Darwin's demon'. *Proc. R. entomol. Soc. (Lond.)* **39**, 59—64.

Leggett, W. C. (1977). The ecology of fish migrations. *Ann. Rev. Ecol. Systemat.* **8**, 285—308.

Leigh, E. G., Jr. (1975). Population fluctuations, community stability, and environmental variability. In *Ecology and evolution of communities* (ed. M. L. Cody and J. M. Diamond), pp. 51—73. Belknap Press, Cambridge, Massachusetts.

Lerner, I. M. (1954). *Genetic homeostasis.* Oliver and Boyd, Edinburgh.

Leuthold, W. (1977). *African ungulates.* Springer-Verlag, Berlin.

—— Sale, J. B. (1973). Movements and patterns of habitat utilization of elephants in Tsavo National Park, Kenya. *E. afr. wildlife J.* **11**, 369—84.

Levin, B. R., Petras, M. L., and Rasmussen, D. I. (1969). The effect of migration on the maintenance of a lethal polymorphism in the house mouse. *Amer. Naturalist* **103**, 647—61.

Levin, D. A. and Kerster, H. W. (1974). Gene flow in seed plants. *Evol. Biology* **7**, 139—220.

—— —— Niedzlek, M. (1971). Pollinator flight directionality and its effect on pollen flow. *Evolution* **25**, 113—18.

Levin, S. A. (1974). Dispersion and population interactions. *Amer. Naturalist* **108**, 207—28.

—— (1976). Population dynamics models in heterogeneous environments. *Ann. Rev. Ecol. Systemat.* **7**, 287—310.

Levins, R. (1968). *Evolution in changing environments.* Princeton University Press, New Jersey.

—— (1969). The effect of random variations of different types on population growth. *Proc. Nat. Acad. Sci.* USA **62**, 1061—5.

—— (1970). Extinction. In *Some mathematical questions in biology* (ed. M. L. Gerstenhaber,) Vol. 2, pp. 75—107. (Series 'Lectures on mathematics in the life sciences'.) American Mathematical Society, Providence, Rhode Island.

—— MacArthur, R. (1966). The maintenance of genetic polymorphism in a spatially heterogeneous environment: variations on a theme by Howard Levene. *Amer. Naturalist* **100**, 585—9.

Lewontin, R. C. (1965). Selection for colonizing ability. In *The genetics of colonizing species* (ed. H. G. Baker and G. L. Stebbins), pp. 74—94. Academic Press, New York.

—— Cohen, D. (1969). On population growth in randomly varying environment. *Proc. Nat. Acad. Sci.* USA **62**, 1056—60.

Li, K. -P., Wong, H. -H., and Woo, W. -S. (1964). Route of the seasonal migration of the Oriental armyworm in the eastern part of China as indicated by a three-year result of releasing and recapturing marked moths. (In Chinese — with summary in English). *Acta phytophyl. Sinica* **3**, 101—10; *Rev. appl. Entomol.* [A] **53**, 391—2. (1965)

Lidicker, W. Z. Jr. (1962). Emigration as a possible mechanism permitting the regulation of population density below carrying capacity. *Amer. Naturalist* **96**, 29—33.

—— (1965). Comparative study of density regulation in confined population of four species of rodents. *Res. pop. Ecol.* **7**, 57—72.

—— (1973). Regulation of numbers in an island population of the California vole; a problem in community dynamics. *Ecol. Monogr.* **43**, 271–302.

—— (1975). The role of dispersal in the demography of small mammals. In *Small mammals: productivity and dynamics of populations.* (ed. K. Petrusewicz, E. B. Golley, and L. Ryszkowski,) pp. 103–28. Cambridge University Press, London.

—— (1978). Regulation of numbers in small mammal populations – historical reflections and a synthesis. In *Populations of small mammals under natural conditions* (ed. D. P. Snyder), pp. 122–41. Special Publ. Ser. Pygmatuning Lab. Ecol., University of Pittsburgh, Pennsylvania.

—— Anderson, P. K. (1962). Colonization of an island by *Microtus californicus,* analysed on the basis of runway transects. *J. animal Ecol.* **31**, 502–17.

Lincoln, F. C. (1952). *Migration of birds.* Doubleday, Garden City, New York.

Lomnicki, A. (1978). Individual differences between animals and the natural regulation of their numbers. *J. animal Ecol.* **47**, 461–75.

—— (1980). Regulation of population density due to individual differences and patchy environments. *Oikos* **35**, 185–93.

Longwell, A. C. and Stiles, S. S. (1973). Gamete cross incompatibility and inbreeding in the commercial American oyster, *Crassostrea virginica* Gmelin. *Cytologia* **38**, 521–33.

MacArthur, R. H. (1959). On the breeding distribution of North American migrants. *Auk* **76**, 218–25.

—— (1972). *Geographical ecology: patterns in the distribution of species.* Harper and Row, New York.

—— Pianka, E. R. (1966). On the optimal use of a patchy environment. *Amer. Naturalist* **100**, 603–9.

—— Wilson, E. O. (1967). *The theory of island biogeography.* Princeton University Press, New Jersey.

Macaulay, E. D. M. (1972). Flight activity of *Plusia gamma* in the laboratory. *Entomologia Experimentalis Applicata* **15**, 387–91.

Mace, G. M. (1979). The evolutionary ecology of small mammals. D. Phil. thesis, University of Sussex, UK.

MacKay, T. F. C. and Doyle, R. W. (1978). An ecological genetic analysis of the settling behaviour of a marine polychaete. I. Probability of settlement and gregarious behaviour. *Heredity* **40**, 1–12.

Maclean, S. F., Jr., Fitzgerald, B. M., and Pitelka, F. A. (1974). Population cycles in arctic lemmings: winter reproduction and predation by weasels. *Arctic Alpine Res.* **6**, 1–12.

MacRoberts, M. H. and MacRoberts, B. R. (1974). Social organization of acorn woodpeckers *Melanerpes formicivorus. Emu* **74**, 310.

Maddock, L. (1979). The "migration" and grazing succession. In *Serengeti: dynamics of an ecosystem* (ed. A. R. E. Sinclair and M. Norton-Griffiths), pp. 104–29. University of Chicago Press.

Mallory, F. F. and Brooks, R. J. (1978). Infanticide and other reproductive strategies in the collared lemming, *Dicrostonyx groenlandicus. Nature* **273**, 144–6.

Manning, J. T. (1976a). Gamete dimorphism and the cost of sexual reproduction: are they separate phenomena? *J. theoret. Biol.* **55**, 393–5.

—— (1976b). Is sex maintained to facilitate or minimize mutational advance? *Heredity* **36**, 351–7.

Manton, S. M. (1977). *The Arthropoda: habits, functional morphology, and evolution.* Clarendon Press, Oxford.

Martin, R. D. (1981). Field studies of primate behaviour. *Symp. Zool. Soc. Lond.* **46**, 287–336.

Masaki, S. and Oyama, N. (1963). Photoperiodic control of growth and wing form on *Nemobius yezoensis* Shiraki. *Kontyu* **31**, 16–26.

Marsden, W. (1964). *The lemming year.* Chatto and Windus, London.

Massey, D. R. (1977). Spatial–temporal changes in the genetic composition of deer mouse populations (*Peromyscus maniculatus*). MA thesis, University of Colorado at Denver (cited from Gaines 1981).

Mathad, S. B. and McFarlane, J. E. (1968). Two effects of photoperiod on wing development in *Gryllodes sigillatus* (Walk). *Can. J. Zool.* **46**, 57–60.

Mather, K. (1943). Polygenic inheritance and natural selection. *Biol. Rev.* **18**, 32–64.

May, R. M. (1973). *Stability and complexity in model ecosystems.* Princeton University Press, Princeton.

—— (1976). Models for single populations. In *Theoretical ecology* (ed. R. M. May), pp. 4–25. Saunders, Philadelphia.

—— (1978). The dynamics and diversity of insect faunas. *Diversity of insect faunas* (ed. L. A. Mound and N. Waloff), pp. 188–204. R. Entomol. Soc. Symp., no. 9. Blackwells, Oxford.

—— Oster, G. F. (1976). Bifurcations and dynamic complexity in simple ecological models. *Amer. Naturalist* **110**, 573–99.

Maynard Smith, J. (1964). Kin selection and group selection. *Nature* **201**, 1145–7.

—— (1966). Sympatric speciation. *Amer. Naturalist* **100**, 637–50.

—— (1968). *Mathematical ideas in biology.* Cambridge University Press.

—— (1972). Game theory and the evolution of fighting. In *On evolution* (ed. J. Maynard Smith). Edinburgh University Press.

—— (1974*a*). *Models in ecology.* Cambridge University Press.

—— (1974*b*). The theory of games and the evolution of animal conflicts. *J. theoret. Biol.* **47**, 209–21.

—— (1976*a*) Group selection. *Quart. Rev. Biol.* **51**, 277–83.

—— (1976*b*). Evolution and the theory of games. *Amer. Scientist,* **64**, 41–5.

—— (1978). *The evolution of sex.* Cambridge University Press.

—— (1979). Game theory and the evolution of behaviour. *Proc. R. Soc. Lond.* **B205**, 475–88.

—— Parker, G. A. (1976). The logic of asymmetric contests. *Animal Behav.* **24**, 159–75.

—— Price, G. R. (1973). The logic of animal conflicts. *Nature* **246**, 15–18.

—— Stenseth, N. C. (1978). On the evolutionary stability of female biased sex ratio in the wood lemming (*Myopus schisticolor*); the effect of inbreeding. *Heredity* **41**, 205–14.

Mayr, E. (1942). *Systematics and the origin of species.* Columbia University Press, New York.

—— (1963). *Animal species and evolution.* Belknap Press, Cambridge, Massachusetts.

—— (1965). Summary. In *The genetics of colonizing species* (ed. H. G. Baker and G. L. Stebbins), pp. 553–62. Academic Press, New York.

—— (1975). The unity of genotype. *Biol. Zbl.* **94**, 377–88.

—— (1976). *Evolution and the diversity of life.* Belknap Press, Cambridge, Massachusetts.

McCart, P. J. (1970). A polymorphic population of *Oncorhynchus nerka* in Babine Lake, BC. Ph.D. Thesis, University of British Columbia.

McClenaghan, L. R., Jr. and Gaines, M. S. (1976). Density-dependent dispersal in *Sigmodon*: a critique. *J. Mammal.* **57**, 758–9.

McFarlane, J. E. (1962). Effect of diet and temperature on wing development of *Gryllodes sigillatus* (Walk) (Orthoptera: Gryllidae), *Ann. Soc. Entomol. Quebec* **7**, 28–33.

McNab, B. K. (1963). Bioenergetics and the determination of home range size. *Amer. Naturalist* **97**, 133–40.

Mech, L. D. (1970). *The wolf*. Natural History Press, Garden City, New York.

Menge, B. A. (1975). Brood or broadcast? The adaptive significance of different reproductive strategies in the two intertidal sea stars, *Leptasterias hexactis* and *Pisaster ochraceus*. *Marine Biol.* **31**, 87–100.

Merrell, D. J. (1970). Migration and gene dispersal in *Rana pipiens*. *Amer. Zool.* **10**, 47–52.

Metzgar, L. H. (1967). An experimental comparison of screech owl predation on resident and transient white-footed mice (*Peromyscus leucopus*). *J. Mammal.* **48**, 387–91.

Michener, G. R. and Michener, D. R. (1977). Population structure and dispersal in Richardson's ground squirrels. *Ecology* **58**, 359–68.

Mihok, S. (1981). Chitty's hypothesis and behaviour in subarctic red-backed voles (*Clethrionomys gapperi*). *Oikos* **36**, 281–95.

Mikkola, K. (1970). The interpretation of long-range migrations of *Spodoptera exigua* Hb. (Lepidoptera:Noctuidae). *J. animal Ecol.* **39**, 593–8.

Milinski, M. and Heller, R. (1978). The influence of a predator on the optimal foraging behaviour of sticklebacks (*Gasterosteus aculeatus* L.). *Nature* **275**, 642–4.

Miller, A. H. (1947). Panmixia and population size with reference to birds. *Evolution* **1**, 186–90.

Milne, A. (1961). Definition of competition among animals. *Symp. Soc. exp. Biol.* **15**, 40–61.

Milne, H. and Robertson, F. W. (1965). Polymorphism in egg albumen protein and behaviour in the eider duck. *Nature, Lond.* **205**, 367–9.

Milton, K. and May, M. L. (1976). Body weight, diet and home range area in primates. *Nature* **259**, 459–62.

Mineau, P. and Cooke, F. (1979). Territoriality in snow geese or the protection of parenthood—Ryder's and Inglis's hypotheses reassessed. *Wildfowl* **30**. 16–19.

Mitchell, B. (1963). Ecology of two carabid beetles, *Bembidion lampros* (Herbst) and *Trechus quadristiriatus* (Schrank). I. Life cycles and feeding behaviour. *J. animal Ecol.* **32**, 289–99.

Mitchell, R. (1970). An analysis of dispersal in mites. *Amer. Naturalist* **104**, 425–31.

Moodie, G. E. E. (1972). Predation, natural selection and adaptation in an unusual three-spine stickleback. *Heredity* **28**, 155–67.

Moreau, R. E. (1966). *The bird faunas of Africa and its islands*. Academic Press, New York.

—— (1972). *The Palaearctic–African bird migration systems*. Academic Press, New York.

Morris, R. D. and Grant, P. R. (1972). Experimental studies of competitive interaction in a two-species system. IV. *Microtus* and *Clethrionomys* species in a single enclosure. *J. animal Ecol.* **41**, 275– 90.

Morrison, J. P. E. (1963). Notes on American Siphonaria. *Bull. Amer. Malac.*

Union (1963), pp. 7—9.

Mueller, H. C., Berger, D. D., and Allez, G. (1977). The periodic invasions of goshawks. *Auk* **94**, 652—63.

Murie, J. O. and Harris, M. A. (1978). Territoriality and dominance in male columbian ground squirrels (*Spermophilus columbianus*). *Can. J. Zool.* **56**, 2402—12.

Murray, B. G., Jr. (1967). Dispersal in vertebrates. *Ecology* **48**, 975—8.

Myers, J. H. and Krebs, C. J. (1971). Genetic, behavioural and reproductive attributes of dispersing field voles. *Microtus pennsylvanicus* and *Microtus ochrogaster*. *Ecol. Monogr.* **41**, 53—78.

Myllymäki, A. (1974). Outbreaks and damage by the field vole, *Microtus agrestis* (L.), in Finland 1945—1973. *EPPO Publ. Ser. C*, pp. 31—24.

—— (1977*a*). Demographic mechanisms in the fluctuating populations of the field vole, *Microtus agrestis*. *Oikos* **29**, 468—93.

—— (1977*b*). Interactions between the field vole, *Microtus agrestis*, and its microtine competitors in Central-Scandinavian populations. *Oikos* **29**, 570—80.

—— (1977*c*). Intraspecific competition and home range dynamics in the field vole, *Microtus agrestis*. *Oikos* **29**, 553—69.

—— Aho, J., Lind, E., and Tast, J. (1962). Behaviour and daily activity of the Norwegian lemming, *Lemmus lemmus* (L.) during autumn migration. *Ann. Zool., Soc. Zool.—Bot. Fenn. Vanamo* **24**, 1—31.

Myrberget, S. (1973). Geographic synchronism of cycles of small rodents in Norway. *Oikos* **24**, 220—4.

Mysterud, I. (1961). Bestandsvariasjoner hos skoglemen. *Fauna* **14**, 106—14. (In Norwegian with English summary).

Nash, L. T. (1976). Troop fission in free-ranging baboons in the Gombe Stream National Park. *Amer. J. phys. Anthropol.* **44**, 63—77.

Naumov, N. P. (1972). *The ecology of animals*. University of Illinois Press.

Neilsen, E. T. (1961). On the habits of the migratory butterfly *Ascia monuste* L. *Biologiske Meddelelser udgivet af Det Kongelige Danske Videnskabernes Selskab* **23**, 1—81.

Newton, I. (1970). Irruptions of crossbills in Europe. In *Animal populations in relation to their food resources* (ed. A. Watson). Blackwells, Oxford.

—— (1972). *Finches*. Collins, London.

—— (1977). Timing and success of breeding in tundra-nesting geese. In *Evolutionary ecology* (ed. B. Stonehouse and C. Perrins). MacMillan, London.

—— (1979). *Population ecology of raptors*. Poyser, Berkhamsted, England.

Nice, M. M. (1937). Studies in the life history of the song sparrow. *Trans. Linn. Soc. NY* **4**, 1—247.

—— (1964). *Studies in the life history of the song sparrow.*, Vol. I. A population study of the song sparrow and other passerines. Dover, New York.

Nix, H. A. (1976). Environmental control of breeding, post-breeding dispersal, and migration of birds in the Australian region. *Proc. 16th Int. ornithol. Cong.* Canberra, pp. 272—306.

Nordeng, H. (1977). A pheromone hypothesis for homeward migration in anadromous salmonids. *Oikos* **28**, 155—9.

Norris, K. S. (1967). Some observations on the migration and orientation of marine mammals. In *Animal orientation and navigation* (ed. R. M. Storm). Oregon State Univ. Press, Corvallis, Oregon.

Northcote, T. G. (1969). Patterns and mechanisms in the lakeward migratory behaviour of juvenile trout. In *Symposium on salmon and trout in streams* (ed. T. G. Northcote), pp. 180–203. University of British Columbia.

Nottebohm, F. (1969). The song of the chingolo, *Zonotrichia capensis,* in Argentina: description and evaluation of a system of dialects. *Condor* 71, 229–315.

Okubo, A. (1980). *Diffusion and ecological problems: mathematical models.* Springer-Verlag, Berlin.

Oliver, C. G. (1979). Genetic differentiation and hybrid viability within and between some Lepidoptera species. *Amer. Naturalist* 114, 681–94. .

Orians, G. H. (1969). On the evolution of mating systems in birds and mammals. *Amer. Naturalist* 103, 589–603.

Orr, R. T. (1970). *Animals in migration.* MacMillan, London.

Packer, C. R. (1979). Inter-troop transfer and inbreeding avoidance in *Papio anubis. Animal Behav.* 27, 1–36.

Palenzona, D. L., Allicchio, R., and Rocchetta, G. (1975). Interaction between artificial and natural selection. *Theoret. appl. Genet.* 46, 223–38.

— — Mochi, M., and Boschieri, E. (1974). Investigation of the founder effect. *Genetica* 45, 1–10.

Palmer, R. S. (1962). *Handbook of North American birds. V. I. Loons through flamingos.* Yale University Press, New Haven, Connecticutt.

Parker, G. A. (1978). Searching for mates. In *Behavioural ecology, an evolutionary approach* (ed. J. R. Krebs and N. B. Davies), pp. 214–44. Blackwell Scientific Publications, Oxford.

— — Stuart, R. A. (1976). Animal behaviour as a strategy optimizer: evolution of resource assessment strategies and optimal emigration thresholds. *Amer. Naturalist* 110, 1055–76.

Parker, G. R. (1972). Biology of the Kaminuriak population of barren-ground caribou. Pt. 1. Total numbers, mortality, recruitment, and seasonal distribution. *Can. Wildlife Serv. Rep.,* Ser. no. 20.

Pashtan, A. and Ritte, U. (1978). Genetic variability in colonizer and non-colonizer *Cerithium* species (Gastropoda: Cerithidae). *2nd Int. Congr. Ecol. INTECOL,* Jerusalem, 1978. Abstracts volume, p. 286.

Pearson, K. and Blakeman, J. (1906). Mathematical contributions to the theory of evolution. XV. A mathematical theory of random migration. *Drapers Company Res. Memoirs, Biomed. Ser.* 1, 1–54.

Pearson, O. P. (1966). The prey of carnivores during one cycle of mouse abundance. *J. animal Ecol.* 35, 217–33.

— — (1971). Additional measurements of the impact of carnivores on California voles (*Microtus californicus*). *J. Mammal.* 52, 41–9.

Pennycuick, L. (1975). Movements of the migratory wildebeest population in the Serengeti area between 1960 and 1973. *East African Wildlife J.* 13, 65–87.

Perdeck, A. C. (1958). Two types of orientation in migrating starlings *Sturnus vulgaris* L. and chaffinches *Fringilla coelebs* L. as revealed by displacement. *Ardea* 46, 1–37.

— — (1964). An experiment on the ending of autumn migration in starlings. *Ardea* 52, 133–9.

Perrins, C. M. (1956). Population studies of the great tit, *Parus major. Proc. Advan. Study Inst.* Dynamics, Numbers Populations, (Oosterbeck, 1970), pp. 524–31.

— — (1970). The timing of birds' breeding seasons. *Ibis* 112, 242–55.

Petrusewicz, K. (1957). Investigation of experimentally induced population

growth. *Ekolgia Polska, Ser. A* **5**, 281–301.
— — (1963). Population growth induced by disturbance in the ecological structure of the population. *Ekologia Polska, Ser. A* **11**, 87–125.
— — (1966). Dynamics, organization and ecological structure of populations. *Ekologia Polska, Ser. A* **14**, 413–36.
Pianka, E. R. (1970). On *r*- and *K*- selection. *Amer. Naturalist* **104**, 592–7.
— — Parker, W. S. (1975). Age-specific reproductive tactics. *Amer. Naturalist* **109**, 453–64.
Pickering, J., Getz, L. L., and Whitt, G. S. (1974). An esterase phenotype correlated with dispersal in *Microtus. Trans. I. State Acad. Sci* **67**, 471–5.
Pielou, E. C. (1969). *An Introduction to mathematical ecology.* John Wiley, New York.
— — (1976). *Population and community ecology.* Gordon and Breach, New York.
— — (1979). *Biogeography.* Wiley and Sons, New York.
Pimm, S. L. and Rosenzweig, M. L. (1981). Competitors and habitat use. *Oikos* **37**, 1–6.
Pitelka, F. A. (1973). Cyclic pattern in lemming populations near Barrow, Alaska. In *Alaskan arctic tundra* (ed. M. E. Britton), pp. 199–215. Arctic Inst. N. Amer., Tech. Paper, no. 25.
Podoler, H. and Rogers, D. (1975). A new method for the identification of key factors from life-table data. *J. animal Ecol.* **44**, 85–114.
Poisson, R. (1924). Contribution a l'etude des Hemipteres aquatiques. *Bull. Biologique Francais Belg.* **58**, 49–305.
Por, F. D. (1978). Lessepsian migration. The influx of Red Sea biota into the Mediterranean by way of the Suez Canal. *Ecological studies*, vol. 23. Springer-Verlag, Berlin.
Porter, W. P., Mitchell, J. W., Beckman, W. A., and Tracy, C. R. (1975). Environmental constraints on some predator–prey interactions. In *Perspectives of biophysical ecology* (ed. D. M. Gates and R. B. Schmerl), Chapter 20, pp. 347–64. Springer-Verlag, Berlin.
Price, M. V. and Waser, N. M. (1979). Pollen dispersal and optimal out-crossing in *Delphinium nelsoni. Nature Lond.* **277**, 294–7.
Provost, M. W. (1953). Motives behind mosquito flights. *Mosquito News* **13**, 106–9.
— — (1960). The dispersal of *Aedes taeniorhynchus.* III. Study methods for migratory exodus. *Mosquito News* **20**, 148–61.
Pusey, A. (1979). Intercommunity transfer of chimpanzees in Gombe National Park. In *The great apes.* (ed. D. A. Hamburg and E. R. McCown), pp. 465–79. Benjamin/Cummings, Menlo Park, California.
Pyke, G. H. (1978*a*). Optimal foraging: movement patterns of bumblebees between inflorescences. *Theoret. pop. Biol.* **13**, 72–97.
— — (1978*b*). Are animals efficient harvesters? *Animal Behav.* **26**, 241– 50.
— — (1978*c*). Optimal foraging in hummingbirds: Testing the marginal value theorem. *Amer. Zool.* **18**, 739–52.
— — (1979). The economics of territory size and time budget in the golden-winged sunbird. *Amer. Naturalist* **114**, 131–45.
— — (1980). Optimal foraging in nectar-feeding animals and coevolution with their plants. In *Foraging behavior: ecological, ethological and psychological approaches* (ed. A. C. Kamil and J. D. Sargent). Garland, New York.
— — (1981). Foraging in honeyeaters: A test of optimal foraging theory. *Anim. Behav.* **29**, 878–88.
— — (1982). Foraging in bumblebees: Rule of departure from an inflorescence.

Can. J. Zool. (In press.)

—— Pulliam, H. R., and Charnov, E. L. (1977). Optimal foraging: a selective review of theory and tests. *Quart. Rev. Biol.* **52**, 137—54.

Rabb, R. L. and Kennedy, G. G. (ed.) 1979. *Movement of highly mobile insects.* North Carolina State University.

Rainey, R. C. (1951). Weather and the movements of locust swarms: a new hypothesis. *Nature, Lond.* **168**, 1057—60.

—— (1963). Meteorology and the migration of desert locusts. Applications of synoptic meteorology in locust control. *Anti-Locust Memoir,* no. 7, 115 pp.

—— (1976). Flight behaviour and features of the atmospheric environment. *R. entomol. Soc. Symp.* **7**, 75—112.

—— (1978). The evolution and ecology of flight: The 'oceanographic' approach. In *The evolution of insect migration and diapause* (ed. H. Dingle), pp. 33—48, Springer, New York.

—— (1979). Interactions between weather systems and populations of locusts and noctuids in Africa. In *Movement of highly mobile insects* (ed. R. L. Rabb and G. G. Kennedy), pp. 109—19. North Carolina State University.

Ralls, K. (1977). Sexual dimorphism in mammals : Avian models and unanswered questions. *Amer. Natur.* **111**, 917—38.

—— Brugger, K., and Ballou, J. (1979). Inbreeding and juvenile mortality in small populations of ungulates. *Science* **206**, 1101—3.

Ralph, C. J. and Pearson, C. A. (1971). Correlation of age, size of territory, plumage and breeding success in white crowned sparrows. *Condor* **73**, 77—80.

Rao, Y. R. (1960). The desert locust in India. Ind. Council Agricult. Res. New Dehli, 721 pp.

Rasmuson, B., Rasmuson, M., and Nygren, J. (1977). Genetically controlled differences between cycling and stable populations of field vole (*Microtus agrestis*). *Hereditas* **87**, 33—41.

Rasmussen, D. I. (1964). Blood group polymorphism and inbreeding in natural populations of the deer mouse, *Peromyscus maniculatus. Evolution* **18**, 219—29.

Raveling, D. G. (1976). Migration reversal : a regular phenomenon of Canada geese. *Science* **193**, 153—4.

Redfield, J. A., Taitt, M. J., and Krebs, C. J. (1978*a*). Experimental alterations of sex-ratios in populations of *Microtus oregoni,* the creeping vole. *J. animal Ecol.* **47**, 55-70.

—— —— —— (1978*b*). Experimental alterations of sex ratios in populations of *Microtus townsendii. Can. J. Zool.* **56**, 17—27.

Remington, C. L. (1968). The population genetics of insect introductions. *Ann. Rev. Entomol.* **13**, 415—26.

Richdale, L. E. (1957). *A population study of penguins.* Oxford University Press, London.

Richter-Dyn, N. and Goel, N. S. (1972). On the extinction of a colonizing species. *Theoret. Pop. Biol.* **3**, 406—33.

Ricklefs, R. E. and Cox, G. W. (1972). Taxon cycles in the West Indian avifauna. *Amer. Naturalist* **106**, 195—219.

Ridley, M. (1978). Paternal care. *Animal Behav.* **26**, 904—32.

Riley, J. R. and Reynolds, D. R. (1979). Radar-based studies of the migratory flight of grasshoppers in the middle Niger area of Mali. *Proc. R. Soc. Lond.* **B 204**, 67—82.

—— —— Farmery, M. J. (1981). Radar observations of *Spodoptera exempta* Kenya, March—April 1979. *Centre for Overseas Pest Research, miscellaneous*

Re. No. 54.

Ritte, U. and Lavie, B. (1977). The genetic basis of dispersal behaviour in the flour beetle *Tribolium castaneum. Can. J. genet. Cytol.* **19**, 717–22.

Rockwell, R. F. and Cooke, F. (1977). Gene flow and local adaptation in a colonially nesting dimorphic bird: the lesser snow goose (*Anser caerulescens caerulescens*). *Amer. Naturalist* **111**, 91–7.

Roff, D. A. (1974*a*). Spatial heterogeneity and persistence of populations. *Oecologia* **15**, 245–58.

—— (1974*b*). The analysis of a population model demonstrating the importance of dispersal in a heterogeneous environment. *Oecologia* **15**, 259–75.

—— (1975). Population stability and the evolution of dispersal in a heterogeneous environment. *Oecologia* **19**, 217–37.

Roffey, J. (1972). Tropical pest investigations. Radar studies of insects. PANS **18**, 303–8.

Rose, R. K. (1979). Levels of wounding in the meadow vole, *Microtus pennsylvanicus. J. Mammal.* **60**, 37–45.

—— Gaines, M. S. (1976). Levels of aggression in fluctuating populations of the prairie vole, *Microtus ochrogaster*, in eastern Kansas. *J. Mammal.* **57**, 43–57.

Rosenzweig, M. L. (1974). On the evolution of habitat selection. *Proc. 1st int. Congr. Ecol.* , pp. 401–404.

—— (1979). Optimal habitat selection in two-species competitive systems. *Fortschr. Zool.* **25**, 283–93.

—— (1981). A theory of habitat selection. *Ecology* **62**, 327–35.

—— Abramsky, Z. (1980). Microtine cycles: the role of habitat heterogeneity. *Oikos* **34**, 141–6.

Ross, R. (1905). An address on the logical basis of the sanitary policy of mosquito reduction. *Brit. med. J.* **1**, 1025–9.

—— (1911). *The prevention of malaria.* John Murray, London.

Roughgarden, J. (1972). Evolution of niche width. *Amer. Naturalist* **106**, 683–718.

—— (1977). Patchiness in the spatial distribution of a population caused by stochastic fluctuations in resources. *Oikos* **29**, 52–9.

—— (1979). *Theory of population genetics and evolutionary ecology: an introduction.* McMillan, New York.

Roitberg, B. D., Myers, J. H., and Frazer, B. D. (1979). The influence of predators on the movement of apterouspea aphids between plants. *J. animal Ecol.* **48**, 111–22.

Royama, T. (1977). Population persistence and density dependence. *Ecol. Monogr.* **47**, 1–35.

Ryman, N., Allendorf, F. W., and Stahl, G. (1979). Reproductive isolation with little genetic divergence in sympatric populations of brown trout (*Salmo trutta*). *Genetics* **92**, 247–62.

Ryszkowski, L., Goszczynski, J., and Truszkowski, J. (1973). Trophic relationships of the common vole in cultivated fields. *Acta theriologica* **18**, 125–65.

Sade, D. S. (1968). Inhibition of son–mother mating among free-ranging rhesus monkeys. *Scient. Psychoanalyst* **12**, 18–27.

Sadlier, R. M. F. S. (1965). The relationship between agonistic behaviour and population changes in the deer mouse, *Peromyscus maniculatus* (Wagner). *J. animal Ecol.* **34**, 331–52.

Saeki, H. (1966*a*). The effect of population density on the occurrence of the macropterous form in a cricket, *Scapsipedus aspersus* Walker (Orthoptera: Gryllidae). *Jap. J. Ecol.* **16**, 1–4.

— — (1966*b*). The effect of day-length on the occurrence of the macropterous form in a cricket *Scapispedus aspersus* Walker (Orthoptera: Gryllidae) *Jap. J. Ecol.* **16**, 49–52.

Safriel, U. N., Gilboa, A., and Felsenburg, T. (1980). Distribution of rocky intertidal mussels in the Red Sea coasts of Sinai, the Suez Canal and the Mediterranean coast of Israel, with special reference to recent colonizers. *J. Biogeog.* **7**, 39–62.

— — Lipkin, Y. (1975). Petterns of colonization of the eastern Mediterranean intertidal zone by Red Sea immigrants. *J. Ecol.* **63**, 61–3.

— — Ritte, U. (1980). Criteria for the identification of potential colonizers. *Biol. J. Linn. Soc.* **13**, 287–97.

Saila, S. B. and Shappy, R. A. (1963). Random movement and orientation in salmon migration. *J. Conseil Int. l.Exploration de la Mer* **28**, 410–17.

Salomonsen, F., (1955). The evolutionary significance of bird migration. *Dan. Bio. Medd.* **22**, 1–62.

Sanders, H. L. (1977). Evolutionary ecology and the deep sea benthos. In *The changing scenes in the natural sciences, 1776–1976* (ed. C. E. Goulden), pp. 223–43. Academy of Natural Sciences, Philadelphia.

Sayer, H. J. (1956). A photographic method for the study of insect migration. *Nature, Lond.* **177**, 226.

— — (1962). The desert locust and tropical convergence. *Nature, Lond.* **194**, 330–6.

Schaeffer, G. W. (1976). Radar observations of insect flight. In *Insect flight* (ed. R. C. Rainey), pp. 157–97. Royal Entomological Society, London.

Schaffer, W. M. and Tamarin, R. H. (1973). Changing reproductive rates and population cycles in lemmings and voles. *Evolution* **27**, 111–24.

Scheltema, R. S. (1971). The dispersal of the larvae of shoal-water benthic invertebrate species over long distances by ocean currents. *Eur. Marine Biol. Symp.* **4**, 7–28.

Schifferli, L. (1973). The effect of egg weight on the subsequent growth of nestling great tits *Parus major*. *Ibis* **115**, 549–88.

Schmidt-Nielsen, K. (1972). Locomotion: energy costs of swimming, flying and running. *Science* **177**, 222–8.

Schoener, T. W. (1968). Sizes of feeding territories among birds. *Ecology* **49**, 123–41.

— — (1969). Optimal size and specialisation in constant and fluctuating environments : an energy–time approach. *Brookhaven Symposia in Biology,* **22**, 103.

Schopf, T. J. M. and Gooch, J. L. (1977). Gene frequencies in a marine ectoproct: a cline in natural populations related to sea temperature. *Evolution* **25**, 286–9.

Schuster, R. H. (1976). Lekking behavior in Kafue lechwe. *Science* **192**, 1240–2.

Selander, R. K. (1970). Behavior and genetic variation in natural populations. *Amer. Zool.* **10**, 53–66.

— — Hudson, R. O. (1976). Animal population structure under close inbreeding: the land snail *Rumina* in southern France. *Amer. Naturalist* **110**, 695–718.

— — Kaufman, D. W. (1973). Self-fertilization and genetic population structure in a colonizing land snail. *Proc. Nat. Acad. Sci.* USA **70**, 1186–90.

— — Yang, S. Y. (1969). Protein polymorphism and genic heterozygosity in a wild population of the house mouse (*Mus musculus*). *Genetics* **63**, 653–7.

Sellier, R. (1954). Recherches sur la morphogenese et le polymorphisme alaires chez les orthopteres Gryllides. *Ann. Sci. Nat. 11th Ser.* **16**, pp. 595–740.

Selye, H. (1950). *The physiology and pathology of exposure to stress; a treatise based on the concepts of the General Adaptation – Syndrome and the diseases of adaptation.* Acta Inc., Montreal.

Serventy, D. L. (1971). Biology of desert birds. In *Avian biology* (ed. D. S. Farner and J. R. King), Vol. 1, pp. 287–339. Academic Press, New York.

Shaw, M. J. P. (1970). Effects of population density on alienicolae of *Aphis fabae* scop. II. The effects of crowding on the expression of migratory urge among alatae in the laboratory. *Ann. appl. Biol.* **65**, 205–12.

—— (1973). Effects of population density on alienicolae of *Aphis fabae* scop. IV. The expression of migratory urge among alatae in the field. *Ann. appl. Biol.* **74**, 1–7.

Sheppard, D. H., Klopfer, P. H., and Oelke, H. (1968). Habitat selection : differences in stereotypy between insular and continental birds. *Wilson Bull.* **80**, 452–7.

Sherman, P. W. (1977). Nepotism and the evolution of alarm calls. *Science* **197**, 1246–53.

—— (1979). The limits of ground squirrel nepotism. In *Sociobiology beyond nature/nurture?* (ed. G. W. Barlow and J. Silverbeg). Westview Press, Boulder, Colorado.

—— (1980). Reproductive competition and infanticide in Belding's ground squirrel and other animals. In *Natural selection and social behaviour : recent research and new theory* (ed. R. D. Alexander and D. W. Tinkle). Chiron Press, New York.

Shields, W. M. (1980). Ground squirrel alarm calls: nepotism or parental care? *Amer. Naturalist* **116**, 599–603.

—— (1982). *Philopatry, inbreeding and the evolution of sex.* State University of New York Press, Albany, New York.

Shine, R. (1978). Propagule size and parental care: The "safe harbor" hypothesis. *J. theoret. Biol* **75**, 417–24.

Simberloff, D. (1976a). Species turnover and equilibrium island biogeography. *Science* **194**, 572–8.

—— (1976b). Experimental zoogeography of islands: effects of island size. *Ecology* **57**, 629–48.

—— (1976c). Trophic structure determination and equilibrium in an arthropod community. *Ecology* **57**, 395–8.

—— (1978). Using island biogeographic distribution to determine if colonization is stochastic. *Amer. Naturalist* **112**, 713–26.

—— Wilson, E. O. (1969). Experimental zoogeography of islands: the colonization of empty islands. *Ecology* **50**, 278–96.

Sinclair, A. R. E. (1975). The resource limitation of trophic levels in tropical grassland ecosystems. *J. animal Ecol* **44**, 497–520.

—— (1977). *The African buffalo.* University of Chicago Press.

—— (1978). Factors affecting the food supply and breeding season of resident birds and movements of Palaearctic migrants in a tropical African savannah. *Ibis* **120**, 480–97.

—— (1979). Dynamics of the Serengeti ecosystems: process and pattern. In *Serengeti: dynamics of an ecosystem* (ed. A. R. E. Sinclair and M. Norton-Griffiths), pp. 1–30. University of Chicago Press.

—— (1982). The adaptations of African ungulates and their effects on community function. In *World ecosystems. Savannas.* (ed. F. Bourlière). Elsevier. (In press).

Skellam, J. G. (1951). Random dispersal in theoretical populations. *Biometrika*

38, 196–218.

Skertchly, S. B. J. (1879). Butterfly swarms. *Nature, Lond.* **20**, 266.

Slade, N. A. and Balph, D. F. (1974). Population ecology of Vinta ground squirrels. *Ecology* **55**, 989–1003.

Smith, J. N. M. and Sweatman, H. P. A. (1974). Food-searching behavior of titmice in patchy environments. *Ecology* **55**, 1216–32.

Smith, M. H., Baccus, R., Chesser, R. K., Johns, P. E., Manlove, M. N., Ryman, N., and Straney, D. O. (1980*a*). Concept of animal populations: a case study of temporal and spatial changes in gene frequency in white-tailed deer. Manuscript.

— — Chesser, R. K., Gaines, M. S., Stenseth, N. C., and Tamarin, R. H. (1980*b*). The evolution of dispersal in small mammals. (In preparation.)

— — Carmon, J. L., and Gentry, J. B. (1972). Pelage color polymorphism in *Peromyscus polionotus. J. Mammal.* **53**, 824–33.

— — Garten, C. T., Jr., and Ramsey, P. R. (1975). Genetic heterozygosity and population dynamics in small mammals. In *Isozymes IV, genetics and evolution* (ed. C. L. Markert). Academic Press, New York.

— — Manlove, M. N., and Joule, J. (1978). Spatial and temporal dynamics of the genetic organization of small mammal populations. In *Populations of small mammals under natural conditions* (ed. D. P. Snyder), pp. 99–113. Special Publ. Ser. Pymatuning Lab. Ecol., University of Pittsburg.

Smith, R. H. (1979). On selection for inbreeding in polygynous animals. *Heredity* **43**, 205–11.

Smith, R. W. and Whittaker, J. B. (1980). Factors affecting *Gastrophysa viridula* Degeer populations (Coleoptera:Chrysomelidae) in different habitats. *J. animal Ecol.* **49**, 537–48.

Smith, W. J. (1977). *The behavior of communicating.* Harvard University Press, Cambridge, Massachusetts.

Sokal, R. R. and Rohlf, F. J. (1969). *Biometry.* Freemans, San Francisco.

Solbreck, C. (1978). Migration, disapause and direct development as alternative life histories in a seed bug, *Neacoryphus bicrucis. Evolution of insect migration and diapause* (ed. H. Dingle), pp. 195–217. Springer-Verlag, New York.

Sorensen, A. E. (1978). Somatic polymorphism and seed dispersal. *Nature, Lond.* **276**, 174–6.

Soule, M. and Stewart, B. R. (1970). The niche variation 'hypothesis: a test and alternative. *Amer. Naturalist* **104**, 87–97.

Southwood, T. R. E. (1961). A hormonal theory of the mechanism of wing polymorphism in Heteroptera. *Proc. R. entomol. Soc. Lond.* **A36**, 63–6.

— — (1962). Migration of terrestrial arthropods in relation to habitat. *Biol. Rev.* **37**, 171–214.

— — (1971). The role and measurement of migration in the population system of an insect pest. *Trop. Sci.* **13**, 275–8.

— — (1977). Habitat, the templet for ecological strategies? *J. animal Ecol* **46**, 337–65.

— — (1978). Escape in space and time – concluding remarks. *Evolution of insect migration and diapause* (ed. H. Dingle), pp. 277–9. Springer-Verlag, New York.

— — (1981) Ecological aspects of insect migration. In *Animal migration* (ed. D. J. Aidley), pp. 196–208. Cambridge University Press.

— — May, R. M., Hassell, M. P., and Conway, G. R. (1974). Ecological strategies and population parameters. *Amer. Naturalist* **108**, 791–804.

Stallcup, J. A. and Woolfenden, G. E. (1978). Family status and contributions to

breeding by Florida scrub jays. *Animal Behav.* **26**, 1144—56.
Stearns, S. C. (1976). Life history tactics: a review of the ideas. *Quart. Rev. Biol.* **51**, 3—47.
—— Crandall, R. E. (1981). Bet-hedging and persistence as adaptations of colonizers. In *Evolution Today: Proceedings of the International Congress of Systematics and evolutionary Biology* (Vancouver, July 1980), (ed. G. G. E. Scudder and J. L. Reveal). pp. 371—83.
Steele, J. H. (1979). Interactions in marine ecosystems. In *Population dynamics* (ed. R. M. Anderson, B. D. Turner, and L. R. Taylor), pp. 343—57. Blackwell Scientific Publications, Oxford.
Stehn, R. and Richmond, M. (1975). Male-induced pregnancy termination in the prairie vole, *Microtus ochrogaster. Science* **187**, 1211—13.
Steinbeck, J. (1962). *Travels with Charley: in search of America.* Heinemann, London.
Stenseth, N. C. (1977a). Evolutionary aspects of demographic cycles: the relevance of some models for microtine fluctuations. *Oikos* **29**, 525—38.
—— (1977b). On the importance of spatio—temporal heterogeneity for the population dynamics of rodents: towards a theoretical foundation of rodent control. *Oikos* **29**, 545—52.
—— (1978a). Demographic strategies in populations of small rodents. *Oecologia* **33**, 149—72.
—— (1978b). Is the female biased sex ratio in wood lemming, *Myopus schisticolor*, maintained by cyclic inbreeding? *Oikos,* **30**, 83—9.
—— (1978c). Do grazers maximize individual plant fitness? *Oikos* **31**, 299—306.
—— (1978d). Energy balance and the Malthusian parameter, m, of grazing small rodents. *Oecologia* **32**, 37—55.
—— (1980). Spatial heterogeneity and population stability: some evolutionary consequences. *Oikos* **35**, 165—84.
—— (1981). On Chitty's theory for fluctuating populations: the importance of polymorphisms in the generation of regular cycles. *J. theoret. Biol.* **90**, 9—36.
—— Framstad, E. (1980). Reproductive effort and optimal reproductive rates in small rodents. *Oikos* **34**, 23—34.
—— Hansson, L. (1981). The importance of population dynamics in heterogenous landscapes: management of vertebrate pests and some other animals. *Aaro-ecosystems* **7**, 187—211.
—— Ugland, K. I. (1982). On the evolution of demographic strategies in populations with equilibrium and cyclic densities. Manuscript.
—— Hansson, L., and Myllymäki, A. (1977a). Population dynamics of the field vole (*Microtus agrestis* (L.)): a model. *EPPO Bull.* **7**, 371—84.
—— —— Ugland, K. I. (1982). On the evolution of reproductive rates in microtime rodents. Manuscript.
—— Framstad, E., Migula, P., Trojan, P., and Wojciechowska-Trojan, B. (1980). Energy models for the common vole, *Microtus arvalis:* energy as a limiting resource for reproductive output. *Oikos* **34**, 1—22.
—— Hansson, L., Myllymäki, A., Andersson, M., and Katila, J. (1977b). General models for the population dynamics of field vole, *Microtus agrestis*, in central Scandinavia. *Oikos* **29**, 616—42.
Stickel, L. (1949). The source of animals moving into a depopulated area. *J. Mammal.* **27**, 301—7.
—— (1968). Home range and travels. In *'Biology of Peromyscus (Rodentia)'* (ed. J. A. King), pp. 373—411. Special Publications No. 2, American Society of Mammalogists.
Stirling, I. (1969). Ecology of the Weddell seal in McMurdo Sound, Antarctica.

Ecology **50**, 574—86.
— — (1975). Factors affecting the evolution of social behaviour in the pinipedia. *Rapp. P. v. Reun. Cons. Int. Explor. Mer.* **169**, 205—12.
Storer, T. I., Evans, F. C., and Palmer, F. G. (1944). Some rodent populations in the Sierra Nevada of California. *Ecol. Monogr.* **14**, 165—92.
Stubbs, M. (1977). Density dependence in the life-cycles of animals and its importance in K and r-strategies. *J. animal Ecol* **46**, 677—88.
Sullivan, T. P. (1977). Demography and dispersal in island and mainland populations of the deer mouse, *Peromyscus maniculatus*. *Ecology* **58**, 964—78.
Svärdson, G. (1949). Competition and habitat selection in birds. *Oikos* **1**, 156—74.
Svendsen, G. E. (1974). Behavioural and environmental factors in the spatial distribution and population dynamics of a yellow-bellied marmot population. *Ecology* **55**, 760—71.
Swarth, H. S. (1920). Revision of the avian genus *Passerella*, with special reference to the distribution and migration of the races in California. *Univ. Cal. Publ. Zool.* **21**, 75—224.
Swingland, I. R. (1977). Reproductive effort and life history strategy of the Aldabran giant tortoise. *Nature, Lond.* **269**, 402—4.
— — Coe, M. (1978). The natural regulation of giant tortoise populations on Aldabra Atoll. Reproduction. *J. Zool. Lond.* **186**, 285—309.
— — Frazier, J. G. (1979). The conflict between feeding and overheating in the Aldabran giant tortoise. In *A handbook on biotelemetry and radio tracking* (ed. C. J. Amlaner and D. W. Macdonald). Pergamon, Oxford and New York.
— — Lessells, C. M. (1979). The natural regulation of giant tortoise populations on Aldabra Atoll. Movement polymorphism, reproductive success and mortality. *J. animal Ecol.* **48**, 639—54.
— — Parker, M. J., and North, P. M. (1981). What determines individual movement patterns in giant tortoises? Second European Chelonian Symposium, Oxford.
Talbot, L. M. and Talbot, M. H. (1963). The wildebeest in western Masailand, East Africa. *Wildlife Monogr.* **12**, 1—88.
Tamarin, R. H. (1977). Dispersal in island and mainland voles. *Ecology* **58**, 1044—54.
— — (1978*a*). Dispersal, population regulation and K-selection in field mice. *Amer. Naturalist* **112**, 545—55.
— — (Ed.) (1978*b*). *Population regulation.* Benchmark Papers in Ecology (Series ed. F. Dowden). Hutchinson and Ross, Stroudsberg, Pennsylvania.
— — (1980*a*). Dispersal and population regulation in rodents. In *Biosocial mechanisms of population regulations* (ed. M. N. Cohen, R. S. Malpass, and H. G. Klein) pp. 117—33. Yale University Press, New Haven, Connecticut.
— — (1980*b*). Animal population regulation through behavioural interactions. (In press in *Mammal. Soc. Special. Publ. Series.*)
Tanaka, S., Matsuka, M., and Sakai, T. (1976). Effects of change in photoperiod on wing form in *Pteronemobius taprobanesis* Walker (Orthoptera: Gryllidae). *J. appl. entomol. Zool.* **11**, 27—32.
Tast, J. (1966). The root vole (*Microtus oeconomous*) as an inhabitant of seasonally flooded land. *Ann. Zool. Fennici* **3**, 127—71.
Tauber, M. J. and Tauber, C. A. (1976). Insect seasonality: diapause maintenance, termination, and post-diapause development. *Ann. Rev. Entomol.* **21**, 81—107.
Taylor, C. R., Schmidt-Nielsen, K., and Raab, S. L. (1970). Scaling of energetic

cost of running to body size in mammals. *Amer. J. Physiol* **219**, 1104–7.

Taylor, L. R. (1957a) (in Hyde, H. A.). Aerobiology. *Nature, Lond.* **179**, 890–2.

—— (1957b). Temperature relations of teneral development and behaviour in *Aphis fabae* Scop. *J. exp. Biol.* **34**, 189–208.

—— (1958). Aphid dispersal and diurnal periodicity. *Proc. Linn. Soc. Lond.* **169**, 67–73.

—— (1960a). Mortality and viability of insect migrants high in the air. *Nature, Lond.* **186**, 410.

—— (1960b). The distribution of insects at low levels in the air. *J. animal Ecol.* **29**, 45–63.

—— (1961). Aggregation, variance and the mean. *Nature, Lond.* **189**, 732–5.

—— (1965a). Flight behaviour and aphid migration. *Proc. N. Central Branch Entomol. Soc. Amer.* **20**, 9–19.

—— (1965b). A natural law for the spatial disposition of insects. *Proc. 12th Int. Congr. Entomol.* 1964, pp. 396–7.

—— (1971). Aggregation as a species characteristic. In *Statistical ecology*, (ed. G. P. Patil, E. C. Pielou, and W. E. Waters) Vol. I pp. 357–77. Pennsylvania University Press, University Park.

—— (1974). Insect migration, flight periodicity and the boundary layer. *J. animal Ecol.* **43**, 225–38.

—— (1975). Longevity, fecundity and size; control of reproductive potential in a polymorphic migrant, *Aphis fabae* Scop. *J. animal Ecol.* **44**, 135–63.

—— (1979). The Rothamsted Insect Survey – an approach to the theory and practice of synoptic pest forecasting in agriculture. *Movement of highly mobile insects* (ed. R. L. Rabb and G. G. Kennedy), pp. 148–85. North Carolina State University.

—— Brown, E. S. (1972). Effects of light-trap design and illumination on samples of moths in the Kenya highlands. *Bull. entomol. Res.* **62**, 91–112.

—— —— Littlewood, S. C. (1979a). The effect of size on the height of flight of migrant moths. *Bull. entomol. Res.* **69**, 605–9.

—— French, R. A., and Macaulay, E. D. M. (1973). Low-altitude migration and diurnal flight periodicity; the importance of *Plusia gamma* L. (Lepidoptera: Plusiidae). *J. animal Ecol.* **42**, 751–60.

—— Taylor, R. A. J. (1977). Aggregation, migration and population mechanics. *Nature, Lond.* **265**, 415–21.

—— —— (1978). The dynamics of spatial behaviour. In *Population control by social behaviour* (ed. F. J. Ebling and D. M. Stoddart), pp. 181–212. Institute of Biology, London.

—— Woiwod, I. P. (1980). Temporal stability as a density-dependent species characteristic. *J. animal Ecol.* **49**, 209–24.

—— —— Perry, J. N. (1978). The density dependence of spatial behaviour and the rarity of randomness. *J. animal Ecol.* **47**, 383–406.

—— —— —— (1980). Variance and the large scale spatial stability of aphids, moths and birds. *J. animal Ecol.* **49**, 831–54.

—— —— Taylor, R. A. J. (1979b). The migratory ambit of the hop aphid and its significance in aphid population dynamics. *J. animal Ecol.* **48**, 955–72.

Taylor, P. D. (1981). Intra-sex and inter-sex sibling interactions as sex ratio determinants. *Nature* **291**, 64–6.

Taylor, R. A. J. (1978). The relationship between density and distance of dispersing insects. *Ecol. Entomol.* **3**, 63–70.

—— (1979a). Simulation studies and analysis of migration dynamics. Unpublished Ph.D. thesis, London University.

—— (1979*b*). A simulation model of locust migratory behaviour. *J. animal Ecol.* **48**, 1–26.

—— (1980). A family of regression equations describing the density distribution of dispersing organisms. *Nature, Lond.* **286**, 53–5.

—— (1981*a*) A behavioural basis for redistribution. I. The Δ-model concept. *J. animal Ecol.* **50**, 573–86.

—— (1981*b*). A behavioural basis for redistribution. II. Simulations of the Δ-model. *J. animal Ecol.* **50**, 587–603.

—— Taylor, L. R. (1979). A behavioural model for the evolution of spatial dynamics. In *Population dynamics*, (ed. R. M. Anderson, B. D. Turner, and L. R. Taylor), pp. 1–27. Blackwells Scientific Publications, Oxford.

Taylor, V. A. (1981). The adaptive and evolutionary significance of wing polymorphism and pathenogenesis in *Ptinella* Motschulsky (Coleoptera : Ptiliidae). *Ecol. Entomol.* **6**, 89–98.

Templeton, A. R. (1979). Genetics of colonization and establishment of exotic species. In *Genetics in relation to insect management* (ed. M. A. Hoy and J. J. McKelvey, Jr.), pp. 41–9. The Rockerfeller Foundation, New York.

—— Carson, H. L., and Sing, C. F. (1976). The population genetics of parthenogenetic strains of *Drosophila mercatorum*. II. The capacity for parthenogenesis in a natural, bisexual population. *Genetics* **82**, 527–42.

Thoday, J. M. (1972). Disruptive selection. *Proc. R. Soc. Lond.* **B182**, 109–43.

Thompson, W. R. (1929). On natural control. *Parasitol.* **21**, 269–81.

Thornhill, R. (1980). Sexual selection within mating swarms of the lovebug. *Plecia nearctica* (Diptera:Bibionidae). *Animal Behav.* **28**, 405–12.

Thorson, G. (1950). Reproductive and larval ecology of marine bottom invertebrates. *Biol. Rev.* **25**, 1–45.

Thorsrud, O. and Stenseth, N. C. (1980). On the evolutionary stability of the female biased sex ratio in the wood lemming (*Myopus schisticolor*) : the effect of metapopulations being divided into small units. (In preparation.)

Tilzey, R. D. J. (1977). Repeat homing of brown trout (*Salmo trutta*) in Lake Eucumbene, New South Wales, Australia. *J. Fish. Res. Bd. Can.* **34**, 1085–94.

Tinkle, D. W. (1965). Population structure and effective size of a lizard population. *Evolution* **19**, 569–73.

—— (1967). Home range, density, dynamics and structure of a Texas population of the lizard, *Uta stansburiana*. In *Lizard ecology: a symposium* (ed. W. M. Milstead). University of Missouri Press, Columbia.

Tomlinson, J. (1966). The advantages of hermaphroditism and parthenogenesis. *J. theoret. Biol.* **11**, 54–8.

Treisman, M. (1978). Bird song dialects, repertoire size and kin association. *Animal Behav.* **26**, 814–17.

Trivers, R. L. (1971). The evolution of reciprocal altruism. *Quart. Rev. Biol.* **46**, 35–57.

—— (1974). Parent–offspring conflict. *Amer. Zool.* **14**, 249–64.

—— Hare, H. (1976). Haplodiploidy and the evolution of the social insects. *Science* **191**, 249–63.

Tucker, V. A. (1969). The energetics of bird flight. *Sci. Amer.* **220**, 70–8.

—— (1970). Energetic costs of locomotion in animals. *Comp. biochem. Physiol.* **34**, 841–6.

Turner, F. B., Jeinrich, R. I., and Weintraub, J. D. (1969). Home ranges and body size of lizards. *Ecology* **50**, 1076–81.

Tutt, J. W. (1902). *The migration and dispersal of insects*. Elliott Stock, London.

Udvardy, M. D. F. (1969). *Dynamic zoogeography, with special reference to land animals.* Van Nostrand Reinhold, New York.
Uexküll, J. (1957). A stroll through the world of animals and men. In *Instinctive behaviour* (ed. C. H. Schiller). Methuen, London.
Ugland, K. I. and Stenseth, N. C. (1982). On the evolution of reproductive rates in populations with equilibrium and cyclic densities. (In preparation).
Urquhart, F. A. and Urquhart, N. R. (1976). The overwintering site of the eastern population of the monarch butterfly. (*Danaus p. plexippus*; Danaidae) in southern Mexico. *J. Lepidopt. Soc.* **30**, 153–8.
—— —— (1978). Autumnal migration routes of the eastern population of the monarch butterfly (*Danaus p. plexippus* L.; Danaidae; Lepidoptera) in North America to the overwintering site in the Neovolcanic Plateau of Mexico. *Can. J. Zool.* **56**, 1758–64.
—— —— (1979). Vernal migration of the monarch butterfly (*Danaus p. plexippus*, Lepidoptera:Danaidae) in North America from the overwintering site in the Neo-volcanic Plateau of Mexico. *Can. Entomol.* **111**, 15–18.
Uvarov, B. (1966). *Grasshoppers and locusts*, Vol. 1. Cambridge University Press, London.
—— (1977). *Grasshoppers and locusts*, Vol. 2. Cambridge University Press, London.
Van Ryzin, M. T. and Fisher, H. I. (1976). The age of Laysan albatrosses, *Diomedea immutabilis*, at first breeding. *Condor* **78**, 1–9.
Van Valen, L. (1965). Morphological variation and width of ecological niche. *Amer. Naturalist* **99**, 377–90.
—— (1971). Group selection and the evolution of dispersal. *Evolution* **25**, 591–8.
Van Vleck, D. B. (1968). Movements of *Microtus pennsylvanicus* in relation to depopulated areas. *J. Mammal.* **49**, 92–103.
Vance, R. R. (1973). On reproductive strategies in marine benthic invertebrates. *Amer. Naturalist* **107**, 339–52.
—— (1980). The effect of dispersal on population size in a temporally varying environment. *Theoret. pop. Biol.* **18**, 343–62.
Varley, G. C. and Gradwell G. R. (1968). Population models for the winter moth. In *Insect abundance* (ed. T. R. E. Southwood), pp. 132–42. R. Entomol. Soc. Symp. no. 4.
Vehrencamp, S. (1979). The role of individual, kin and group selection in the evolution of sociality. In *Handbook of behavioural neurobiology*, Volume III. *Social behaviour and communication* (ed. P. Marler and J. G. Vandenbergh), pp. 351–94. Plenum Press, New York.
Vepsäläinen, K. (1973). The distribution and habitats of *Gerris* Fabr. species (Heteroptera Gerridae) in Finland. *Ann. Zool. Fenn.* **10**, 419–44.
—— (1974*a*). The wing lengths, reproductive stages and habitats of Hungarian Gerris Fabr. species (Heteroptera, Gerridae). *Ann. Acad. Sci. Fenn.* **AIV**.
—— (1974*b*). The life cycles and wing lengths of Finnish *Gerris* Fabr. species (Heteroptera, Gerridae). *Acta Zool. Fenn.* **141**, 1–73.
—— (1978). Wing dimorphism and diapause in *Gerris*: Determination and adaptive significance. In *Evolution of insect migration and diapause* (ed. H. Dingle), pp. 218–53. Springer-Verlag, New York.
Vermeij, G. J. (1978). *Biogeography and adaptation.* Harvard University Press, Cambridge, Massachusetts.
Verner, L. (1979). The significance of dispersal in fluctuating populations of *Microtus ochrogaster* and *Microtus pennsylvanicus*.) Unpublished Ph.D.

thesis, University of Illinois, Urbana.

Vuilleumier, F. and Matteo, M. B. (1972). Esterase polymorphism in European and American populations of the periwinckle, *Littorina littorea* (Gastropoda). *Experientia* **28**, 1241–2.

Wade, M. J. (1979). The evolution of social interactions by family selection. *Amer. Naturalist* **113**, 399–417.

Waldman, B. and Adler, K. (1979). Toad tadpoles associate preferentially with siblings. *Nature* **282**, 611–13.

Wallace, A. R. (1876). *The geographical distribution of animals* (2 vols.). Macmillan, London.

Wallace, B. (1968). *Topics in population genetics.* W. W. Norton, New York.

Waloff, Z. (1946). Seasonal breeding and migrations of the desert locust (*Schistocerca gregaria* Forskål) in eastern Africa. *Anti-Locust Memoir* **1**, 74 pp.

—— (1958). The behaviour of locusts in migrating swarms (Abstract) *Proc. 10th Int. Cong. Entomol.* **2**, 567–9.

—— (1966). The upsurges and recessions of the desert locust plague: an historical survey. *Anti-Locust Memoir* **8**, 118 pp.

—— (1972). Orientation of flying locusts, *Schistocerca gregaria* (Forsk.), in migrating swarms. *Bull. entomol. Res.* **62**, 1–72.

Wangersky, P. J. (1978). Lotka–Volterra population models. *Ann. Rev. Ecol. Systemat.* **9**, 189–218.

Ward, P. (1969). The annual cycle of the yellow-vented Bulbul *Pycnonotus goiavier* in a humid equatorial environment. *J. Zool., Lond.* **157**, 25–45.

—— (1971). The migration patterns of *Quelea quelea* in Africa. *Ibis* **113**, 275–97.

Ware, D. M. (1975). Growth, metabolism, and optimal swimming speed of a pelagic fish. *J. Fish. Res. Bd. Can.* **32**, 33–41.

Warham, J. (1964). Breeding behavior in Procellariiformes. In Biologie antarctique. *Proc. 1st Symp. Antarctic Biol.* (ed. R. M. Carrick, M. W. Holdgate, and J. Prevost), pp. 389–94. Paris.

Waser, P. M. and Wiley, R. H. (1979). Mechanisms and evolution of spacing in animals. In *Handbook of behavioral neurobiology* Vol. 3, Social behavior and communication (ed. P. Marler and J. G. Vandenbergh), pp. 159–223. Plenum, New York.

Watson, A. and Moss, R. (1970). Dominance, spacing behaviour and aggression in relation to population limitation en vertebrates. In *Animal population in relation to their food resources* (ed. A. Watson), pp. 167–218. Blackwells, Oxford.

—— —— (1979). Population cycles in the Tetraonidae. *Ornis Fenn.* **56**, 87–109.

Wellings, P. W., Leather, S. R., and Dixon, A. F. G. (1980). Seasonal variation in reproductive potential: a programmed feature of aphid life cycles. *J. animal Ecol.* **49**, 975–85.

Wellington, W. G. (1977). Returning the insect to insect ecology: some consequences for pest management. *Environ. Entomol.* **6**, 1–8.

Wells, K. D. and Wells, R. A. (1976). Patterns of movement in a population of the slimy salamander, *Plethodon glutinosus*, with observations on aggregations. *Herpetologica* **32**, 156–62.

Werren, J. H. (1980). Sex ratio adaptations to local mate competition in a parasitic wasp. *Science* **208**, 1157–9.

White, M. J. D. (1978). *Modes of speciation.* Freeman, San Francisco.

Whitham, T. G. (1977). Coevolution of foraging in Bombus and nectar dispensing in Chilopsis: A last dreg theory. *Science* **197**, 593–6.

Whittaker, J. B., Ellistone, J. and Patrick, C. K. (1979). The dynamics of a chrysomelid beetle, *Gastrophysa viridula* in a hazardous natural habitat. *J. animal Ecol.* **48**, 973–86.

Wiens, J. A. (1966). On group selection and Wynne-Edwards hypothesis. *Amer. Scientist* **54**, 273–87.

—— 1976. Population responses to patchy environments. *Ann. Rev. Ecol. Systemat.* **7**, 81–120.

Wigglesworth, V. B. (1963). The origin of flight in insects. *Proc. R. Entomol. Soc. Lond.* [C] **28**, 23–32.

—— (1976). The evolution of insect flight. Insect flight (ed. R. C. Rainey), pp. 255–69. *R. Entomol. Soc. Symp.* **7**. Blackwells, Oxford.

Wigan, L. G. (1944). Balance and potence in natural populations. *J. Genet.* **46**, 150–60.

Wilbur, H. M. (1977). Propagule size, number, and dispersion pattern in *Ambystoma* and *Asclepias. Amer. Naturalist* **111**, 43–68.

—— Tinkle, D. W., and Collins, J. P. (1974). Environmental certainty, trophic level, and resource availability in life history evolution. *Amer. Naturalist* **108**, 805–17.

Williams, C. B. (1917). Some notes on butterfly migration in British Guiana. *Trans. Entomol. Soc. Lond.* **1917**, 154–64.

—— (1926). Voluntary or involuntary migration of butterflies. *The Entomologist* **59**, 281–8.

—— (1930). *The migration of butterflies.* Oliver and Boyd, Edinburgh.

—— (1949). Migration in Lepidoptera and the problem of orientation. *Proc. R. entomol. Soc. Lond.* [C] **13**, 70–84.

—— (1951). Seasonal changes in flight direction of migrant butterflies in the British Isles. *J. animal Ecol.* **20**, 180–90.

—— (1958). *Insect migration.* Collins, London.

—— (1976). The migrations of the hesperid butterfly. *Andronymus neander* Plötz. in Africa. *Ecol. Entomol.* **1**, 213–20.

—— Cockbill, G. F., Gibbs, M. E., and Downes, J. A. (1942). Studies in the migration of Lepidoptera. *Trans. R. entomol. Soc. Lond.* **92**, 101–283.

—— Common, I. F. B., French, R. A., Muspratt, V., and Williams, M. C. (1956). Observations on the migration of insects in the Pyrenees in the autumn of 1953. *Trans. R. entomol. Soc. Lond.* **108**, 385–407.

Williams, G. C. (1966a). *Adaption and natural selection.* Princeton University Press, New Jersey.

—— (1966b). Natural selection, the cost of reproduction, and a refinement of Lack's principle. *Amer. Naturalist* **100**, 687–90.

—— (1975). *Sex and evolution.* Princeton University Press, New Jersey.

—— (1979). The question of adaptive sex ratio in outcrossed vertebrates. *Proc. R. Soc.* **B205**, 567–80.

Williamson, M. H. (1957). An elementary theory of interspecific competition. *Nature, Lond.* **180**, 422–5.

Willis, E. O. (1967). The behaviour of bicolored antbirds. *Univ. Cal. Publ. Zool.* **79**, 1–132.

—— (1968). Studies of limulated and Salvin's antbirds. *Condor* **70**, 128–48.

Wilson, D. S. (1980). *The natural selection of populations and communities.* Benjamin/Cummings, Menlo Park, California.

Wilson, E. O. (1973). Group selection and its significance for ecology. *BioSci.* **23**, 631–8.

—— (1975). *Sociobiology.* Belknap Press, Cambridge, Massachusetts.

—— Simberloff, D. S. (1969). Experimental zoogeography of islands: defaunation and monitoring techniques. *Ecology* **50**, 267–78.

Wilson, F. (1965). Biological control and the genetics of colonizing species. In *The genetics of colonizing species*. (ed. H. G. Baker and G. L. Stebbins), pp. 307–25. Academic Press, New York.

Wiltshire, E. P. (1946). Studies in the geography of Lepidoptera III. Some Middle East migrants, their phenology and ecology. *Trans. R. entomol. Soc. Lond.* **96**, 163–82.

Windberg, L. A. and Keith, L. B. (1976). Experimental analysis of dispersal in snowshoe hare populations. *Can. J. Zool.* **54**, 2061–81.

Wittenberger, J. F. (1979). The evolution of mating systems in birds and mammals. In *Handbook of behavioural neurobiology*, Vol. 3. Social behaviour and communication. (ed. P. Marler and J. G. Vandenbergh), pp. 271–349. Plenum, New York.

Wium-Andersen, G. (1970). Hemoglobin and protein variation in three species of *Littorina*. *Ophelia* **8**, 267–73.

Wolf, L. L., Hainsworth, F. R., and Gill, F. B. (1975). Foraging efficiencies and time budgets in nectar-feeding birds. *Ecology* **56**, 117–28.

Woolfenden, G. E. (1975). Florida scrub jay helpers at the nest. *Auk* **92**, 1–15.

—— (1976). Co-operative breeding in American birds. *Proc. 16th int. ornithol. Cong.*, Canberra, pp. 674–84.

—— Fitzpatrick, J. W. (1977). Dominance in the Florida scrub jay. *Condor* **79**, 1–12.

—— —— (1978). The inheritance of territory in group-breeding birds. *Biosci.* **28**, 104–8.

Wootton, R. J. (1976). *The biology of the sticklebacks*. Academic Press, London.

Wright, S. (1922). Coefficients of inbreeding and relationship. *Amer. Naturalist* **56**, 330–8.

—— (1931). Evolution in mendelian populations. *Genetics* **16**, 97–159.

—— (1932). The role of mutation, inbreeding, cross-breeding and selection in evolution. *Proc. 6th int. Cong. Genetics*, pp. 356–66.

—— (1940). Breeding structure of populations in relation to speciation. *Amer. Naturalist* **74**, 232–48.

—— (1943). Isolation by distance. *Genetics* **28**, 114–38.

—— (1946). Isolation by distance under diverse systems of mating. *Genetics* **31**, 39–59.

—— (1956). Modes of selection. *Amer. Naturalist* **90**, 7–24.

—— (1969). *Evolution and the genetics of populations*, Vol. 2. The theory of gene frequencies. University of Chicago Press.

—— (1977). *Evolution and the genetics of populations*, Vol. 3. Experimental results and evolutionary deductions. University of Chicago Press.

—— (1978). *Evolution and the genetics of populations*, Vol. 4. Variability within and among natural populations. University of Chicago Press.

Wynne-Edwards, V. C. (1962). *Animal dispersion in relation to social behaviour*. Oliver and Boyd, Edinburgh.

—— (1978). Intrinsic population control: an introduction. In *Population control by social behaviour* (ed. F. J. Ebling and D. M. Stoddart), pp. 1–22. Institute of Biology, London.

Yiftakh, Z., Gilboa, A., and Safriel, U. N. (1978). Can an Indo-Pacific mytilid successfully colonize a Mediterranean mytilid bed? *2nd int. Cong. Ecol. INTECOL*, Jerusalem, 1978. Abstracts volume, p. 419.

Young, E. C. (1965*a*). Flight muscle polymorphism in British Corixidae: ecological observations. *J. animal Ecol.* **34**, 353—96.
— — (1965*b*). The incidence of flight polymorphism in British Corixidae and description of the morphs. *J. Zool.* **146**, 567—76.
Zach, R. and Falls, J. B. (1976*a*). Ovenbird (Aves: Parulidae) hunting behavior in a patchy environment: an experimental study. *Can. J. Zool.* **54**, 1863—79.
— — — — (1976*b*). Do ovenbirds (Aves: Parulidae) hunt by expectation? *Can. J. Zool.* **54**, 1894—903.
Zwickel, F. C., Redfield, J. A., and Kristensen, J. (1977). Demography, behaviour and genetics of a colonizing population of Blue Grouse. *Can. J. Zool.* **55**, 1948—57.

Scientific names index

English names index

Subject index

sink 99
dynamic motion, models for 202–7

emigration 250–5, 258
  evolution of nomadism 250–1
  in unpredictable environments, 251–5
environment, spatial and temporal variation in 56–7
environmental hostility, survival strategies 171–6
environmental stochasticity 223
environmental templet 182
evolutionarily stable strategies (ESS) 76, 79, 103–5, 106, 108

Finland 109
flight
  linear, Williams' theory of 193, 194
  oogenesis syndrome 172, 197
  'paranotal theory' in insects 168
Florida 194, 228
flying, or walking, alternatives 167–71
food resources, aerial 33
foraging
  by bumblebees 10–12
  by chickadees 12–15
  and dispersal 2–3
  by great tits 9–10, 17–18
  by honeyeaters 24–6
  by hummingbirds 21–3
  by ovenbirds 17–19
  by parasitic wasps 15–17
  and predation 3–4
  predictive theory of behaviour 7
  by spotted flycatchers 19–21
  and territories 2

Gadgil's model, of asexual reproduction 68–9, 70, 71, 73
Galapagos Islands 185
gene frequency, in eider duck 111
General Adaptation Syndrome 210
genetics
  basis, of DDT resistance, 151
  consequences of habitat selection 113–15
  correlates of colonizing ability 225–38
  cost of meiosis 140
  dispersal 66–7
  homogenization 138
  polymorphism 104, 105, 107, 108
  in Finland 109
genomes 139, 140
genomic consequences of incest 141
genotypes 105
  unity of 148
giving-up time 13, 19, 20

habitat
  colonization of recreated 228
  colonization of virgin 227–8
  dispersal in stable 56
  donor or reception 83
  populations in 97
  rodent 65
  selection, genetic consequences and life histories 113–15
  suitability, in evolution of dispersal 67–8, 172
  transition 65
Hawaii, and North Pacific 147
herbivores 34
hermaphroditism 226
heterozygote decay 138
home-range 32
  and metabolic needs in vertebrates 36–46
  methods 37–40
  sex differences in size 43–4
  size 4–5, 33
  and average daily metabolic needs (ADMN) 50, 51
  metabolic needs and diet in primates 46–53
  relationship between 52
  review and critique of 34–6

Iceland 227
immigration, lack of 99
inbreeding 57, 157
  avoidance 123
  intensities 142
  and relative fecundity 142
  optimal levels of 136–43
  consequences of 137–9
  definition of 136–7
  pedigree characteristics of 137
  selective value of 139–40
incest 157
  among siblings 61
  genomic consequences of 141
infanticide 129, 130
inflorescence 29
  nectar threshold on 23, 25, 26
insect
  Carboniferous Palaeopteran 169
  diversity of types 169–70
  evaluation of flight in 168
  in Quaternary era 169
  evolution of wings in 170–1
  flight, 'paranotal theory' in 168
  migration, hypotheses for 160–1
  in Permian eras 169
  Tutt's review of 189, 190, 191, 192, 193